Qiang Wang and Ziyi Zhong (Eds.)
Environmental Functional Nanomaterials

Also of interest

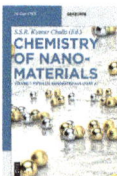

Chemistry of Nanomaterials
Volume 1: Metallic Nanomaterials (Part A)
Kumar (Ed.), 2018
ISBN 978-3-11-034003-7, e-ISBN 978-3-11-034510-0

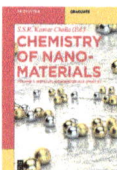

Chemistry of Nanomaterials
Volume 1: Metallic Nanomaterials (Part B)
Kumar (Ed.), 2018
ISBN 978-3-11-063660-4, e-ISBN 978-3-11-063666-6

Advanced Materials
van de Ven, Soldera (Eds.), 2019
ISBN 978-3-11-053765-9, e-ISBN 978-3-11-053773-4

Nanoscience and Nanotechnology
Advances and Developments in Nano-sized Materials
Van de Voorde (Ed.), 2018
ISBN 978-3-11-054720-7, e-ISBN 978-3-11-054722-1

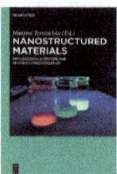

Nanostructured Materials
Applications, Synthesis and In-Situ Characterization.
Terraschke (Ed.), 2019
ISBN 978-3-11-045829-9, e-ISBN 978-3-11-045909-8

Environmental Functional Nanomaterials

Edited by
Qiang Wang and Ziyi Zhong

DE GRUYTER

Editors
Prof. Qiang Wang
Environmental Functional Nanomaterials (EFN) Lab
College of Environmental Science and Engineering
Beijing Forestry University
35 Qinghua East Road
100083 Beijing
China
qiangwang@bjfu.edu.cn

Prof. Ziyi Zhong
College of Engineering
Guangdong Technion Israel Institute of Technology (GTIIT)
241 Da Xue Road, Shantou, Guangdong, 515063
China
ziyi.zhong@gtiit.edu.cn

&

Technion–Israel Institute of Technology (IIT)
Haifa, 32 000
Israel
ziyi.zhong@technion.ac.il

ISBN 978-3-11-054405-3
e-ISBN (PDF) 978-3-11-054418-3
e-ISBN (EPUB) 978-3-11-054436-7

Library of Congress Control Number: 2019947324

Bibliographic information published by the Deutsche Nationalbibliothek
The Deutsche Nationalbibliothek lists this publication in the Deutsche Nationalbibliografie;
detailed bibliographic data are available on the Internet at http://dnb.dnb.de.

© 2020 Walter de Gruyter GmbH, Berlin/Boston
Cover image: ktsimage/iStock/Getty Images Plus
Typesetting: Integra Software Services Pvt. Ltd.
Printing and binding: CPI books GmbH, Leck

www.degruyter.com

Preface

With the advancement of industrialization worldwide and rapid population growth, a series of environmental problems have emerged or become much more severe than before. These problems include air pollution due to the emission of NO_x, volatile organic compounds (VOCs), and heavy metals such as Hg; water contamination by toxic organic and inorganic compounds discharged by some chemical and pharmaceutical plants; generation of huge amounts of electronic-wastes due to wide use of electronic devices, and so on. In recent years, the proper disposal of municipal and industrial solid wastes has also become a main concern in many cities or countries. To ensure sustainable development, many governments are adopting or implementing more stringent environmental regulations, while both academia and industry are making great effort to find technical solutions to the environmental problems. From the wide range of literature and practical applications available, it seems that functionalized materials, particularly those in the form of nanomaterials, are playing important roles in dealing with environmental pollutants. The application of these nanomaterials (catalysts, adsorbents, etc.) can not only improve treatment efficiency, but also lower energy requirements. Considering the importance and the rapid development in this area, we believe that it is necessary to summarize the representative and important achievements in one book.

To serve the aforementioned purpose, we have organized seven chapters focusing on the topics of wastewater treatment, air pollution control, e-waste treatment, and environmental reclamation by application of various nanomaterials. For example, Chapter 1 deals with the treatment of organics in wastewater. Besides reviewing the noncatalytic methods, the catalytic oxidation of organic pollutants in wastewater by air is also specially discussed and reviewed. Chapters 2 and 3 are about catalytic removal of VOCs. In particular, Chapter 2 describes the catalytic oxidation of a number of VOCs such as benzene, formaldehyde, and toluene, and briefly highlights the oxidation mechanisms. And Chapter 3 reviews the functional catalysts for catalytic removal of formaldehyde from air, but special attention is paid to the catalysts workable at low reaction temperatures, for example, at room temperature. Chapters 4 and 5 are about catalytic removal of NO_x. In particular, Chapter 4 presents the latest research developments in the catalytic oxidation of NO to NO_2 on various materials, including Pt-based catalysts, metal oxides, perovskites, zeolite, and carbon-based materials for the oxidation of NO to NO_2, a main reaction used in the different industrial NO_x emission control technologies. And Chapter 5 shows the recent progress on the selective catalytic reduction of nitrogen oxides by NH_3. Chapter 6 describes the application of photocatalytic approach for removal of gaseous elemental mercury from emissions. In the final Chapter 7, a concise and comprehensive picture of the recovery of metals from e-waste by covering theories, mechanisms, materials synthesis, and processes is provided. In particular, the recent progress on the selective extraction of precious metals from e-waste is discussed in detail.

https://doi.org/10.1515/9783110544183-202

We hope this book will be relevant to a wide range of readers such as high-school students, undergraduates, postgraduates, and researchers. This book aims to not only provide them with common knowledge in this area, but also raise their awareness of environmental protection issues, and promote research work in both academia and industry in related fields.

The editors would like to thank all the contributors to this book, including Prof. Jiang Wu (Shanghai University of Electric Power, China), Dr Shuande Li and Yunfa Chen (Institute of Process Engineering, CAS, China), Prof. Qingling Liu (Tianjin University, China), Dr Cheng Zhang in Beijing Forestry University, Dr Poernomo Gunawan in Nanyang Technological University, Singapore, Dr Wen Xiao in National University of Singapore, Dr Ubong Jerome Etim (Guangdong Technion Israel Institute of Technology (GTIIT), China) and Prof. Bai Peng from China University of Petroleum (East China), and all the students and researcher involved in each chapters. Without their hard work and professionalism, it is impossible to have this book written and published.

Finally, we are very grateful to Ms Vivien Schubert for all the support during the process of organizing and editing this book, and de Gruyter Publisher for publishing this book.

Prof. Qiang Wang,
College of Environmental Science and Engineering,
Beijing Forestry University,
China

Prof. Ziyi Zhong,
College of Engineering,
Guangdong Technion Israel Institute of Technology (GTIIT),
China

&

Technion–Israel Institute of Technology (IIT),
Israel

Contents

Qingling Liu and Mingyu Guo

Jiang Wu and Fengguo Tian

List of Contributors

Peng Bai
State Key Laboratory of Heavy Oil Processing
Key Laboratory of Catalysis CPNC
College of Chemical Engineering
China University of Petroleum (East China)
Qingdao 266580, China
baipeng@upc.edu.cn

Yunfa Chen
State Key Laboratory of Multi-phase Complex
Systems
Institute of Process Engineering
Chinese Academy of Sciences
Beijing 100190, China
yfchen@ipe.ac.cn

Ubong Jerome Etim
College of Engineering
Guangdong Technion Israel Institute of
Technology (GTIIT)
241 Da Xue Road (Office A503)
Jinping District
Shantou 515063, China

Yanshan Gao
College of Environmental Science and
Engineering
Beijing Forestry University
35 Qinghua East Road
Haidian District
Beijing 100083, China

Poernomo Gunawan
School of Chemical & Biomedical Engineering
Nanyang Technological University Singapore
62 Nanyang Drive
Singapore 637459
pgunawan@ntu.edu.sg

Mingyu Guo
Tianjin Key Laboratory of Indoor Air
Environmental Quality Control
School of Environmental Science and
Engineering
Tianjin University
Tianjin, China

Liang Huang
College of Environmental Science and
Engineering
Beijing Forestry University
35 Qinghua East Road
Haidian District
Beijing 100083, China

Shuangde Li
State Key Laboratory of Multi-phase Complex
Systems
Institute of Process Engineering
Chinese Academy of Sciences
Beijing 100190, China
sdli@ipe.ac.cn

Qingling Liu
Tianjin Key Laboratory of Indoor Air
Environmental Quality Control
School of Environmental Science and
Engineering
Tianjin University
Tianjin, China
liuql@tju.edu.cn

Fengguo Tian
College of Environmental Science and
Engineering
Donghua University
Shanghai 201620, China

Qiang Wang
College of Environmental Science and
Engineering
Beijing Forestry University
35 Qinghua East Road
Haidian District
Beijing 100083, China
qiangwang@bjfu.edu.cn

Jiang Wu
College of Energy and Mechanical
Engineering
Shanghai University of Electric Power
Shanghai 200090, China
wjcfd2002@163.com

https://doi.org/10.1515/9783110544183-204

Wen Xiao
Department of Materials Science and
Engineering
National University of Singapore
4 Engineering Drive 3
Singapore 117583
wen.xiao@outlook.com

Zifeng Yan
State Key Laboratory of Heavy Oil Processing
Key Laboratory of Catalysis CPNC
College of Chemical Engineering
China University of Petroleum (East China)
Qingdao 266580, China

Cheng Zhang
College of Environmental Science and
Engineering

Beijing Forestry University
35 Qinghua East Road
Haidian District
Beijing 100083, China

Ziyi Zhong
College of Engineering,
Guangdong Technion Israel Institute of
Technology (GTIIT),
241 Da Xue Road, Shantou, Guangdong,
515063, China
ziyi.zhong@gtiit.edu.cn

&

Technion–Israel Institute of Technology (IIT)
Haifa, 32 000
Israel
ziyi.zhong@technion.ac.il

Poernomo Gunawan and Ziyi Zhong

1 Noncatalytic and catalytic oxidation of organic pollutants in wastewater by air

1.1 Introduction

With the ever-increasing worldwide population and demand for various goods and commodities, rapid economic development gives rise to increasing volume of wastewater globally produced by various industrial sectors, such as chemical, pharmaceutical, petrochemical, food, oil refineries, dyes and textiles, and agricultural activities. Consequently, they pose a greater challenge to the environmental remediation and protection, which calls for more effective and efficient technology for the wastewater treatment in order to meet more stringent discharge standards. Among different types of pollutants generated by industry, organic wastes account for the majority, which often contain highly toxic and hazardous compounds, such as aromatic and phenolic hydrocarbons, chlorinated and halogenated solvents, organic dyes and azo derivatives, pesticides, herbicides, and antibiotics [1]. Each of these organic pollutant families requires specific elimination methods, which should be judiciously selected to obtain minimum traces of residual organic contents as the main criterion (Figure 1.1).

Traditionally, wastewater treatment processes, such as biological, thermal, and physicochemical methods, are widely used for the removal of pollutants. Activated sludge method, a conventional biological treatment, normally requires relatively long residence time (25–60 h) and large-scale treatment plants. Furthermore, it is prone to contaminant fluctuation and unable to handle nonbiodegradable and toxic compounds as they are deemed poisonous to the microorganisms [2, 3]. Thermal method, such as incineration at temperature above 1,000 °C is usually appropriate for effluents with chemical oxygen demand (COD) larger than 100 g L^{-1} but is highly energy intensive and generates hazardous emissions of NO_x, SO_x, dioxin, and furan [4]. Physical techniques such as flocculation, precipitation, adsorption, and membrane separation require posttreatment and often generate a large amount of sludge [5], while chemical methods using strong oxidizing agents such as chlorine [6] and potassium permanganate [7] are expensive and environmentally unfriendly although they can be carried out at ambient condition. Given the aforementioned limitations, it is apparent that more effective and eco-friendly processes for industrial wastewaters need to be further developed.

Poernomo Gunawan, School of Chemical & Biomedical Engineering, Nanyang Technological University Singapore, Singapore
Ziyi Zhong, College of Engineering, Guangdong Technion Israel Institute of Technology (GTIIT), Shantou, Guangdong Province, China; Technion–Israel Institute of Technology (IIT) Haifa, Israel

https://doi.org/10.1515/9783110544183-001

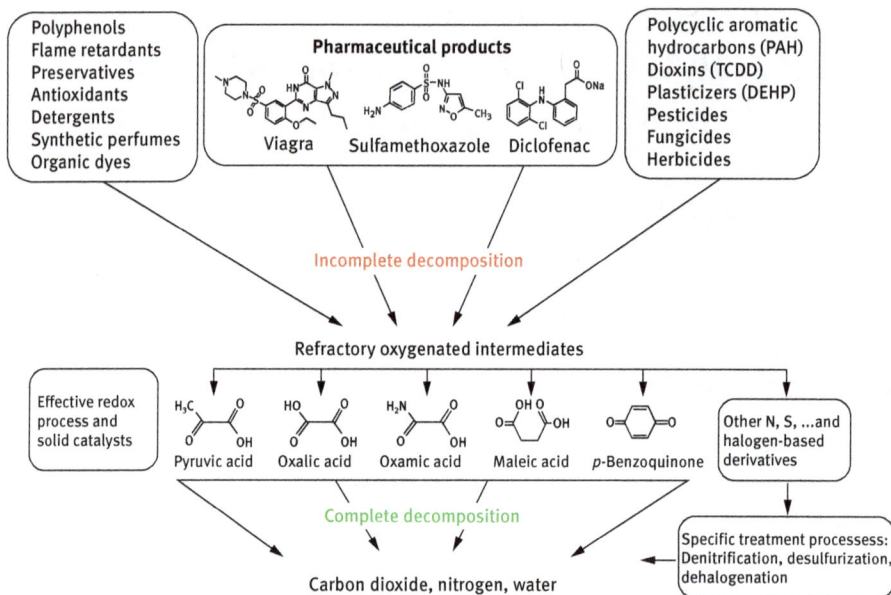

Figure 1.1: Major organic pollutants and their removal strategies. Reprinted from ref. [1] with permission from Elsevier.

Advanced oxidation processes (AOPs), which was first defined by Glaze et al. in 1987 [8], are alternatives for conventional wastewater treatment techniques and are considered low or even nonwaste generation technologies. As a result, they have been widely investigated and well reviewed in several review articles [1, 9–13]. The main advantage of AOPs is their high reactivity and nonselectivity that result in the mineralization of organic contaminants in the wastewater. This is possible due to the generation of short-lived free radical species with high oxidation power and reactivity, such as hydroxyl (OH·) and sulfate ($SO_4^{·-}$) free radicals [10]. However, some recalcitrant organic compounds such as acetic, maleic, and oxalic acids cannot be oxidized by OH radicals [14]. In general, several ways to perform AOPs are as follows [15]:

a) Homogeneous systems without irradiation: O_3/H_2O_2, O_3/OH^-, H_2O_2/Fe^{2+} (Fenton process)

b) Homogeneous systems with irradiation: O_3/UV, H_2O_2/UV, $O_3/H_2O_2/UV$, photo-Fenton, electron beam, ultrasound, and vacuum-UV

c) Heterogeneous systems with irradiation: $TiO_2/O_3/UV$

d) Heterogeneous systems without irradiation: electro-Fenton

Depending on the COD loading, oxidations of the organic pollutants can be carried out by AOPs at ambient conditions for low-loaded wastewaters, or at elevated temperature and pressure for high COD loading of approximately 200 g L^{-1} COD [16]. The nonambient oxidation methods can be classified as wet air oxidation (WAO), supercritical WAO, or other hydrothermal oxidation processes that involve high temperature and pressure, which can be above 150 °C and 1 MPa, respectively [9]. Figure 1.2 depicts the suitability of different water treatment techniques in relation to the wastewater COD loading.

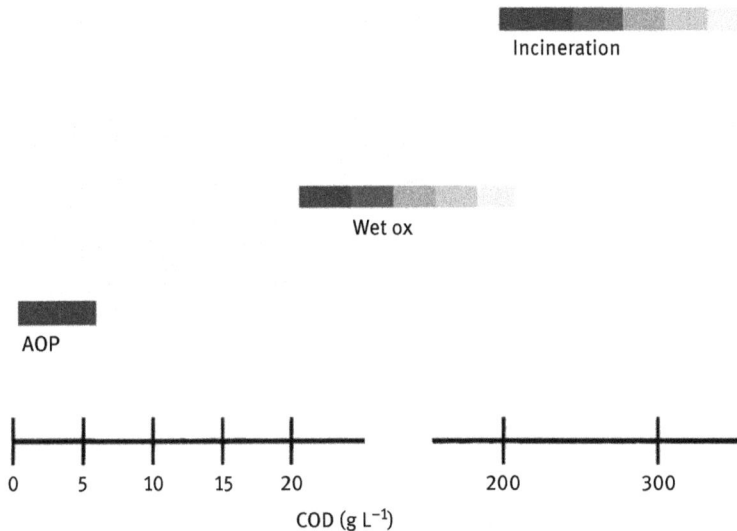

Incineration

Wet ox

AOP

COD (g L^{-1})

Figure 1.2: Suitability of water treatment technologies according to COD contents. Reprinted from ref. [16] with the permission from Elsevier.

1.2 Wet air oxidation

WAO process was first patented by Zimmerman [17, 18] over 50 years ago and ever since it has been commercially used around the world to treat industrial effluents and sludge containing a high content of organic matters (COD 10–100 g L^{-1}) or toxic contaminants [19]. The WAO process can be defined as the oxidation of organic and inorganic substances in an aqueous solution or suspension by means of oxygen or air as oxidant at elevated temperatures and pressures either in the presence or absence of catalysts [20]. WAO destroys toxic materials in industrial wastewater by breaking down complex molecular structures into innocuous components such as water and carbon dioxide. In contrast to other thermal processes, it does not produce NO_x, SO_2, HCl, dioxins, furans, and fly ash, making this process extremely clean. It is reported that the WAO process is capable of achieving a high degree of conversion of toxic organics with more than 99% destruction rate; however, some materials cannot be completely oxidized to carbon dioxide and water, but to some small oxygenated compounds, which typically represent a quarter of the initial mass of organic matter [20], such as small carboxylic acids (e.g., acetic acid and propionic acid), methanol, ethanol, and acetaldehyde. Sulfur and halogens in the wastewater are converted to sulfate and halides, respectively, whereas phosphorus to phosphates. Organic nitrogen may be converted to ammonia, nitrate, or nitrogen gas [21].

Due to its effectiveness, there had been construction of several large WAO plants for the treatment of municipal wastewater sludge by the Zimpro Company in the early 1960s. To date, many of WAO plants are in operation worldwide, treating wastewaters from petrochemical, chemical, and pharmaceutical industries as well as residual sludge from biological treatment plants [21].

In the past three decades, there have also been extensive research activities and publications in this area, and a number of comprehensive reviews are available for readers' references, as summarized in Table 1.1.

Table 1.1: Reviews on wet air oxidation (WAO) process [2, 4, 19, 21–31].

Author	Year	Highlights	Ref. no.
Mishra et al.	1995	Comprehensive review on mechanism, kinetics, and structure-oxidizability correlation for WAO of various organic compounds as well as WAO industrial applications	[22]
Debellefontaine et al.	1996	Kinetic study and mechanism of oxygen transfer and balance during WAO	[4]
		Wet peroxide oxidation of phenol	
Imamura	1999	Relationship between reactivity and oxygen content in the organic molecules	[31]
		Cu salts and H_2O_2 as homogeneous catalysts	
		Oxides and supported metal oxides as solid catalysts	
Luck	1999	Technical features of commercial noncatalytic WAO	[19]
		Development of CWAO catalysts and processes	
Kolaczkowski et al.	1999	Kinetics and mass transfer study of WAO and various industrial designs and applications	[23]
Debellefontaine et al.	1999	Design and modeling of bubble column reactor for WAO	[24]
Debellefontaine et al.	2000	Kinetics and reactor design and industrial applications in Europe	[25]

Table 1.1 (continued)

Author	Year	Highlights	Ref. no.
Bhargava et al.	2006	Fundamental chemistry of WAO/CWAO for various organic compounds	[26]
		Catalyst developments for CWAO	
		Engineering aspects and reactor design of WAO/CWAO	
Levec et al.	2007	Development, mechanistic, and kinetic study of CWAO process	[21]
Hii et al.	2014	Hydrothermal technologies as pretreatment method	[27]
		Roles of WAO and thermal hydrolysis in sludge treatment	
Fu et al.	2014	WAO for treatment of dyes' wastewaters	[28]
Arena et al.	2015	Reaction mechanism and kinetics of homogeneous and heterogeneous CWAO	[2]
		Recent advances on design and development of supported noble metal and oxide catalysts.	
Jing et al.	2016	Catalyst development, reaction mechanism, and kinetics of CWAO	[29]
Luan et al.	2017	WAO for treatment of phenolic compounds and dyes	[30]

Figure 1.3 illustrates a typical WAO treatment with a bubble column reactor. The wastewater is first pumped to reach the operating pressure of the reactor and mixed with compressed air. This mixture of gas and liquid is pumped to the bubble column reactor in which the oxidation takes place. Preheating of the feed stream may be necessary to raise the temperature of wastewater, which can be done by exchanging heat with the hot treated effluent. The reactor is typically adiabatic and the feed temperature is normally adjusted such that the exothermic heat of the reaction increases the mixture temperature to the operating temperature. The reactor effluent can be subsequently cooled with cooling water before being sent to the separator to separate the gas and liquid phases. The gases are then treated to further reduce the concentration of organic matter. The liquid phase can be disposed of directly or most often subjected to further biological treatment, depending on the oxidation level [22].

In general, WAO takes place at high reaction temperatures (473–593 K) and pressures (20–200 bar). The residence times of the liquid phase in the reactor range from 15 to 120 min, with typical COD removal of about 75–90% [21]. As the oxidation of the organic compounds involves three-phase reaction, there are two main stages: (1) physical stage, during which oxygen is transferred from gas to the liquid phase, and (2) chemical stage, which involves the reaction between the dissolved oxygen (or an active species formed by oxygen) and the organic compound [26].

Transfer of oxygen from the gas phase to the liquid phase has been described in detail by Debellefontaine et al. [4, 24], and the rate of oxygen transfer can be enhanced by improving mixing efficiency through careful reactor design and geometry [26]. Overall, the extent of oxidation is dependent on several important factors, such as temperature, oxygen partial pressure, residence time, pH, and the nature of the pollutants. At high temperatures and pressures, solubility of oxygen gas in aqueous solutions increases, thus providing a strong driving force for oxidation. In addition, elevated pressures are also required to keep water in the liquid state while high temperature is necessary to achieve high reaction rate and thus greater extent of oxidation. On the other hand, pH of the solution will determine the formation of different types of free radical species and intermediates involved in the oxidation process, thus affecting the effectiveness of organic oxidation.

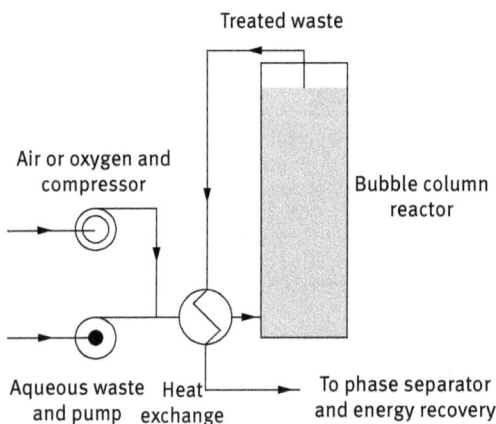

Figure 1.3: Schematic diagram of WAO process flow. Reprinted from ref. [25] with the permission from Elsevier.

In order to further improve WAO performance, a better understanding of the mechanism and the types of reactions involved during the oxidation period is beneficial. The general consensus is that the chemical reactions of WAO mostly take place via free-radical chemical reactions [32–34], although very few direct experimental evidences of the free radicals are available. Bhargava et al. discussed in detail numerous possible free-radical reactions from each of the three main types of reactions (initiation, propagation, and termination) that occur during WAO of various organic compounds as observed and proposed by many researchers in this field [26].

In general, noncatalytic WAO is exothermic and the reaction rate expression can be described by the following kinetic model [23]:

$$r = Ae^{\left(-\frac{E_A}{RT}\right)} \left(C_{org}\right)^m \left(C_{O_2,L}\right)^n$$

where r is the reaction rate, A the pre-exponential factor, E_A the activation energy, R the universal gas constant, T the reaction temperature, C_{org} the organic compound concentration in the bulk liquid, and $C_{O_2,L}$ the dissolved oxygen concentration in the bulk liquid. The superscripts m and n are the orders of reaction, with m is usually 1, whereas n is typically ~0.4. The typical temperature dependence of organic removal profile is illustrated graphically in Figure 1.4.

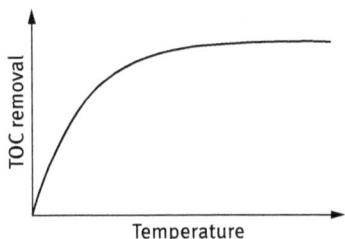

Figure 1.4: TOC removal profile with regard to temperature during WAO process. Reprinted with permission from [26]. Copyright (2006) American Chemical Society.

For more complex mixture of organic compounds, more detailed kinetic models are required to explain the effects of the main reaction parameters on the rate of WAO. To simplify, the organic decomposition can be assumed to occur via two pathways: (i) deep and complete direct oxidation to carbon dioxide and water, and (ii) through the formation of intermediate compounds, which further undergo hydrolysis to form relatively refractory species such as acetic acid before eventually being oxidized to carbon dioxide [26, 27]. Based on this simplified scheme, several kinetic models have been proposed, such as those developed by Li et al. [35] who considered the generation of acetic acid intermediate to be the rate-limiting step, as presented in Figure 1.5. This model is widely used as it shows a good fit with experimental data. Zhang et al. [36] proposed lumped kinetic model by considering all of reactants concentrations in wastewater and those of the intermediates formed during the reaction as a lumped concentration of total organic carbon (TOC), whereas Imteaz et al. [37] simplified the

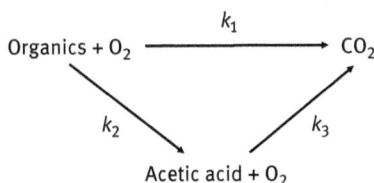

$$\frac{[\text{Org} + \text{acetic acid}]}{[\text{Org} + \text{acetic acid}]_0} = \frac{k_2}{k_1 + k_2 - k_3}e^{-k_3 t} + \frac{k_1 - k_3}{k_1 + k_2 - k_3}e^{-(k_1 + k_2)t}$$

Figure 1.5: Generalized kinetic model for wet air oxidation [35].

reaction scheme in terms of solubilization and oxidation of all COD in wastewater. Here, it is assumed that any solid COD must first solubilize via hydrolysis prior to contact with the oxidants and final oxidation.

1.3 Catalytic wet air oxidation

As mentioned previously, a significant drawback of the WAO process is incomplete mineralization of organics, since some low-molecular-weight oxygenated compounds, which are initially present in the wastewater or generated during the oxidation process, are resistant to further transformation to carbon dioxide [21]. For example, the removal of acetic acid is usually negligible at temperature less than 300 °C. Furthermore, organic nitrogen compounds are mostly transformed to ammonia, which is stable at WAO operating conditions. In this case, WAO process can be very expensive to achieve deep oxidation of all organic compounds. For this reason, WAO can be alternatively considered as a pretreatment step of wastewaters, which requires additional treatment, typically in a conventional biological treatment plant. In addition, high temperatures and pressures are usually required to achieve a high degree of oxidation within a reasonable length of time.

The performance of WAO can be considerably improved by introducing catalysts, either homogeneous or heterogeneous catalysts. Compared to conventional WAO, catalytic wet air oxidation (CWAO) has lower energy requirements. With the presence of a catalyst, one can use less severe reaction conditions to reduce organic compounds to the same degree as that of noncatalytic process. Moreover, recalcitrant acetic acid and ammonia will also be more susceptible to further oxidation [21].

With these motivations, CWAO has been investigated extensively in the past decades and it is believed as one of the most economical and environmentally friendly AOPs for treatment of refractory organic pollutants in industrial wastewaters. It is estimated that by employing CWAO, the operating cost of noncatalytic wet oxidation process can be reduced by 50% due to much milder operating conditions and shorter residence time [38]. The development of commercial CWAO processes started as early as the mid-1950s in the United States [39]. Several Japanese companies developed heterogeneous catalysts based on precious metals deposited on titania or titania–zirconia oxides. Whereas in Europe, homogeneous catalysts, such as soluble transition metal catalysts based on iron or copper salts, have been successfully commercialized in the past two decades (e.g., Ciba-Geigy, LOPROX, WPO, ORCAN, and ATHOS processes) [21]. Table 1.2 summarizes some examples of homogenous catalysts used for removal of various organic compounds and industrial effluents.

Despite their high effectiveness and efficiency, homogeneous catalysts require an additional separation step to remove or recover the metal ions from the treated

Table 1.2: Catalytic wet air oxidation using homogeneous catalysts.

Catalyst	Organic type	Reaction conditions	TOC removal	Ref.
$CuSO_4 + H_2O_2$	Phenol	448 K, 0.4 MPa O_2	100% in 20 min	[113]
$MnSO_4 + H_2O_2$	Phenol	448 K, 0.4 MPa O_2	100% in 30 min	[113]
$Cu(NO_3)_2$	Phenol	323 K, 1.16 MPa O_2	90% in 60 min	[114]
Fe-salt	Phenol	423 K, 5.1 MPa O_2	100% in 50 min	[115]
$FeSO_4$	Phenolic compounds	413 K, 1 MPa O_2	90% in 200 min	[116]
$Cu(NO_3)_2$	Phenol	473 K, 1 MPa O_2	100% in 60 min	[117]
$Ni(NO_3)_2$	Phenol	473 K, 1 MPa O_2	100% in 60 min	[117]

effluent, hence increases operational costs. This issue can be overcome by developing heterogeneous catalysts. Various solid catalysts for CWAO processes have been developed and are widely available, primarily based on either metal oxides [40–68], supported precious metals [69–93], or carbonaceous materials [12, 94–110]. They have been applied for the removal of organic pollutants, in particular phenols and substituted phenols, carboxylic acids, ammonia, dyes, and some N-containing compounds such as aniline and dyes, which have been comprehensively reviewed by Kim et al. [111] and Stüber et al. [112]. Different from homogeneous catalytic reaction where the dissolved oxygen can be readily used for oxidation, heterogeneous catalytic reaction requires adsorption and activation of molecular oxygen from the liquid phase onto the solid catalyst as illustrated in Figure 1.6.

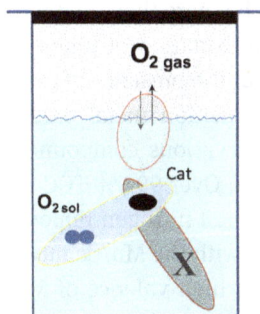

Figure 1.6: Mass transfer of oxygen from gas phase to the active site of the solid catalyst. X indicates the organic molecule to be oxidized. Reprinted from ref. [56] with permission from Elsevier.

1.3.1 Metal oxide catalysts for CWAO

In general, catalytic property of metal oxides is primarily attributed to redox cycle between two oxidation states, which makes transition metals suitable as oxidation catalysts. Among the family of metal oxides, Cu-based catalysts are commonly used for CWAO due to their high activity. Kim and Ihm studied the CWAO of phenol over transition metal (Cu, Ni, Co, Fe, and Mn) oxide catalysts supported on Al_2O_3 [45] and found that CuO_x/Al_2O_3 catalyst showed the highest catalytic activity due to its highest surface reducibility. Fortuny et al. conducted a lifetime test of a mixed oxide containing 2% CoO, Fe_2O_3, MnO, or ZnO and 10% CuO supported on γ-alumina [83]. Despite high initial phenol removal of 99% by Cu/Mn oxide in the first 6 h, its activity considerably decreased to 20% after 120 h of reaction. The decreasing activity of the catalyst was assumed to be due to the dissolution of the metal oxides from the catalyst surface under hot acidic environment. Another study by Zapico et al. also demonstrated high catalytic performance of CuO_x/Al_2O_3 catalyst in terms of both the catalytic activity and the mineralization selectivity [66]. However, it was found that the leached copper ions also contributed significantly to the high conversion of phenol, as homogeneous reaction is considerably faster than the heterogeneous one. These evidences show that more robust and stable metal oxide catalysts need to be developed, and attempts made to improve the stability of CWAO catalysts include exploring various supports for copper catalysts as well as finding novel alternative oxides. A selection of transition metal oxide catalysts that have been developed in the past decade is presented in Table 1.3.

Imamura first explored the activity of various metal oxides on the oxidation of refractory acetic acid and found that Bi/γ-Al_2O_3, Co/Bi, Co/Bi/γ-Al_2O_3, Sn/Bi, and Zn/Bi were active [31]. High activity of Bi-based catalyst could be due to the basic nature of Bi, which has high affinity and adsorption toward carboxylic acids. The results show that Co/Bi catalyst with 16 mol% Bi (Co/Bi 5/1) has the highest activity. At 1 MPa of O_2 and 248 °C, the catalyst exhibits high initial TOC removal (ppm/min) of acetic acid, about seven times that in the absence of catalyst. In particular, when propionic acid is used as organic substrate, this value could be more than 20 times. On the contrary, the effect of catalyst is incremental on the oxidation of phenolic compounds and formic acid, as they decompose easily without the presence of catalyst. Further experiments with various lanthanides and transition metal oxides showed that Mn/Ce oxide was highly active for oxidation of various compounds, including N-containing species, such as amine and ammonia. Over 99.5% TOC removal was observed for acetic acid at 247 °C, whereas 70% total nitrogen removal was observed for ammonia at 263 °C on Mn/Ce oxide catalyst with the Mn/Ce molar ratio of 7/3. The presence of ceria was proposed to maintain high valence of Mn, which increases its oxidation ability.

Further literature studies show that manganese–cerium mixed oxides have been identified as one of the most effective CWAO catalysts for the oxidation of various organic compounds, such as ammonia [130], pyridine [131], phenol [122, 123, 132], acrylic acid [127], and ethylene glycol [62]. The primary role of ceria is likely to be a good medium for oxygen activation and transport, which is associated with the formation of oxygen defect sites and oxygen activation. On the other hand, the incorporation of Mn into ceria lattice was found to greatly improve the oxygen storage

Table 1.3: Development of transition metal oxide catalysts for CWAO of organic compounds.

Catalyst	Organic type	Reaction condition	Organic removal	TOC removal	Ref.
$Ce_{0.75}Fe_{0.25}O_2$	4-Chlorophenol	80 °C, atmospheric pressure	100%	60%	[118]
$Ce_{0.75}Fe_{0.25}O_2$	2,4-Dichlorophenol	80 °C, atmospheric pressure	100%	74.30%	[118]
Mo-Cu-Fe-O composite	Organic dyes	ambient condition	>90% in 1 h	N.A.	[119]
MnO_x–CeO_2 nanorods	Phenol	140 °C, 2 MPa air	100% in 1 h	98% in 1 h	[120]
5%Cu/activated carbon	Phenolics	150 °C, 0.8 MPa O_2, pH ~13	90% in 4 h	60% in 4 h	[121]
1.64%Cu/SBA-15	Aniline	70 °C, 0.05 mol L^{-1} H_2O_2, pH 8	100% in 3 h	66.9% in 3 h	[67]
6%Fe/zeolite-13X	1,4-Dioxane	400 °C, O2/dioxane = 8 (mol)	97% at GHSV 24,000 h^{-1}	95% at GHSV 24,000 h^{-1}	[58]
$MnCeO_x$ (Mn/Ce = 1)	Acetic acid	150 °C, 0.9 MPa O_2 (pH 3–7)	100% in 6 h (Cat/acid = 5)	100% in 6 h (Cat/acid = 5)	[122]
$MnCeO_x$ (Mn/Ce = 1)	Formic acid	150 °C, 0.9 MPa O_2 (pH 3–7)	100% in 1 h (Cat/acid = 5)	100% in 1 h (Cat/acid = 5)	[122]
MnO_2/CeO_2	Methylamine	190 °C, 5 MPa	100% in 2 h	75% in 2 h	[54]
$Ce_{0.85}Zr_{0.15}O_2$	Bisphenol A	160 °C, 2 MPa air	76% in 3 h	75% in 3 h	[55]
7%Ni/MgAlOₓ	Crystal violet	180 °C, 5 MPa air	85% at 350 h on stream	64% at 350 h on stream	[52]
$MnCeO_x$ (Mn/Ce = 1)	Phenol	150 °C, 0.9 MPa O_2 (pH 4–6)	100% in 1 h (Cat/Ph = 2)	90% in 1 h (Cat/Ph = 2)	[123]
$CuO_x/Ce_{0.65}Zr_{0.35}O_2$	Phenol	150 °C, 2 MPa air	100% in 1 h	70% in 1 h	[48]
$MnO_x/Ce_{0.65}Zr_{0.35}O_2$	Phenol	150 °C, 2 MPa air	85% in 1 h	80% in 1 h	[48]
CeO_2-TiO_2 (Ce/Ti = 1)	Phenol	150 °C, 3 MPa air	100% COD in 2 h (batch)	77% in 2 h (batch)	[124]
$LaFeO_3$	Salicylic acid	140 °C, 0.5 MPa O_2	87% in 2 h	84% in 2 h	[47]
CeO_2-TiO_2 (Ce/Ti = 1)	Acetic acid	230 °C, 5 MPa	N.A.	64% COD in 3 h	[125]
Cu/MCM-41	Aniline	200 °C, 0.69 MPa O_2	96% in 2 h	76% in 2 h	[63]
$5%Cu/Al_2O_3$	Phenol	150 °C, 2 MPa air	100% in 1.5 h	80% in 1.5 h	[45]
$MnCeO_x$	Ethylene glycol	200 °C, 1.5 MPa O_2	100% in 1 h	99% in 1 h	[62]
$CeO_2/γ$-Al_2O_3	Phenol	180 °C, 1.5 MPa O_2	100% in 2 h	80% in 2 h	[126]
$MnCeO_x$ (70% Mn)	Acrylic acid	200 °C, 1.5 MPa O_2 (pH 4–5)	100% in 2 h	97.7% in 2 h	[127]
5%Cu/activated carbon	Phenol	180 °C, 3.6 MPa (pH 5–6)	100% in 1 h	N.A.	[60]
$16%Fe/CeO_2$	p-Coumaric acid	130 °C, 2 MPa air	N.A.	90% in 10 min	[41]
K-Mn-Ce oxide	Phenol	110 °C, 0.5 MPa O_2	100% in 20 min	99.2% in 20 min	[128, 129]

N.A., data is not available.

capacity (OSC) and the oxygen mobility on the catalyst surface, as well as tuning the acid–base properties of the mixed oxides [56, 133]. Ma et al. reported that the synergy between Mn and Ce could further improve the catalyst redox properties and produce a larger amount of active oxygen species [120]. Figure 1.7 shows the synthesized $MnCeO_x$ catalyst with nanocubes and nanorods morphology. It was observed that CeO_2 nanorods displayed superior catalytic activity than the cubic counterparts. This is because the former contains more surface active oxygen than the latter; therefore, more oxygen vacancies are formed on the CeO_2 nanorods' surface. The presence of oxygen vacancies would improve the reducibility of the catalyst surface and increase the productivity of active oxygen species [120]. Moreover, the incorporation of Mn into CeO_2 lattice can enhance the phenol adsorption ability and thus further promotes the oxidizing ability of the catalyst.

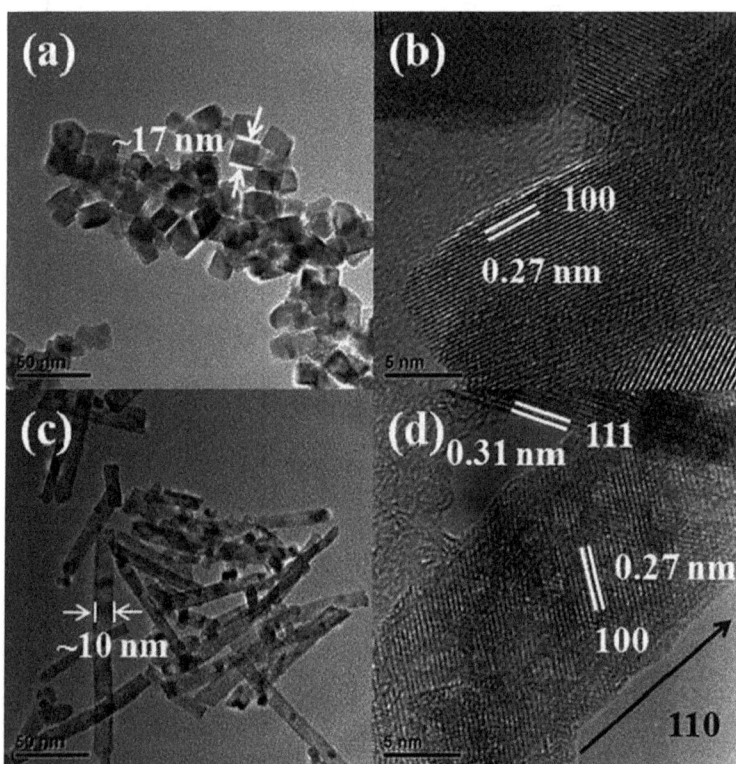

Figure 1.7: TEM images of $MnCeO_x$ nanocubes (a and b) and nanorods (c and d). Reprinted from ref. [120] – Published by The Royal Society of Chemistry.

Recent works on the oxidation of phenol by Arena et al. over Ce–Mn oxide showed that the textural and redox properties of the catalyst are important aspects for CWAO performance [122, 123, 132] and can influence both the reaction kinetics and mineralization of phenol. In the study conducted by Arena et al. [122], a novel redox-precipitation route was used for catalyst preparation, which resulted in the formation of highly dispersed MnCeO$_x$ catalysts with improved physicochemical properties. Irrespective of the Mn/Ce ratio, this preparation route ensured a large surface area, a uniform pore size distribution, and a monolayer active phase dispersion, strongly promoting the catalytic performance for the CWAO of phenol. The performance of the catalyst was evaluated and a systematic study on the reaction kinetic and mechanism was presented, demonstrating the superior activity of MnCeO$_x$. At 150 °C and 0.9 MPa of O$_2$, CuCeO$_x$ could achieve a phenol disappearance after 3 h whereas it took only 20 min over MnCeO$_x$ [132]. On the other hand, the TOC removal of the former was only approximately 70% while the latter was over 90% within the same period of reaction, indicating high CWAO efficiency toward phenol and C$_1$–C$_2$ intermediate organic acids in terms of adsorption and mineralization rates. Figure 1.8 shows the differences of homogeneous and heterogeneous CWAO processes by the activity–selectivity relationships [2]. The linear correlation of phenol and TOC conversion reflects the selective characteristic of the surface adsorption on the solid catalyst, while the exponential trend shows the unselective behavior of homogeneous systems. Mechanistic and kinetic study on the catalyst suggests that CWAO of phenol proceeds via a Langmuir–Hinselwood mechanism, involving adsorption as first reaction step and surface oxidation as rate determining step, while adsorption and mineralization steps of acidic substrates proceed at a comparable rate [122].

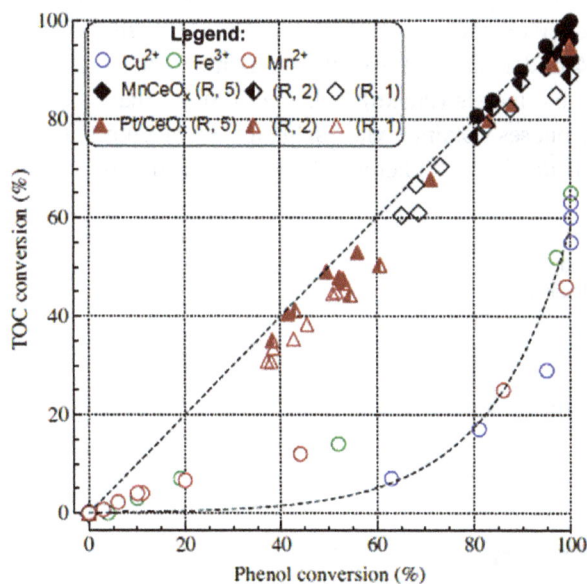

Figure 1.8: TOC versus phenol conversion over homogeneous and heterogeneous oxide catalysts. Reprinted from ref. [2] with permission from Elsevier.

Posada et al. studied CWAO process for oil field produced water containing refractory organic pollutants with TOC content around 1,200–1,400 mg L^{-1} [68]. The oxidation was carried out using 4 wt% Cu supported on Mn–CeO$_x$ catalyst. Approximately 80% of TOC reduction was achieved at 433 K and oxygen partial pressure of 1 MPa with no leaching of Cu or Mn was detected. It is believed that the relative abundance of the lattice oxygen species in the mixed Mn–Ce oxide is responsible for the catalytic activity due to the activation of molecular oxygen via oxygen transfer mechanism. Further study on the adsorption of O$_2$ gas on the mixed oxide surface suggested the formation of superoxide (O$_2^-$) and peroxide (O$_2^{2-}$) intermediate species, which are more oxidative than molecular O$_2$, as depicted in Scheme 1.1. The FTIR analysis indicated that the superoxide species was formed on the CeO$_2$ surface defects (the coordinately unsaturated Ce^{4+} ions) immediately after the introduction of O$_2$ and was successively converted into O$^-$ and O^{2-} by accepting electrons from the oxide surface:

$$O_{2\,ads} \rightarrow O_{2\,ads}^- \rightarrow O_{2\,ads}^{2-} \rightarrow 2O_{ads}^- \rightarrow 2O_{lattice}^{2-}$$

Scheme 1.1: Oxygen adsorption mechanism on Mn–Ce oxide surface. Adapted from ref. [68] with permission from Springer Nature.

In general, CWAO over solid catalysts can result in three different routes as depicted in Scheme 1.2: (i) partial oxidation to soluble oxygenates, (ii) deposition (ca. reactive adsorption) of carbonaceous products on the catalyst surface, and (iii) complete mineralization to CO$_2$ [2]. Abecassis-Wolfovich et al. prepared Mn–Ce-based catalysts with Mn/Ce atomic ratio of 6:4 under different activation conditions and promoted with alkali metals (K, Cs) or noble metals (Pt, Ru) [43]. It was found that catalysts containing mixed Mn$_3$O$_4$–CeO$_2$ phases obtained via calcination in vacuum (350 °C) and promoted with alkali metals displayed higher activity and adsorption of organic

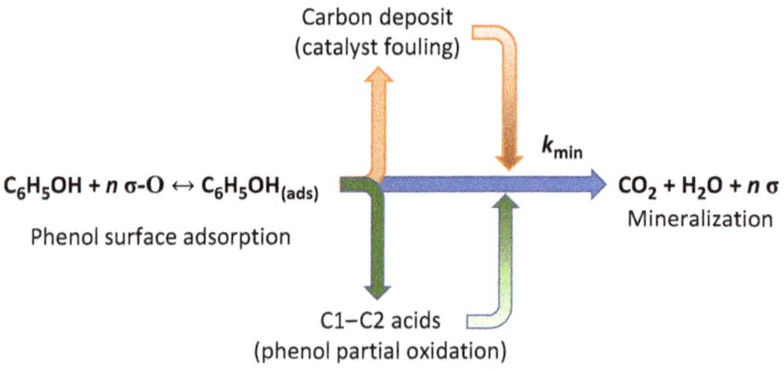

Scheme 1.2: Reaction scheme for CWAO of phenol over solid catalyst. Adapted from ref. [2] with permission from Elsevier.

deposits on their surface than those promoted with noble metals. At temperature as low as 100 °C and 10 bar of O_2 partial pressure, Mn–Ce–Cs oxide exhibits 82% phenol conversion and 75% TOC removal with around 90% of carbon being deposited on the catalyst surface. It was claimed that low-temperature oxidative regeneration of the catalysts at 270 °C showed full recovery of its activity and adsorption capacity in three consecutive runs, superior to activated carbon regeneration, which is usually conducted at higher temperatures and results in partial loss of adsorption ability.

Another work by Larachi et al. reported that addition of K onto the co-precipitated $MnCeO_x$ (atomic Mn/Ce = 1) enhanced the catalytic activity and stability, resulting in complete removal of phenol and over 95% TOC conversion after 10–20 min of reaction at 110 °C and 0.5 MPa of O_2 [128]. Santiago et al. also investigated the effect of potassium impregnation (up to 10%) and showed that MnO_2–CeO_2 impregnated with 1 wt% of K showed the highest efficiency, achieving 90% TOC conversion within 60 min at 130 °C and 20.4 atm [134]. It was observed that further increasing the amount of impregnated potassium adversely affected the phenol conversion due to the decrease in surface area. It was proposed that potassium increases the production of peroxide species on the catalyst surface due to its electron-donating ability, which promotes deep oxidation and mineralization [128, 129].

Apart from Mn, composites of CeO_2 with other various transition metal oxides have also been reported, such as CeO_2/Al_2O_3 [65, 126], CeO_2/ZrO_2 [56, 135], and CeO_2/TiO_2 [50, 124, 125, 136]. The combination of ceria with other metal oxides modulates the redox and acid–base properties of the catalysts. For example, although zirconia is more acidic than ceria, their mixed oxides possess more basic sites, which are stronger than those in pure ceria [56]. In addition, the mixed oxide also exhibits higher OSC [137]. It was observed that both redox and acid–base properties can affect the selectivity of the reaction by changing the adsorption mode of the organic molecules on the catalyst surface.

Chen et al. investigated the activities of CeO_2 catalysts supported on several oxides, such as γ-Al_2O_3, SiO_3, $AlPO_4$-5, and TiO_2 for the CWAO of phenol [126], and found that $CeO_2/γ$-Al_2O_3 catalyst had the most abundant exchangeable structure oxygen species and the highest reducibility as verified by O_2-TPD and H_2-TPR, respectively, which consequently showed the highest activity. With the optimal 20 wt% of Ce, about 100% phenol conversion and 80% COD removal were achieved after 2 h reaction at 180 °C and 1.5 MPa O_2 partial pressure. Increasing Ce content above 20 wt% resulted in the decreasing COD removal due to the formation of larger CeO_2 crystals. In the subsequent work, various promoters such as Cu, Mn, Co, Cr, Fe, or Ni were added into $CeO_2/γ$-Al_2O_3 via incipient wetness impregnation, and it was found that Cu-promoted catalyst rendered the most effective performance [65]. At 180 °C, the catalyst with 5 wt% Cu-15 wt% Ce exhibited complete removal of phenol after 1 h of reaction and 95% removal of COD after 2 h, higher than that of $CeO_2/γ$-Al_2O_3. On the contrary, only 30% phenol conversion could be achieved by Mn-promoted catalyst.

Despite its superior performance, it was found that the leached Cu ions contributed for the majority of phenol mineralization, hence caused a severe decline in activity and posed additional secondary treatment for Cu recovery.

Yang et al. reported that the CeO_2/TiO_2 mixed oxides prepared by co-precipitation method showed higher activity than that by pure CeO_2 and TiO_2 [124]. The mixed oxide with Ce/Ti molar ratio of 1:1 exhibited the highest activity for the CWAO of phenol with over 91% COD and 80% TOC removal at 140 °C and air total pressure of 3.5 MPa. The catalytic activity was maintained throughout 100 h of continuous reaction during which the metal leaching was negligible. Wang et al. applied CeO_2/TiO_2 catalyst for the oxidation of fomesafen-containing wastewater, a type of herbicide [136]. The effects of temperature, oxygen partial pressure, and initial pH on the COD removal efficiency were investigated, and the kinetics of fomesafen oxidation was obtained, The results showed that with Ti/Ce of 3:1, the COD removal rate reached 78% at 250 °C, 3 MPa oxygen partial pressure, and 3 h reaction time. The ratio of biochemical oxygen demand (BOD5) to COD increased from 0.03 to 0.32 after 3 h, which indicated that CWAO can effectively improve the biodegradability of wastewater of fomesafen production.

CeO_2-ZrO_2 mixed oxide ($Ce_{0.65}Zr_{0.35}O_2$) prepared by Kim et al. via supercritical synthesis was used as support for various transition metals, such as Mn, Fe, Co, Ni, and Cu, in view of higher thermal stability, better OSC, and higher surface area, which makes a promising support for wet oxidation of phenol due to its excellent redox properties [48]. $CuO_x/Ce_{0.65}Zr_{0.35}O_2$ was found to be the most effective catalyst with both high catalytic activity and CO_2 selectivity due to its highly dispersed Cu^{2+} clusters. However, the leached copper ions (58.3 mg L^{-1}) contributed to the higher conversion of phenol. Substitution of Cu with Mn considerably decreased metal leaching while yielding a high conversion of phenol and TOC. However, the oxidized phenol mainly formed carbon deposits on the surface of catalyst, resulting in catalyst deactivation.

Parvas et al. also investigated the oxidation of phenol over copper catalysts supported on CeO_2-ZrO_2 that was prepared via sonochemistry method [135]. This preparation method produced small copper particles and excellent dispersion over the support surface. With 20 wt% of Cu on CeO_2-ZrO_2, the reaction was carried out under atmospheric pressure of oxygen at 160 °C in a stirred batch reactor, after which about 100 wt% phenol is removed within 180 min. According to Balcaen et al. [138], the addition of Cu on ceria increases the number of active surface lattice oxygen species, which leads to higher phenol conversion. The oxidation process was proposed to occur by reacting the surface lattice oxygen species ($O_{L,S}$) and/or the weakly adsorbed O_{ads} species with phenol molecules. Meanwhile, the role of zirconia is to facilitate the reduction of Ce^{4+} to Ce^{3+} and to preserve the oxygen vacancies as the source of OSC.

The reaction mechanism of CWAO of phenol is illustrated in Figure 1.9 and involves the following reaction steps [135]:

1. Formation of adsorbed O_{ads} species by dissociative adsorption–desorption of gas phase O_2
2. Replenishment of the bulk lattice oxygen ($O_{L,B}$) by O_{ads} species
3. Diffusion of $O_{L,B}$ to the surface, forming active surface lattice oxygen species ($O_{L,S}$)
4. Oxidation of phenol by $O_{L,S}$ to CO_2 and H_2O
5. Spillover of the generated CO_2 from CuO and CeO_2 to CeO_2–ZrO_2 support
6. Desorption of CO_2 from CeO_2–ZrO_2 support
7. Alternatively, oxidation of phenol could also occur by reaction with weakly adsorbed O_{ads} to form CO_2

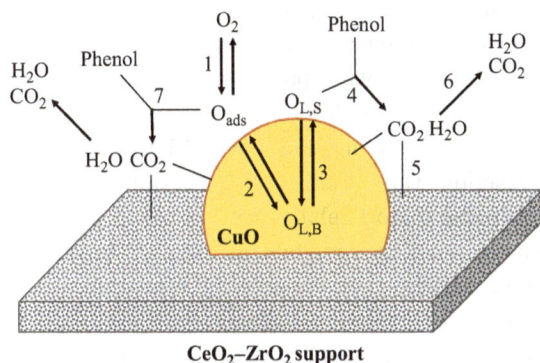

Figure 1.9: Proposed reaction mechanism of CWAO of phenol over Cu/CeO_2–ZrO_2 catalyst. Redrawn with reference to [135].

Further exploration on mixed oxides leads to mesoporous materials such as MCM-41 [63], SBA-15 [67], zeolites [58], and pillared clay [5, 139]. They could be a promising catalyst support for CWAO due to their uniform pore structure and large surface area. Copper catalysts supported on mesoporous SBA-15 were evaluated in the oxidation of aniline with H_2O_2 as oxidizing agent [67]. The effects of reaction conditions such as initial pH of the reaction mixture and reaction temperature on the aniline and TOC conversion were investigated. It was observed that weakly alkaline condition was favorable for aniline and TOC removal. In addition, the leaching Cu^{2+} ions could also be decreased from 45.0% to 3.66% when the initial pH of solution was increased from 5 to 10. At 70 °C and initial pH of 8.0, 100% aniline conversion and 66.9% TOC removal can be achieved after 180 min in the presence of 0.05 mol L^{-1} H_2O_2 as an oxidant.

In another work, Gomes et al. reported the modification of mesoporous MCM-41 with various transition metals (Cr, Cu, and V) via hydrothermal synthesis, and the catalytic activity was evaluated at 200 °C and 6.9 bar O_2 partial pressure [63]. Among the transition metal, Cu-MCM-41 showed a significant catalytic activity with aniline conversion of 96% and CO_2 selectivity of 76% after 2 h reaction. Although leaching of some Cu was observed, the catalyst showed a good performance and stability after three consecutive recycles.

Ramakrishna et al. prepared Zeolite-13X-supported Fe (Fe/zeolite-13X) catalysts with various Fe contents and they were subjected to oxidation of 1,4-dioxane using air as the oxidant in the temperature range of 100–400 °C [58]. The effect of various process parameters such as reaction temperature and metal loading on dioxane removal efficiency was investigated. It was found that catalyst with 6 wt% Fe exhibited the highest catalytic activity at 400 °C with 97% dioxane conversion and 95% CO_x selectivity (CO and CO_2). The material analysis reveals that the redox-active site density increases with increasing Fe loading up to 6 wt%, which favors complete oxidation of dioxane to carbon oxides. Above 6 wt% loading, aggregation of iron oxide species was observed, hence, decreasing the activity of the catalyst.

A novel hybrid material, $Na_2Mo_4O_{13}/\alpha\text{-}MoO_3$, was reported to be highly effective for CWAO of organic dyes (GTL) under ambient conditions [140]. With a catalyst to dye mass ratio as low as 0.83, complete oxidation of cationic red GTL was achieved at 1 bar and 30 °C within 40 min of reaction. The material characterizations showed that the presence of $Na_2Mo_4O_{13}$ within the $\alpha\text{-}MoO_3$ phase was crucial to generate more interfaces and O^{2-} ions in the oxygen-deficient regions. As these O^{2-} ions have higher mobility than the typical lattice oxygen, they promote the formation of OH radicals that actively take part in the catalytic oxidation reactions.

In general, with many combinations of metal oxide composition, mixed oxides remain potential catalysts for further development and optimization to achieve more efficient organic removal as well as optimum operating conditions. It can be concluded that their catalytic activity is mainly dependent on the redox property and OSC, which determine the rate of lattice oxygen activation on the oxide surface. However, the issues of the metal dissolution and surface accumulation of carbon-containing species remain a drawback that needs to be resolved.

1.3.2 Supported noble metal catalysts for CWAO

Noble metal-based catalysts, on the other hand, despite being much more expensive than the transition metal oxides, are generally more stable and more active. They are traditionally supported on metal oxides, such as Al_2O_3, CeO_2, TiO_2, ZrO_2, or SiO_2, as well as on activated carbon. The metal contents typically range from 0.1 to 5.0 wt%. The role of the oxide supports is important as they provide large surface area for better dispersion of metal nanoparticles as well as oxygen source for the oxidation reaction. In particular, due to their high capability for C–C bond cleavage, they are effective in the degradation of refractory intermediates. Among the frequently studied noble metals (Ir, Pd, Pt, Rh, and Ru), platinum, ruthenium, and palladium are reported to be the most effective ones and the discussion will be focused more on these catalysts. For the readers' reference, a summary of supported noble metal catalysts is presented in Table 1.4.

Monteros et al. studied the catalytic performances of Ru (1.25 wt%) and Pt (2.5 wt%) supported on TiO_2-(x wt%)CeO_2 for CWAO of phenol at 160 °C and 20 bar of oxygen pressure [79]. The amount of ceria was varied to study the effect of OSC on phenol conversion, of which maximum OSC was observed for catalyst with 50 wt% ceria. The stability of the catalysts was confirmed since no leaching of Ru and Pt was detected. The catalytic results revealed different behaviors of the two metals: 100% of phenol conversion could be achieved in 2 h on Ru/CeO_2 and 30 min on Pt/CeO_2 with ceria content between 0 and 25 wt%. Figure 1.10 shows the carbon distribution after 3 h reaction. The amounts of adsorbed compounds and insoluble polymers in solution are important to determine the effectiveness of the catalyst in mineralization of the organic pollutant. It is apparent that the platinum catalysts (Figure 1.10, bottom) are more efficient than their ruthenium counterparts, given the higher CO_2 selectivity for the former. Contrary to Ru catalysts, the loading of ceria has a direct effect on the catalytic performance of Pt catalysts. The best performance was obtained on Pt catalyst containing 0–10 wt% CeO_2, resulting in 88% of CO_2 selectivity. Interestingly, at higher CeO_2 loadings, the formation of polymers in solution and the carbon deposit became more prominent, which is detrimental to catalytic performance.

The surface acidic properties of both Ru and Pt catalysts were determined through pyridine adsorption experiment, which allows a clear identification of Brønsted and Lewis acid sites. The surface acidity measured for all of the samples showed primarily Lewis acid sites as presented in Figure 1.10. It was observed that increasing ceria loading from 0 to 25 wt% decreases the number of Lewis acid sites and CO_2 selectivity. This indicates that Lewis acid sites could minimize the formation of polymers and carbon deposit and instead promote CO_2 formation. However, with further increase in ceria loading to 50 wt%, the positive effect of increasing Lewis acid sites was counteracted by increase in OSC value, which results in a significant formation of carbon deposit. This could be explained by different phenol orientation on the catalyst surface during oxidation reaction (Figure 1.11). The orientation would be governed by the redox and acid–base properties of the catalyst [56, 79]. High OSC would favor *para*-orientation of phenol, leading to *p*-benzoquinone and eventually to the formation of polymeric products and accumulation of adsorbed carbon species [79]. On the other hand, Lewis acid sites could activate the electronic doublet of oxygen of the hydroxyl function and thus enhance the *ortho*-oxidation of phenol leading to catechol formation and finally to CO_2.

Yang et al. used $Ce_xZr_{1-x}O_2$ mixed oxides as support for Ru and Pt for oxidation of succinic acid at 150 °C and 50 bar total pressure [81]. The catalytic activity of these solids was demonstrated to be strongly influenced by the Ce content. The higher the Ce content in the mixed oxide support, the higher the succinic acid removal rate would be. In particular, pure CeO_2 and $Ce_{0.9}Zr_{0.1}O_2$ catalyzed 100%

Table 1.4: Development of supported noble metal catalysts for CWAO of organic compounds.

Catalyst	Organic type	Reaction condition	Organic removal	TOC removal	Ref.
Ru/Ti$_{0.9}$Zr$_{0.1}$O$_2$	Aniline	180 °C and 1.5 MPa O$_2$	100% in 5 h	88% in 5 h	[141]
Ru/CeO$_2$ (added alcohol)	Butyric acid	180 °C and 0.8 MPa O$_2$	N.A.	90.13% COD in 5 h	[82]
1.25% Ru/CeO$_2$	Phenol	160 °C and 2 MPa O$_2$	100% in 2 h	45% CO$_2$ selectivity	[79]
2.5% Pt/TiO$_2$–(10%)CeO$_2$	Phenol	160 °C and 2 MPa O$_2$	100% in 0.5 h	88% CO$_2$ selectivity	[79]
0.5% Pt/Ce$_{0.9}$Zr$_{0.1}$O$_2$	Succinic acid	190 °C and 5 MPa	100% in 6 h	65% in 6 h	[81]
0.5% Ru/Ce$_{0.9}$Zr$_{0.1}$O$_2$	Succinic acid	190 °C and 5 MPa	100% in 6 h	93% in 6 h	[81]
5%(Pt–Ru)/TiO$_2$ (Pt/Ru = 1)	Methylamine	200 °C	100% at LHSV 5.4 h^{-1}	100% at LHSV 5.4 h^{-1}	[80]
1.25% Ru/Zr$_{0.1}$Ce$_{0.9}$O$_2$	Phenol	160 °C and 2 MPa O$_2$	100% in 3 h	80% in 3 h	[92]
1.5% Pt/Ce0.5Zr.5O2	Formic acid	40 °C	> 90% in 8 h	N.A.	[53]
0.08% Pt/Ce$_{0.9}$Zr$_{0.1}$O$_2$	Formic acid	53 °C and 0.1 MPa	100% in 6 h	N.A.	[88]
Pt/CeO2	Phenol	150 °C, 0.9 MPa O$_2$ (pH 4–6)	55% in 6 h (Cat/Ph = 2)	45% in 6 h (Cat/Ph = 2)	[123]
Rh/γ-Al$_2$O$_3$–(1%)CeO$_2$	Ethyl tert-butyl ether	100 °C and 1 MPa O$_2$	85% in 1 h	79% in 1 h	[87]
1% Pt/TiO$_2$-C	Methanol	220 °C and 4 MPa	92% at 180 h on stream	N.A.	[85]
6% Pt/CeO$_2$	p-Coumaric acid	80 °C and 2 MPa air	> 95% in 2 h	> 80% in 2 h	[142]
5% Pt/CeO$_2$	Stearic acid	200 °C and 2 MPa O$_2$	N.A.	95% in 3 h	[143]
3% Ru/mesoporous TiO$_2$	Succinic acid	190 °C and 5 MPa air	100% in 4 h	> 90% in 4 h	[73]
2.1% Pt/mesoporous TiO$_2$	p-Coumaric acid	190 °C and 5 MPa air	100% in 5 h	> 90% in 5 h	[73]
Pt/graphite	Phenol	165 °C and 1.7 MPa	>99% after 6 ks	98% after 6 ks	[144]
Pt/graphite	Maleic acid	150 °C and 1.7 MPa	>95% after 6 ks	>90% after 6 ks	[144]
5% Ru/CeO$_2$	Aniline	200 °C and 2 MPa O$_2$	100% in 3 h	85% in 3 h	[74]
1% Pt/C (wet impregnation)	Butyric acid	200 °C, 0.69 MPa O$_2$	75% in 8 h	N.A.	[145]
1% Pt/C (wet impregnation)	Propionic acid	200 °C, 0.69 MPa O$_2$	96% in 8 h	N.A.	[145]
1% Pt/C (wet impregnation)	Acetic acid	200 °C, 0.69 MPa O$_2$	93% in 8 h	N.A.	[145]
3% Ru/TiO$_2$	Kraft bleach plant effluent	190 °C, 5.5 MPa air	N.A.	>95% in 8 h	[146]
5% Ir/C	Butyric acid	200 °C, 0.69 MPa O$_2$	43% in 2 h	N.A.	[95]
5% Ir/C	Isobutyric acid	200 °C, 0.69 MPa O$_2$	52% in 2 h	N.A.	[95]

N.A., data is not available.

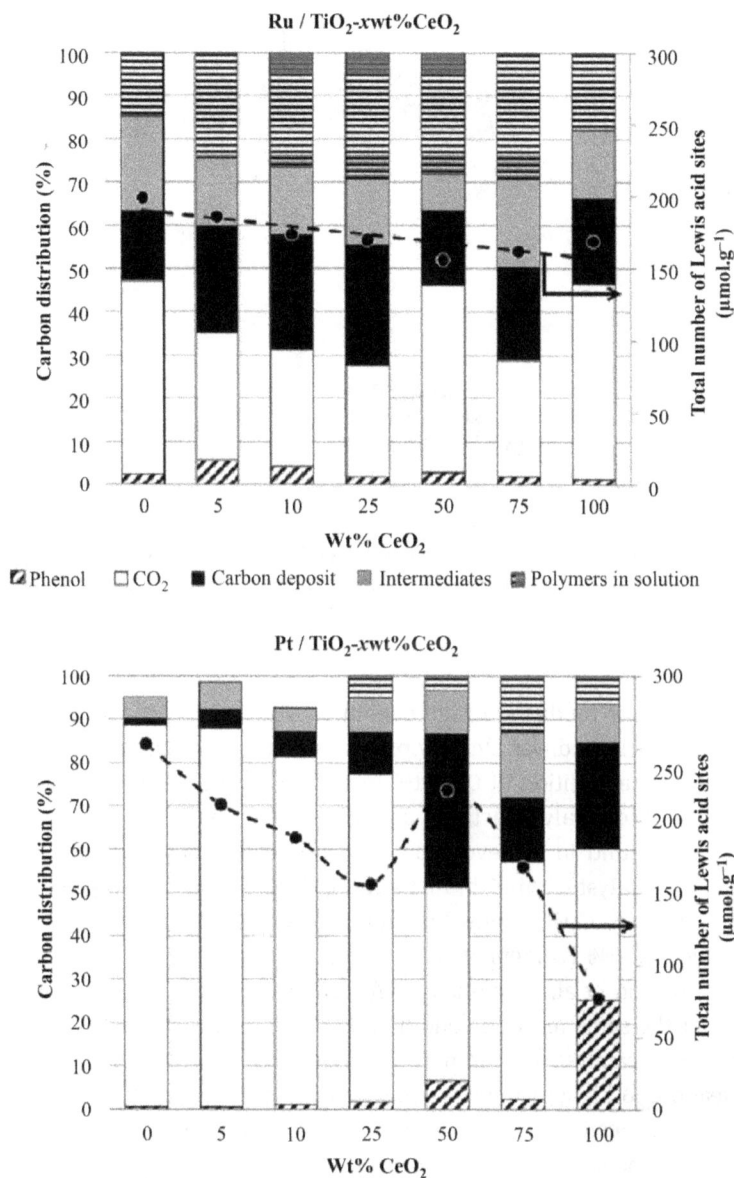

Figure 1.10: Carbon distribution after 3 h of phenol oxidation and total number of Lewis acid sites for Ru (top) and Pt (bottom) supported on $TiO_2-(x\ wt\%)CeO_2$. Reprinted from ref. [79] with the permission from Elsevier.

succinic acid conversion within 5 h. The evolution of the TOC content in the reaction mixture showed that succinic acid was not completely mineralized to CO_2 and H_2O. The main reaction intermediates were identified to be acetic and acrylic acids.

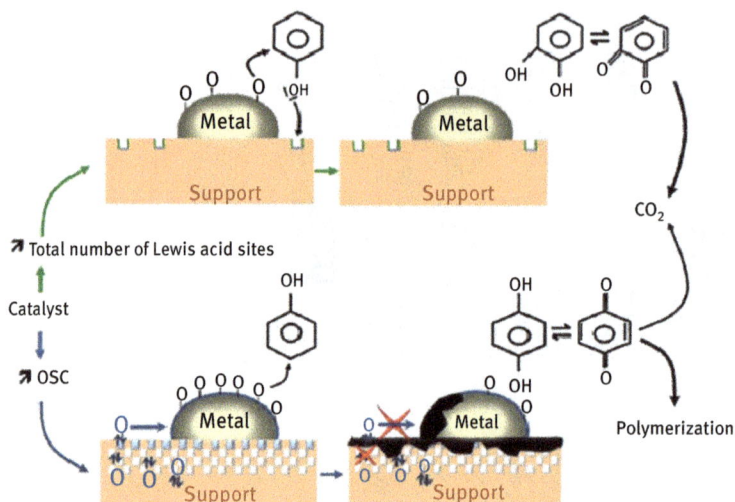

Figure 1.11: Conflicting effect of OSC and Lewis acid sites on the orientation of phenol on supported noble metal catalyst surface during oxidation reaction. Reprinted from ref. [79] with the permission from Elsevier.

Acrylic acid was transiently produced in the beginning of the reaction and rapidly removed. In contrast, acetic acid was steadily produced in the reaction mixture and remained recalcitrant. The addition of 0.5 wt% of Pt and Ru on the mixed oxides significantly increased the catalytic activity as well as TOC abatement. The supported Pt catalyst was found to achieve 100% succinic acid removal within 3 h, whereas supported Ru catalyst within 2 h. In addition, the C–C bond cleavage capability of Ru catalyst increased TOC removal to around 90% mineralization, whereas Pt catalyst only achieved 70% TOC removal.

As comparison, Besson et al. performed CWAO of succinic acid over gold supported on TiO_2 under the same reaction condition [84]. The results showed that the catalytic activity of gold catalyst was strongly dependent on the gold particle size, with smaller Au leading to higher turnover frequency and succinic acid conversion. The main by-product was acetic acid and a trace amount of acrylic acid. Substituting Au with Ru significantly increased the catalytic activity.

Ru is considered one of the most effective metals for CWAO of nitrogen-containing pollutants. However, the complexation reaction that occurs between the nitrogen atom and Ru may lead to Ru leaching, which is the main cause of catalytic deactivation, as reported by Grosjean et al. [147]. In order to improve the catalyst activity and stability, Song et al. prepared Ru catalyst (2 wt%) supported on TiO_2 doped with Zr or Ce by the impregnation method [141]. The catalyst was then used for CWAO of aniline. The amount of Zr or Ce was varied to obtain the maximum aniline conversion. The reaction was performed at 180 °C and 1.5 MPa total oxygen pressure, and it was found that $Ru/Ti_{0.9}Zr_{0.1}O_2$ was the most active with complete aniline conversion and over 88% of COD removal achieved within 5 h reaction. Comparing different supports, the catalytic activity was in the following order: $Ru/Ti_{0.9}Zr_{0.1}O_2 > Ru/Ti_{0.9}Ce_{0.1}O_2 > Ru/TiO_2$. The results indicated that doping of Zr into TiO_2 lattice could increase the specific surface area and dispersion of the metal. In addition, higher amount of surface oxygen species and strong interaction between Ru and the support increased the redox property of $Ru/Ti_{0.9}Zr_{0.1}O_2$, hence, explains the high degradation of aniline. Despite high conversion and COD removal, nitrogen abatement was relatively low, as only about 58% of total nitrogen content was converted to N_2, whereas the remainder was in the form of ammonium or nitrate ions. The proposed reaction pathway of the CWAO of aniline is presented in Figure 1.12.

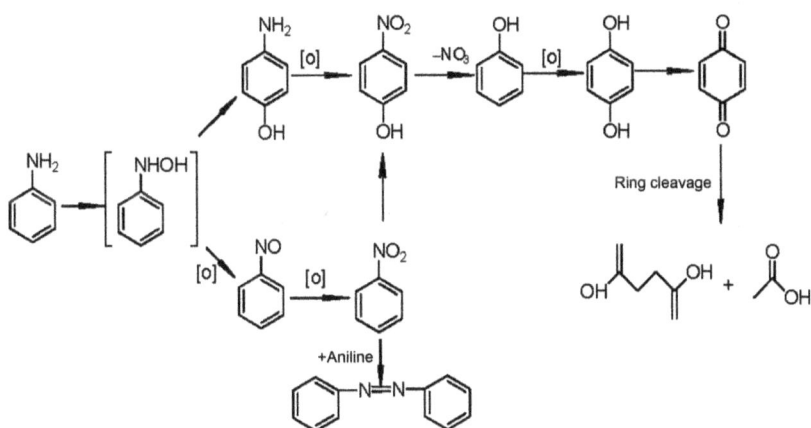

Figure 1.12: Reaction pathway of CWAO of aniline over supported Ru catalyst. Reprinted from ref. [141] with permission from Elsevier.

1.3.3 Carbon-based catalysts for CWAO

In comparison with transition metal oxide and noble metal catalysts, carbon-based catalyst would avoid the use and leaching of metal species during the preparation and reaction processes. Carbon materials also provide a cheaper alternative due to their low-cost production as well as better stability due to their resistance to acidic and basic operating conditions. A broad variety of carbons, such as activated carbon, carbon black powder, graphite, and more recently, carbon nanotubes and graphene, are commercially available. Depending on the preparation method, carbonaceous materials usually exhibit large surface area (e.g., up to 2,500 m^2/g for activated carbon) and pore distribution, giving rise to remarkable adsorption capacities for various chemical species [112]. Apart from large surface area, the adsorption performance of carbon materials is also strongly dependent on the types of surface functional group that can interact with adsorbed species, which conventionally mainly involve oxygen and hydroxyl groups.

Besides being an excellent adsorbent, carbons also exhibit catalytic properties. A review by Coughlin suggested that their catalytic activity correlates to their surface chemistry and electronic structures, which can be tuned by introducing functional groups/atoms such as O, N, P, and B into the carbon surface [148]. Some examples of chemical reactions catalyzed by carbons are hydrogenation, oxidation–reduction, polymerization, and halogenation. Table 1.5 shows some representative examples using various carbon materials for CWAO processes.

Table 1.5: Development of carbon catalysts for CWAO of organic compounds.

Catalyst	Organic type	Reaction condition	Organic removal	TOC removal	Ref.
N-doped CNTs	Oxalic acid	140 °C, 0.7 MPa O_2	100% in 0.25 h	N.A.	[103]
N-doped CNTs	Phenol	160 °C, 0.7 MPa O_2	97% in 2 h	N.A.	[103]
P-doped carbon black	Phenol	110 °C, 0.4 MPA, with H_2O_2	>95% in 6 h	>70% in 6 h	[102]
B-doped carbon black	Phenol	110 °C, 0.4 MPA, with H_2O_2	>95% in 6 h	>70% in 6 h	[102]
N-doped carbon black	Phenol	110 °C, 0.4 MPA, with H_2O_2	>95% in 6 h	>70% in 6 h	[102]
N-doped activated carbon	Phenol	160 °C, 0.2 MPa O_2	80% in 1 h	60% in 1 h	[104]
Graphene oxide	Phenol	155 °C, 1.8 MPa O_2	100% in 1 h	80% in 1 h	[99]
Activated carbon	Methylamine	195 °C, 1.6 MPa O_2	45% in 2 h	N.A.	[149]
Activated carbon	Dimethylamine	195 °C, 1.6 MPa O_2	35% in 2 h	N.A.	[149]

N.A., data is not available.

Fortuny et al. [94] conducted continuous oxidation of phenol in a fixed bed reactor at 4.7 MPa of air pressure and at 140 °C operating in trickle flow regime. Either active carbon or a commercial supported copper was used as catalyst and their performance was compared in terms of phenol conversion after 240 h reaction. The results showed that the active carbon significantly catalyzed higher phenol conversion than the copper counterpart, despite declining activity over time from the initial 100% phenol conversion to 48% due to reduction of surface area and combustion. It was observed that the presence of phenol prevents rapid combustion of the active carbon, suggesting that it would be suitable for treating high loading of phenol waste.

Yang et al. modified the surface of activated carbon by nitrogen doping via high-temperature amination [104]. The catalytic performance was investigated for the CWAO of phenol at 120 and 160 °C and 0.2 MPa oxygen. It was found that carbon aminated at 850 °C exhibited the highest activity with around 78% and 60% of phenol and COD removal, respectively, at 160 C reaction temperature. This is significantly higher than that obtained by the parent-activated carbon (ca. 10% phenol removal). Further analysis suggested that the catalytic activity was linearly dependent on the concentration of pyridinic and pyrrolic groups on the catalyst surface, indicating the predominant role of these two nitrogen-containing functional groups in the CWAO of phenol. The positive effect of nitrogen could be attributed to (i) the increased conductivity of the activated carbon, which promotes electron transfer during the reaction and (ii) the changes in the acid–base properties of the carbon surface [150, 151].

In another work by Diaz de Tuesta et al., nonporous carbon black was doped with P, B, and N for the catalytic wet peroxide oxidation (CWPO) of phenol and kinetic model was developed [102]. Figure 1.13 shows that all P-, B-, and N-doped carbon exhibit high conversion of phenol and TOC after 6 h of reaction at 110 °C and 4 bar. In particular, boron-doped carbon black showed the best performance in terms of both activity and reusability. The surface acidity seemed to play an important role in increasing the activity of the doped carbons as the most acidic catalysts (e.g., those doped with H_3PO_4, H_3BO_3, and pyridine) exhibit more efficient consumption of H_2O_2 and consequently give the highest activity.

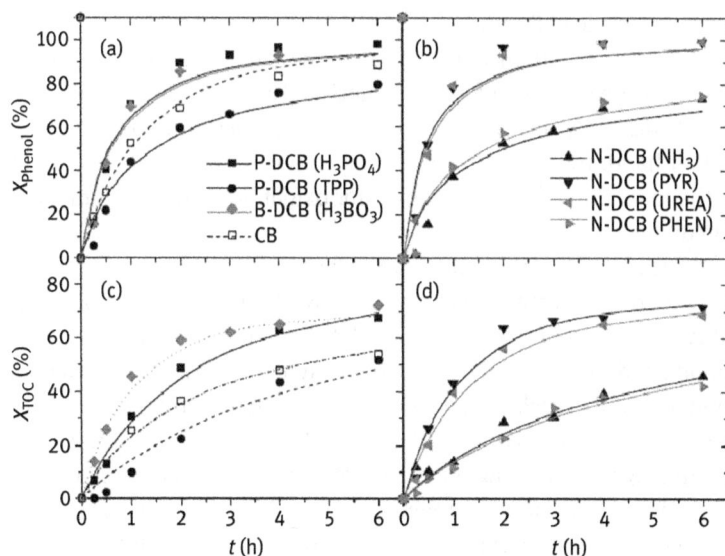

Figure 1.13: Experimental (symbols) and predicted (curves) conversion of phenol (top) and TOC (bottom) obtained upon CWPO of phenol over carbon black doped with P, B, and N. Reprinted from ref. [102] with permission from Elsevier.

Aguilar et al. used the oxidized and reduced forms of activated carbon as catalysts for the oxidation of methylamine and dimethylamine at 195 °C and 16 bar O_2 [149]. Due to their strong basic nature, it is expected that the adsorption of both amines takes place mainly over carboxylic, lactonic, and anhydride surface groups of activated carbons; hence, the presence of these functional groups would also increase the catalytic activity of the activated carbon. However, strong adsorption on the carboxylic sites may cause steric hindrance to the oxygen adsorption, thus inhibiting the reaction, which is evident by a lower conversion obtained for dimethylamine compared to that of methylamine. Therefore, it is proposed that activated carbons with higher catalytic activity are those that have smaller amount of carboxylic-like functional group and larger amount of quinonic groups. Overall, selective or total oxidation of the amines to nitrogen and carbon dioxide can be described by the following global reactions:

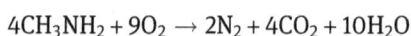

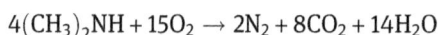

$$4CH_3NH_2 + 9O_2 \rightarrow 2N_2 + 4CO_2 + 10H_2O$$

$$4(CH_3)_2NH + 15O_2 \rightarrow 2N_2 + 8CO_2 + 14H_2O$$

Under the experimental conditions, there was no production of carbon monoxide, showing that the methyl groups were completely oxidized to CO_2 and H_2O.

A type of porous graphite-like materials of the Sibunit family was used for CWAO of aniline and phenol [152]. Despite having lower surface area, Sibunit was superior compared to conventional active carbons in terms of stability toward aggressive and oxidative media. At 433 K and 1.0 MPa O_2, the oxidation of aniline showed a considerable intrinsic catalytic activity, with up to 97% aniline conversion could be achieved, depending on the characteristic of Sibunit used in the reaction. Its activity was found to be comparable or even higher than that of supported oxide catalysts and zeolite catalyst Fe–ZSM, which were used as reference materials. However, the CO_2 selectivity was considerably lower, only about 36%, indicating large accumulation of intermediates. On the other hand, the oxidation of phenol exhibited higher conversion. It was observed that the induction period was longer for oxidation of aniline compared to that for oxidation of phenol.

The presence of unsaturated carbon atoms or defects on the surface of CNTs allows the introduction of other elements, such as O, S, N, B into the carbon surface, which could modify its electronic properties and hence their catalytic activity. Rocha et al. performed N-doping on CNTs by using different N-precursors, such as melamine, urea, and ammonia [103]. Subsequently, the catalytic activity was evaluated for CWAO of oxalic acid. The parent CNT achieved around 75% of oxalic acid degradation in 60 min, significantly higher than that of the noncatalytic oxidation process. However, the incorporation of nitrogen onto the CNT surface led to a drastic decrease in time required to fully mineralize the carboxylic acid, with complete oxalic acid conversion within less than 15 min.

As a building block of CNTs, graphene, a single-layer graphite with a large surface area and excellent mobility of charge carriers, also exhibits the outstanding electrical, optical, electrochemical, and mechanical properties [99]. Graphene is usually prepared by reducing the exfoliated graphene oxide (GO), a readily available and inexpensive intermediate, which contains abundant surface functional groups, including hydroxyl and epoxide groups on their basal planes and carbonyl and carboxyl groups at the sheet edges. In catalysis, GO has been used as a catalyst support (TiO_2/GO, ZnO/GO, etc.) and was found to improve the catalytic decomposition of organic pollutants such as diphenhydramine pharmaceutical, methyl orange dye, and methylene blue. In the absence of metal, different types of GO, such as reduced GO (RGO) and chemically modified graphene, could also catalyze chemical reactions, such as oxidation of various alcohols and *cis*-stilbene, and the hydration of various alkynes.

Yang et al. carried out CWAO of phenol over GO, RGO, and N-modified GO at 155 °C and total oxygen pressure of 1.8 MPa [99]. Among the tested graphenes, GO was the most active catalyst, resulting in 100% phenol removal after 40 min of reaction, as shown in Figure 1.14. The catalyst also showed good mineralization by achieving 80% of TOC removal after 90 min. Through recycling experiments, it was observed that the catalytic performance of GO gradually decreased with the increase in recycling number due to the decrease in the surface area and the amount of oxygen-containing functional groups (e.g., $C-OH, COOH, C-O-C$), as well as the aggregation of the GO during the reaction. Therefore, the presence of oxygen-containing functional groups played an important role in the catalytic activity of the GO. On the other hand, N-modified GO, having the lowest amount of oxygenated functional groups and a certain amount of N-containing functional groups, had no significant catalytic activity in the CWAO of phenol, as indicated by less than 10% phenol conversion within the same reaction time.

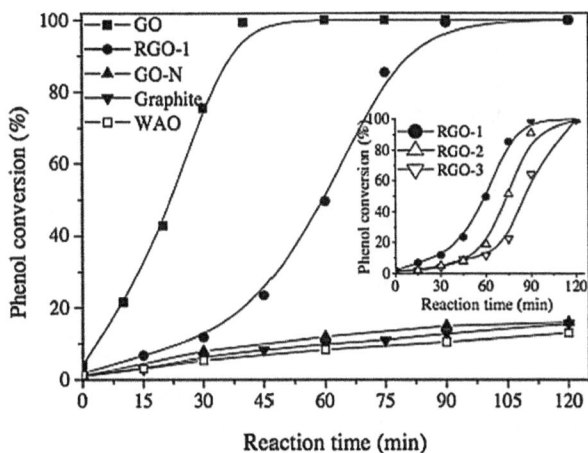

Figure 1.14: Catalytic activity of various types of graphene on the CWAO of phenol. Reprinted from ref. [99] with permission from Elsevier.

1.4 Catalyst deactivation

One of the major limitations of heterogeneous catalysts is the relatively rapid deactivation under CWAO operating condition, which could be due to one or several reasons of the following: (i) leaching and sintering of the active metal species, (ii) loss of surface area of the supporting material, (iii) possible poisoning of the active sites by intermediates, and (iv) deposition of organic or inorganic compounds (coking) on the catalyst surface [21]. Metal oxide catalysts, in particular Cu, as we know it, are more vulnerable to metal leaching under acidic condition [60, 66, 67]. However, it can be controlled to a large extent by an appropriate choice of catalytic metal or metal oxide phase as well as pH of the solution. In contrast, supported precious metal catalysts are less prone to leaching of active species, but coke deposition remains a problem.

The deposition of carbonaceous materials on the catalyst surface would obstruct the access of reactant and dissolved oxygen to the active sites of catalyst, hence decreasing the catalyst activity. Although moderate CWAO conditions are economically advantageous, low temperatures tend to promote stronger adsorptive interactions on catalyst surface, and consequently, stimulate the formation of undesired high-molecular-weight polymeric products [153, 154]. For example, Pintar et al. reported that the deactivation of Ru/TiO_2 catalysts observed at temperatures below 190 °C was attributed to the strong adsorption of partially oxidized intermediates on the catalyst surface, which can be avoided by conducting the CWAO at sufficiently high temperatures [155]. For the CWAO of phenol, Kim et al. found that a large fraction of decomposed phenol was deposited as coke over MnO_x supported on $Ce_{0.65}Zr_{0.35}O_2$ [48]. In another paper, they also reported that among the transition metal oxide (Mn, Fe, Co, Ni, and Cu) supported on Al_2O_3, Mn catalyst showed the highest amount of carbon deposits [45]. The formation/migration of carbon deposits over Pt supported on Al_2O_3 and CeO_2 was studied by Lee et al., who suggested that carbon was primarily formed on Pt particles which subsequently migrated onto the oxide supports [156]. Nousir et al. also conducted phenol oxidation over Pt/CeO_2 catalysts in a stirred batch reactor and observed the deactivation due to the formation of carbonates and polymeric carbon species [157]. They found that the introduction of zirconium into ceria lattice decreased the formation of carbonaceous materials, which correlated to the decrease in catalyst surface area. On the other hand, during the CWAO of acetic acid over Pt/CeO_2, Mikulová et al. suggested that the formation of carbonate species in the lattice of ceria-based support was responsible for the inhibition of oxygen transfer, thus limiting the catalyst performance [86].

The chemical analysis showed that the carbonaceous deposits were mostly of aromatic compounds and contained some oxygen functional groups such as carboxylic acids and alcohols. More detailed study by Keav et al. indicated that the coke formation during the CWAO of phenol consists of several polycyclic aromatic compounds of chromenone, xanthenone, and fluorenone families, as a result of the condensation of phenol with partial oxidation products [158].

Many attempts have been made to prevent or reduce the carbonaceous deposits via the modification of catalyst composition, the introduction of promoter, the development of preparation method and pretreatment procedure, as well as the control of operating conditions. For example, promotion of MnO_2–CeO_2 catalyst with Pt and/or Ag appreciably reduced the amount of coke deposits due to the low-temperature redox properties attained by metal doping, and correspondingly enhanced the mineralization selectivity [159]. Hussain et al. reported that doping of potassium increased the catalytic activity of Mn–Ce–O catalyst for the CWAO of phenol and decreased the formation of carbonaceous deposits [129]. Furthermore, K-doped catalyst also exhibited a high resistance to metal leaching. As mentioned previously, the role of potassium is associated with the electron-donating ability of K_2O in activating oxygen to form peroxide, O_2^{2-}, which rapidly increases the oxidation efficiency and thereby reducing the formation of coke deposits [128, 129].

Several methods have been used to regenerate the solid catalysts deactivated by coke deposition. Thermal treatment in an oxidizing environment is the most common method. Usually, the spent catalysts can recover their initial activity after regeneration at 400 °C under air stream [160]. Rinsing with organic solvent like acetone was also reportedly effective to remove the coke deposits from catalyst surface but it may not be environmentally friendly [65]. Keav et al. proposed an easy-to-perform ex situ reactivation procedure by combustion with diluted oxygen at moderate temperature, which can achieve total carbon degradation [78].

Metal leaching is another important issue. In general, acidic intermediates such as short-chain carboxylic acids are formed during the CWAO of organic compounds, thus decreases the pH value of the solution [64, 132, 161, 162]. Under this hot acidic condition, metal ions easily leached out from the catalyst, resulting in the decrease of catalytic activity [162].

In particular, copper-based catalysts are susceptible to leaching in spite of their good catalytic activity [61, 64, 66, 67]. Santos et al. found that the rate of copper leaching increased noticeably when the pH of the solution was less than 4 [64]. Otherwise, it was negligible in the presence of bicarbonate buffer solution where slightly alkaline pH was maintained in the range of 7–8 [161]. In contrast, the CWAO reaction rate in the basic condition is generally lower than that in acidic condition [121, 163]. The dissolved copper ions from the solid catalyst therefore change the heterogeneous catalysis into the homogeneous reaction, and consequently affecting the oxidation rate of organic compounds and the oxidation routes [161, 164]. Arena et al. observed that the length of induction period in the CWAO of phenol corresponds to the leaching of copper ions [164]. Initially, phenol conversion and the extent of copper leaching were low while the initial pH value hardly changed. Thereafter, the extent of copper leaching rapidly increased due to the decrease in the pH value caused by the generation of acidic intermediates. Both phenol and TOC conversion were increased with copper leaching. The parallel evolution of soluble Cu ions and phenol conversion apparently showed a direct relationship between dissolution and activity of the Cu catalysts. This evidence indicates that CWAO of organic compounds over the Cu catalysts proceeds via a homogeneous–heterogeneous path promoted by the dissolved copper ions.

In order to minimize metal leaching, development of a novel catalyst structure could be beneficial. Spinel-type catalysts are found to be highly resistant to dissolution in acidic conditions [165, 166]. Alejandre et al. evaluated the catalytic activity and the stability of Cu–Al mixed oxide and found that Cu ion leaching from spinel $CuAl_2O_4$ was minimal [166]. Xu et al. prepared $Cu_{(0.5-x)}Fe_xZn_{0.5}Al_2O_4$ spinel catalysts for the CWAO of phenol and reported that the solubility of Cu decreased when $x \leq 0.25$ due to the reduction in the fraction of octahedral Cu exposed to reactants [167]. They also showed that increasing Al content in the $ZnFe_{(2-x)}Al_xO_4$ spinel transformed the aggregated iron oxide clusters to Fe^{3+} species in octahedral sites that are very resistant to dissolution under acidic condition [168, 169].

1.5 Future outlook and challenges

We have discussed environmentally friendly methods for oxidation of organic pollutants in wastewater by air, namely WAO and CWAO. WAO has long been commercialized for despite being too costly due to high utility demand. CWAO is a proven promising technique for the abatement of refractory organic pollutants at moderate conditions with high pollutant removal efficiency. Extensive studies on the CWAO have employed various model pollutants, such as phenolic compounds, carboxylic acids, and nitrogen-containing compounds, as well as industrial effluents. In particular, phenol is often used as a model compound because it is a common intermediate produced in the oxidation of large organic compounds.

To achieve high pollutant removal, it is crucial to develop and apply highly efficient catalysts. Different types of solid catalysts such as transition metal oxides, supported noble metals, various carbon materials, or their combinations have been discussed. It is evident that, in general, catalytic activity and stability are dependent on the combination and composition of the active component and support, catalyst preparation and pretreatment method, the nature of pollutants, and the reaction conditions. Further studies are necessary to develop inexpensive but more active and stable catalysts, which can be scaled up for industrial application. One possible direction is to develop efficient and robust catalysts that can work under very mild reaction conditions. In addition, catalyst processing and handling by wash-coating of powder catalysts on certain supports such as resin, glass beads, or other ceramic supports can improve the separation and recycle efficiency. Furthermore, the catalyst deactivation issue should be well addressed. Factors leading to the catalyst deactivation include leaching and sintering of the active metal species, loss of surface area, poisoning of the active sites by intermediates, and coking. Last but not least, attention also needs to be given to the development of more detailed kinetic models and effective reactor designs besides the conventional packed-bed and slurry reactors, which will be of great importance to practical applications.

References

[1]   Shahidi, D., Roy, R., and Azzouz, A. Advances in catalytic oxidation of organic pollutants –
       Prospects for thorough mineralization by natural clay catalysts. Appl. Catal. B-Environ. 2015,
       174, 277–292.
[2]   Arena, F., Di Chio, R., Gumina, B., Spadaro, L., Trunfio, G. Recent advances on wet air
       oxidation catalysts for treatment of industrial wastewaters. Inorg. Chim. Acta 2015, 431,
       101–109.
[3]   Gottlieb, A., Shaw, C., Smith, A., Wheatley, A., Forsythe, S. The toxicity of textile reactive azo
       dyes after hydrolysis and decolourisation. J. Biotechnol. 2003, 101, 49–56.
[4]   Debellefontaine, H., Chakchouk, M. Foussard, J. N., Tissot, D., Striolo, P. Treatment of
       organic aqueous wastes: Wet air oxidation and wet peroxide oxidation(R). Environ. Pollut.
       1996, 92, 155–164.
[5]   Danis, T. G., Albanis, T. A., Petrakis, D. E., Pomonis, P. J. Removal of chlorinated phenols
       from aqueous solutions by adsorption on alumina pillared clays and mesoporous alumina
       aluminum phosphates. Water Res. 1998, 32, 295–302.
[6]   White, G.C. White's Handbook of Chlorination and Alternative Disinfectants. 2010: Hoboken,
       N.J.: Wiley, ©2010. 5th ed. / Black & Veatch Corporation.
[7]   Welch, W.A. Potassium permanganate in water treatment. J. Am. Water Works Assoc. 1963,
       55, 735–741.
[8]   Glaze, W.H., Kang, J.W., and Chapin, D.H. The chemistry of water-treatment processes
       involving ozone, hydrogen-peroxide and ultraviolet-radiation. Ozone-Science & Engineering
       1987, 9, 335–352.
[9]   Kusic, N.K.H. AOP as an effective tool for the minimization of hazardous organic pollutants in
       colored wastewater; Chemical and photochemical processes, in Hazardous Materials and
       Wastewater: Treatment, Removal and Analysis, Lewinsky, A.A., Editor. 2007, Nova Science
       Publishers, Inc: New York, pp. 149–199.
[10]   Deng, Y. and Zhao, R. Advanced Oxidation Processes (AOPs) in wastewater treatment.
       Current Pollution Reports 2015, 1, 167–176.
[11]   Boczkaj, G. and Fernandes, A. Wastewater treatment by means of advanced oxidation
       processes at basic pH conditions: a review. Chem. Eng. J. 2017, 320, 608–633.
[12]   Nidheesh, P.V. Graphene-based materials supported advanced oxidation processes for water
       and wastewater treatment: a review. Environ. Sci. Pollut. Res. 2017, 24, 27047–27069.
[13]   Salimi, M., Esrafili, A., Gholami, M., Jafari, A. J., Kalantary, R. R., Farzadkia, M., et al.
       Contaminants of emerging concern: a review of new approach in AOP technologies. Environ.
       Monit. Assess. 2017, 189, 414.
[14]   Bigda, R.J. Consider Fenton's chemistry for wastewater treatment. Chem. Eng. Prog. 1995, 91,
       62–66.
[15]   Kasprzyk-Hordern, B., Ziolek, M., and Nawrocki, J. Catalytic ozonation and methods of
       enhancing molecular ozone reactions in water treatment. Appl. Catal. B-Environ. 2003, 46,
       639–669.
[16]   Andreozzi, R., Caprio, V., Insola, A., Marotta, R. Advanced oxidation processes (AOP) for
       water purification and recovery. Catal. Today 1999, 53, 51–59.

[17] Zimmermann, F.J. New waste disposal process. Chem. Eng, 1958, 25, 117–120.

[18] Zimmermann, F.J. and Diddams, D.G. The Zimmermann process and its application to the pulp and paper industry. Tappi 1960, 43, 710–715.

[19] Luck, F. Wet air oxidation: past, present and future. Catal. Today 1999, 53, 81–91.

[20] Zou L.Y., Li Y., Hung Y.T. Wet air oxidation for waste treatment, in Advanced Physicochemical Treatment Technologies, Wang L.K., Hung YT., Shammas N.K. Editors. 2007, Humana Press, pp. 575–610.

[21] Levec, J. and Pintar, A. Catalytic wet-air oxidation processes: a review. Catal. Today 2007, 124, 172–184.

[22] Mishra, V.S., Mahajani, V.V., and Joshi, J.B. Wet air oxidation. Ind. Eng. Chem. Res. 1995, 34, 2–48.

[23] Kolaczkowski, S. T., Plucinski, P., Beltran, F. J., Rivas, F. J., McLurgh, D. B. Wet air oxidation: a review of process technologies and aspects in reactor design. Chem. Eng. J. 1999, 73, 143–160.

[24] Debellefontaine, H., Crispel, S., Reilhac, P., Perie, F., Foussard, J. N. Wet air oxidation (WAO) for the treatment of industrial wastewater and domestic sludge. Design of bubble column reactors. Chem. Eng. Sci. 1999, 54, 4953–4959.

[25] Debellefontaine, H. and Foussard, J.N. Wet air oxidation for the treatment of industrial wastes. Chemical aspects, reactor design and industrial applications in Europe. Waste Manage. (Oxford) 2000, 20, 15–25.

[26] Bhargava, S. K., Tardio, J., Prasad, J., Foger, K., Akolekar, D. B., Grocott, S. C. Wet oxidation and catalytic wet oxidation. Ind. Eng. Chem. Res. 2006, 45, 1221–1258.

[27] Hii, K., Baroutian, S., Parthasarathy, R., Gapes, D. J., Eshtiaghi, N. A review of wet air oxidation and thermal hydrolysis technologies in sludge treatment. Bioresour. Technol. 2014, 155, 289–299.

[28] Fu, J. and Kyzas, G.Z. Wet air oxidation for the decolorization of dye wastewater: an overview of the last two decades. Chin. J. Catal. 2014, 35, 1–7.

[29] Jing, G.L., Luan, M.M., and Chen, T.T. Progress of catalytic wet air oxidation technology. Arabian J. Chem. 2016, 9, S1208–S1213.

[30] Luan, M. M., Jing, G. L., Piao, Y. J., Liu, D. B., Jin, L. F. Treatment of refractory organic pollutants in industrial wastewater by wet air oxidation. Arabian J. Chem. 2017, 10, S769–S776.

[31] Imamura, S. Catalytic and noncatalytic wet oxidation. Ind. Eng. Chem. Res. 1999, 38, 1743–1753.

[32] Ingale, M. N., Joshi, J. B., Mahajani, V. V., Gada, M. K. Waste treatment of an aqueous waste stream from a cyclohexane oxidation unit: a case study. Process Saf. Environ. Prot. 1996, 74, 265–272.

[33] Mantzavinos, D., Lauer, E., Hellenbrand, R., Livingston, A. G., Metcalfe I. S. Wet oxidation as a pretreatment method for wastewaters contaminated by bioresistant organics. Water Sci. Technol. 1997, 36, 109–116.

[34] Rivas, F. J., Kolaczkowski, S. T., Beltran, F. J., McLurgh, D. B. Development of a model for the wet air oxidation of phenol based on a free radical mechanism. Chem. Eng. Sci. 1998, 53, 2575–2586.

[35] Li, L.X., Chen, P.S., and Gloyna, E.F. Generalized kinetic-model for wet oxidation of organic-compounds. AIChE J. 1991, 37, 1687–1697.

[36] Zhang, Q. and Chuang, K.T. Lumped kinetic model for catalytic wet oxidation of organic compounds in industrial wastewater. AIChE J. 1999, 45, 145–150.

[37] Imteaz, M.A. and Shanableh, A. Kinetic model for the water oxidation method for treating wastewater sludges. Dev. Chem. Eng. Min. Process. 2004, 12, 515–530.

[38] Levec, J. Wet oxidation processes for treating industrial wastewaters. Chem. Biochem. Eng. Q. 1997, 11, 47–58.
[39] V. Moses, Douglas, A. Smith, Elgene. Waste Disposal Process, Office, U.S.P., Editor. 1954, Du Pont: United States.
[40] Hocevar, S., Batista, J., and Levec, J. Wet oxidation of phenol on Ce1-xCuxO2-delta catalyst. J. Catal. 1999, 184, 39–48.
[41] Neri, G., Pistone, A., Milone, C., Galvagno, S. Wet air oxidation of p-coumaric acid over promoted ceria catalysts. Appl. Catal. B-Environ. 2002, 38, 321–329.
[42] Hung, C.M., Lou, J.C., and Lin, C.H. Wet air oxidation of aqueous ammonia solutions catalyzed by composite metal oxide. Environ. Eng. Sci. 2003, 20, 547–556.
[43] Abecassis-Wolfovich, M., Landau, M. V., Brenner, A., Herskowitz, M. Catalytic wet oxidation of phenol with Mn-Ce-based oxide catalysts: impact of reactive adsorption on TOC removal. Ind. Eng. Chem. Res. 2004, 43, 5089–5097.
[44] Wan, J. F., Feng, Y. J., Cai, W. M., Yang, S. X., Sun, X. J. Kinetics study on catalytic wet air oxidation of phenol by several metal oxide catalysts. J. Environ. Sci. 2004, 16, 556–558.
[45] Kim, S.K. and Ihm, S.K. Nature of carbonaceous deposits on the alumina supported transition metal oxide catalysts in the wet air oxidation of phenol. Top. Catal. 2005, 33, 171–179.
[46] Kim, M.H. and Choo, K.H. Low-temperature continuous wet oxidation of trichloroethylene over CoOx/TiO2 catalysts. Catal. Commun. 2007, 8, 462–466.
[47] Yang, M., Xu, A. H., Du, H. Z., Sun, C. L., Li, C. Removal of salicylic acid on perovskite-type oxide LaFeO3 catalyst in catalytic wet air oxidation process. J. Hazard. Mater. 2007, 139, 86–92.
[48] Kim, K.H., Kim, J.R., and Ihm, S.K. Wet oxidation of phenol over transition metal oxide catalysts supported on Ce0.65Zr0.35O2 prepared by continuous hydrothermal synthesis in supercritical water. J. Hazard. Mater. 2009, 167, 1158–1162.
[49] Zhang, L. H., Li, F., Evans, D. G., Duan, X. Cu-Zn-(Mn)-(Fe)-Al layered double hydroxides and their mixed metal oxides: physicochemical and catalytic properties in wet hydrogen peroxide oxidation of phenol. Ind. Eng. Chem. Res. 2010, 49, 5959–5968.
[50] Zhao, B. X., Shi, B. C., Zhang, X. L., Cao, X., Zhang, Y. Z. Catalytic wet hydrogen peroxide oxidation of H-acid in aqueous solution with TiO2-CeO2 and Fe/TiO2-CeO2 catalysts. Desalination 2011, 268, 55–59.
[51] Ovejero, G., Rodriguez, A., Vallet, A., Gomez, P., Garcia, J. Catalytic wet air oxidation with Ni- and Fe-doped mixed oxides derived from hydrotalcites. Water Sci. Technol. 2011, 63, 2381–2387.
[52] Ovejero, G., Rodriguez, A., Vallet, A., Garcia, J. Ni supported on Mg-Al oxides for continuous catalytic wet air oxidation of Crystal Violet. Appl. Catal. B-Environ. 2012, 125, 166–171.
[53] Cau, C., Guari, Y., Chave, T., Larionova, J., Nikitenko, S. I. Thermal and sonochemical synthesis of porous (Ce,Zr)O-2 mixed oxides from metal beta-diketonate precursors and their catalytic activity in wet air oxidation process of formic acid. Ultrason. Sonochem. 2014, 21, 1366–1373.
[54] Schmit, F., Bois, L., Chassagneux, F., Descorme, C. Catalytic wet air oxidation of methylamine over supported manganese dioxide catalysts. Catal. Today 2015, 258, 570–575.
[55] Heponiemi, A., Azalim, S., Hu, T., Lassi, U. Cerium Oxide Based Catalysts for Wet Air Oxidation of Bisphenol A. Top. Catal. 2015, 58, 1043–1052.
[56] Lafaye, G., Barbier, J., and Duprez, D. Impact of cerium-based support oxides in catalytic wet air oxidation: conflicting role of redox and acid-base properties. Catal. Today 2015, 253, 89–98.

[57] Nousir, S., Maache, R., Azalim, S., Agnaou, M., Brahmi, R., Bensitel, M. Synthesis and investigation of the physico-chemical properties of catalysts based on mixed oxides CexZr1-xO2. Arabian J. Chem. 2015, 8, 222–227.

[58] Ramakrishna, C., Krishna, R., Gopi, T., Swetha, G., Saini, B., Shekar, S. C., et al. Complete oxidation of 1,4-dioxane over zeolite-13X-supported Fe catalysts in the presence of air. Chin. J. Catal. 2016, 37, 240–249.

[59] Nze, V.M.M., Fontaine, C., and Barbier, J. Preparation and characterization of MgAlCe mixed oxides for catalytic oxidation of acetic acid. C.R. Chim. 2017, 20, 67–77.

[60] Alvarez, P.M., McLurgh, D., and Plucinski, P. Copper oxide mounted on activated carbon as catalyst for wet air oxidation of aqueous phenol. 2. Catalyst stability. Ind. Eng. Chem. Res. 2002, 41, 2153–2158.

[61] Dou, H. R., Zhu, J. D., Chen, Y. J., Wu, M., Sun, C. L. The leaching and control of active component Cu in copper-based catalysts for catalytic wet air oxidation. Chin. J. Catal. 2003, 24, 328–332.

[62] Silva, A.M.T., Oliveira, A.C.M., and Quinta-Ferreira, R.M. Catalytic wet oxidation of ethylene glycol: kinetics of reaction on a Mn-Ce-O catalyst. Chem. Eng. Sci. 2004, 59, 5291–5299.

[63] Gomes, H. T., Selvam, P., Dapurkar, S. E., Figueiredo, J. L., Faria, J. L. Transition metal (Cu, Cr, and V) modified MCM-41 for the catalytic wet air oxidation of aniline. Microporous Mesoporous Mater. 2005, 86, 287–294.

[64] Santos, A., Yustos, P., Quintanilla, A., Ruiz, G., Garcia-Ochoa, F. Study of the copper leaching in the wet oxidation of phenol with CuO-based catalysts: causes and effects. Appl. Catal. B-Environ. 2005, 61, 323–333.

[65] Chen, I. P., Lin, S. S., Wang, C. H., Chang, S. H. CWAO of phenol using CeO2/gamma-Al2O3 with promoter – Effectiveness of promoter addition and catalyst regeneration. Chemosphere 2007, 66, 172–178.

[66] Zapico, R. R., Marin, P., Diez, F. V., Ordonez, S. Assessment of phenol wet oxidation on CuO/gamma-Al2O3 catalysts: competition between heterogeneous and leached-copper homogeneous reaction paths. J. Environ. Chem. Eng. 2017, 5, 2570–2578.

[67] Kong, L. M., Zhou, X., Yao, Y., Jian, P. M., Diao, G. W. Catalytic wet peroxide oxidation of aniline in wastewater using copper modified SBA-15 as catalyst. Environ. Technol. 2016, 37, 422–429.

[68] Posada, D., Betancourt, P., Fuentes, K., Marrero, S., Liendo, F., Brito, J. L. Catalytic wet air oxidation of oilfield produced wastewater containing refractory organic pollutants over copper/cerium-manganese oxide. React. Kinet. Mech. Catal. 2014, 112, 347–360.

[69] Gallezot, P., Chaumet, S., Perrard, A., Isnard, P. Catalytic wet air oxidation of acetic acid on carbon-supported ruthenium catalysts. J. Catal. 1997, 168, 104–109.

[70] Barbier-Jr, J., Delanoe, F., Jabouille, F., Duprez, D., Blanchard, G., Isnard, P. Total oxidation of acetic acid in aqueous solutions over noble metal catalysts. J. Catal. 1998, 177, 378–385.

[71] Pintar, A., Besson, M., and Gallezot, P. Catalytic wet air oxidation of Kraft bleach plant effluents in a trickle-bed reactor over a Ru/TiO2 catalyst. Appl. Catal. B-Environ. 2001, 31, 275–290.

[72] Pintar, A., Besson, M., and Gallezot, P. Catalytic wet air oxidation of Kraft bleaching plant effluents in the presence of titania and zirconia supported ruthenium. Appl. Catal. B-Environ. 2001, 30, 123–139.

[73] Perkas, N., Minh, D. P., Gallezot, P., Gedanken, A., Besson, M. Platinum and ruthenium catalysts on mesoporous titanium and zirconium oxides for the catalytic wet air oxidation of model compounds. Appl. Catal. B-Environ. 2005, 59, 121–130.

[74] Barbier-Jr, J., Oliviero, L., Renard, B., Duprez, D. Role of ceria-supported noble metal catalysts (Ru, Pd, Pt) in wet air oxidation of nitrogen and oxygen containing compounds. Top. Catal. 2005, 33, 77–86.

[75] Besson, M. and Gallezot, P. Stability of ruthenium catalysts supported on TiO2 or ZrO2 in catalytic wet air oxidation. Top. Catal. 2005, 33, 101–108.

[76] Wang, J. B., Zhu, W. P., Wang, W., Yang, S. X., Zhou, Y. R. Activity and stability of pelletized ruthenium catalysts in wet air oxidation. Chin. J. Catal. 2007, 28, 521–527.

[77] Gaalova, J., Barbier-Jr, J., and Rossignol, S. Ruthenium versus platinum on cerium materials in wet air oxidation of acetic acid. J. Hazard. Mater. 2010, 181, 633–639.

[78] Keav, S., Martin, A., Barbier-Jr, J., Duprez, D. Deactivation and reactivation of noble metal catalysts tested in the Catalytic Wet Air Oxidation of phenol. Catal. Today 2010, 151, 143–147.

[79] de los Monteros, A. E., Lafaye, G., Cervantes, A., Del Angel, G., Barbier, J., Torres, G. Catalytic wet air oxidation of phenol over metal catalyst (Ru,Pt) supported on TiO2-CeO2 oxides. Catal. Today 2015, 258, 564–569.

[80] Song, A.Y. and Lu, G.X. Catalytic wet oxidation of aqueous methylamine: comparative study on the catalytic performance of platinum-ruthenium, platinum, and ruthenium catalysts supported on titania. Environ. Technol. 2015, 36, 1160–1166.

[81] Yang, S.X., Besson, M., and Descorme, C. Catalytic wet air oxidation of succinic acid over Ru and Pt catalysts supported on CexZr1-xO2 mixed oxides. Appl. Catal. B-Environ. 2015, 165, 1–9.

[82] Wang, Y. M., Yu, C. Y., Meng, X., Zhao, P. Q., Chou, L. J. The ethanol mediated-CeO2-supported low loading ruthenium catalysts for the catalytic wet air oxidation of butyric acid. RSC Adv. 2017, 7, 39796–39802.

[83] Fortuny, A., Bengoa, C., Font, J., Fabregat, A. Bimetallic catalysts for continuous catalytic wet air oxidation of phenol. J. Hazard. Mater. 1999, 64, 181–193.

[84] Besson, M., Kallel, A., Gallezot, P., Zanella, R., Louis, C. Gold catalysts supported on titanium oxide for catalytic wet air oxidation of succinic acid. Catal. Commun. 2003, 4, 471–476.

[85] Shih, C.C. and Chang, J.R. Pt/C stabilization for catalytic wet-air oxidation: use of grafted TiO2. J. Catal. 2006, 240, 137–150.

[86] Mikulova, J., Barbier-Jr, J., Rossignol, S., Mesnard, D., Duprez, D., Kappenstein, C. Wet air oxidation of acetic acid over platinum catalysts supported on cerium-based materials: influence of metal and oxide crystallite size. J. Catal. 2007, 251, 172–181.

[87] Cuauhtemoc, I., Del Angel, G., Torres, G., Bertin, V. Catalytic wet air oxidation of gasoline oxygenates using Rh/gamma-Al2O3 and Rh/gamma-Al2O3-CeO2 catalysts. Catal. Today 2008, 133, 588–593.

[88] Yang, S.X., Besson, M., and Descorme, C. Catalytic wet air oxidation of formic acid over Pt/CexZr1-xO2 catalysts at low temperature and atmospheric pressure. Appl. Catal. B-Environ. 2010, 100, 282–288.

[89] Cuauhtemoc, I., Del Angel, G., Torres, G., Angeles-Chavez, C., Navarrete, J., Padilla, J. M. Enhancement of catalytic wet air oxidation of tert-amyl methyl ether by the addition of Sn and CeO2 to Rh/Al2O3 catalysts. Catal. Today 2011, 166, 180–187.

[90] Nunez, F., Del Angel, G., Tzompantzi, F., Navarrete, J. Catalytic Wet-Air Oxidation of p-Cresol on Ag/Al2O3-ZrO2 Catalysts. Ind. Eng. Chem. Res. 2011, 50, 2495–2500.

[91] Soukup, K., Topka, P., Hejtmanek, V., Petras, D., Vales, V., Solcova, O. Noble metal catalysts supported on nanofibrous polymeric membranes for environmental applications. Catal. Today 2014, 236, 3–11.

[92] Keav, S., de los Monteros, A. E., Barbier, J., Duprez, D. Wet Air Oxidation of phenol over Pt and Ru catalysts supported on cerium-based oxides: resistance to fouling and kinetic modelling. Appl. Catal. B-Environ. 2014, 150, 402–410.

[93] Wei, H. Z., Wang, Y. M., Yu, Y., Gu, B., Zhao, Y., Yang, X., et al. Effect of TiO2 on Ru/ZrO2 catalysts in the catalytic wet air oxidation of isothiazolone. Catal. Sci. Technol. 2015, 5, 1693–1703.

[94] Fortuny, A., Font, J., and Fabregat, A. Wet air oxidation of phenol using active carbon as catalyst. Appl. Catal. B-Environ. 1998, 19, 165–173.

[95] Gomes, H. T., Figueiredo, J. L., Faria, J. L., Serp, P., Kalck, P. Carbon-supported iridium catalysts in the catalytic wet air oxidation of carboxylic acids: kinetics and mechanistic interpretation. J. Mol. Catal. A: Chem. 2002, 182, 47–60.

[96] Cao, S. L., Chen, G. H., Hu, X. J., Yue, P. L. Catalytic wet air oxidation of wastewater containing ammonia and phenol over activated carbon supported Pt catalysts. Catal. Today 2003, 88, 37–47.

[97] Rodriguez, A., Ovejero, G., Romero, M. D., Diaz, C., Barreiro, M., Garcia, J. Catalytic wet air oxidation of textile industrial wastewater using metal supported on carbon nanofibers. J. Supercrit. Fluids 2008, 46, 163–172.

[98] Hung, C.M. Activity of Cu-activated carbon fiber catalyst in wet oxidation of ammonia solution. J. Hazard. Mater. 2009, 166, 1314–1320.

[99] Yang, S. X., Cui, Y. H., Sun, Y., Yang, H. W. Graphene oxide as an effective catalyst for wet air oxidation of phenol. J. Hazard. Mater. 2014, 280, 55–62.

[100] Tu, Y. T., Xiong, Y., Tian, S. H., Kong, L. J., Descorme, C. Catalytic wet air oxidation of 2-chlorophenol over sewage sludge-derived carbon-based catalysts. J. Hazard. Mater. 2014, 276, 88–96.

[101] Yadav, B.R. and Garg, A. Catalytic oxidation of pulping effluent by activated carbon-supported heterogeneous catalysts. Environ. Technol. 2016, 37, 1018–1025.

[102] de Tuesta, J. L. D., Quintanilla, A., Casas, J. A., Rodriguez, J. J. P-, B- and N-doped carbon black for the catalytic wet peroxide oxidation of phenol: activity, stability and kinetic studies. Catal. Commun. 2017, 102, 131–135.

[103] Rocha, R. P., Soares, O. S. G. P., Goncalves, A. G., Orfao, J. J. M., Pereira, M. F. R., Figueiredo, J. L. Different methodologies for synthesis of nitrogen doped carbon nanotubes and their use in catalytic wet air oxidation. Appl. Catal. A-Gen. 2017, 548, 62–70.

[104] Yang, G., Chen, H. L., Qin, H. D., Zhang, X. M., Feng, Y. J. Effect of nitrogen doping on the catalytic activity of activated carbon and distribution of oxidation products in catalytic wet oxidation of phenol. Can. J. Chem. Eng. 2017, 95, 1518–1525.

[105] Mena, I. F., Diaz, E., Rodriguez, J. J., Mohedano, A. F. CWPO of bisphenol A with iron catalysts supported on microporous carbons from grape seeds activation. Chem. Eng. J. 2017, 318, 153–160.

[106] Wang, Y. M., Wei, H. Z., Zhao, Y., Sun, W. J., Sun, C. L. Low temperature modified sludge-derived carbon catalysts for efficient catalytic wet peroxide oxidation of m-cresol. Green Chem. 2017, 19, 1362–1370.

[107] Chicinas, R. P., Cotet, L. C., Maicaneanu, A., Vasilescu, M., Vulpoi, A. Preparation, characterization, and testing of metal-doped carbon xerogels as catalyst for phenol CWAO. Environ. Sci. Pollut. Res. 2017, 24, 2980–2986.

[108] Yadav, B.R. and Garg, A. Performance assessment of activated carbon supported catalyst during catalytic wet oxidation of simulated pulping effluents generated from wood and bagasse based pulp and paper mills. RSC Adv. 2017, 7, 9754–9763.

[109] Zhao, Q. X., Mao, Q. M., Zhou, Y. Y., Wei, J. H., Liu, X. C., Yang, J. Y., et al. Metal-free carbon materials-catalyzed sulfate radical-based advanced oxidation processes: a review on heterogeneous catalysts and applications. Chemosphere 2017, 189, 224–238.

[110] Primo, A., Parvulescu, V., and Garcia, H. Graphenes as Metal-Free Catalysts with Engineered Active Sites. J. Phys. Chem. Lett. 2017, 8, 264–278.

[111] Kim, K.H. and Ihm, S.K. Heterogeneous catalytic wet air oxidation of refractory organic pollutants in industrial wastewaters: a review. J. Hazard. Mater. 2011, 186, 16–34.

[112] Stuber, F., Font, J., Fortuny, A., Bengoa, C., Eftaxias, A., Fabregat, A. Carbon materials and catalytic wet air oxidation of organic pollutants in wastewater. Top. Catal. 2005, 33, 3–50.

[113] Rivas, F. J., Kolaczkowski, S. T., Beltran, F. J., McLurgh, D. B. Hydrogen peroxide promoted wet air oxidation of phenol: influence of operating conditions and homogeneous metal catalysts. J. Chem. Technol. Biotechnol. 1999, 74, 390–398.

[114] Wu, Q., Hu, X.J., and Yue, P.L. Kinetics study on catalytic wet air oxidation of phenol. Chem. Eng. Sci. 2003, 58, 923–928.

[115] Vicente, J., Rosal, R., and Diaz, M. Catalytic wet oxidation of phenol with homogeneous iron salts. J. Chem. Technol. Biotechnol. 2005, 80, 1031–1035.

[116] Collado, S., Quero, D., Laca, A., Diaz, M. Fe2+-catalyzed wet oxidation of phenolic acids under different pH values. Ind. Eng. Chem. Res. 2010, 49, 12405–12413.

[117] Fu, D. M., Zhang, F. F., Wang, L. Z., Yang, F., Liang, X. M. Simultaneous removal of nitrobenzene and phenol by homogenous catalytic wet air oxidation. Chin. J. Catal. 2015, 36, 952–956.

[118] Kurian, M., Remya, V.R., and Kunjachan, C. Catalytic wet oxidation of chlorinated organics at mild conditions over iron doped nanoceria. Catal. Commun. 2017, 99, 75–78.

[119] Xu, Y., Shao, H. A., Ge, F., Liu, Y. Novel-structured Mo-Cu-Fe-O composite for catalytic air oxidation of dye-containing wastewater under ambient temperature and pressure. Chin. J. Catal. 2017, 38, 1719–1725.

[120] Ma, C. J., Wen, Y. Y., Yue, Q. Q., Li, A. Q., Fu, J. L., Zhang, N. W. Oxygen-vacancy- promoted catalytic wet air oxidation of phenol from MnOx-CeO2. RSC Adv. 2017, 7, 27079–27088.

[121] Mohite, R.G. and Garg, A. Performance of heterogeneous catalytic wet oxidation for the removal of phenolic compounds: catalyst characterization and effect of pH, temperature, metal leaching and non-oxidative hydrothermal reaction. J. Environ. Chem. Eng. 2017, 5, 468–478.

[122] Arena, F., Italiano, C., Ferrante, G. D., Trunfio, G., Spadaro, L. A mechanistic assessment of the wet air oxidation activity of MnCeOx catalyst toward toxic and refractory organic pollutants. Appl. Catal. B-Environ. 2014, 144, 292–299.

[123] Arena, F., Italiano, C., and Spadaro, L. Efficiency and reactivity pattern of ceria-based noble metal and transition metal-oxide catalysts in the wet air oxidation of phenol. Appl. Catal. B-Environ. 2012, 115, 336–345.

[124] Yang, S. X., Zhu, W. P., Wang, J. B., Chen, Z. X. Catalytic wet air oxidation of phenol over CeO2-TiO2 catalyst in the batch reactor and the packed-bed reactor. J. Hazard. Mater. 2008, 153, 1248–1253.

[125] Yang, S. X., Zhu, W. P., Jiang, Z. P., Chen, Z. X., Wang, J. B. The surface properties and the activities in catalytic wet air oxidation over CeO2-TiO2 catalysts. Appl. Surf. Sci. 2006, 252, 8499–8505.

[126] Chen, I. P., Lin, S. S., Wang, C. H., Chang, L., Chang, J. S. Preparing and characterizing an optimal supported ceria catalyst for the catalytic wet air oxidation of phenol. Appl. Catal. B-Environ. 2004, 50, 49–58.

[127] Silva, A.M.T., Marques, R.R.N., and Quinta-Ferreira, R.M. Catalysts based in cerium oxide for wet oxidation of acrylic acid in the prevention of environmental risks. Appl. Catal. B-Environ. 2004, 47, 269–279.

[128] Hussain, S.T., Sayari, A., and Larachi, F. Enhancing the stability of Mn-Ce-OWETOX catalysts using potassium. Appl. Catal. B-Environ. 2001, 34, 1–9.

[129] Hussain, S.T., Sayari, A., and Larachi, F. Novel K-doped Mn-Ce-O wet oxidation catalysts with enhanced stability. J. Catal. 2001, 201, 153–157.

[130] Ding, Z. Y., Li, L. X., Wade, D., Gloyna, E. F. Supercritical water oxidation of NH3 over a MnO2/CeO2 catalyst. Ind. Eng. Chem. Res. 1998, 37, 1707–1716.

[131] Aki, S. and Abraham, M.A. Catalytic supercritical water oxidation of pyridine: comparison of catalysts. Ind. Eng. Chem. Res. 1999, 38, 358–367.

[132] Arena, F., Italian, C., Raneri, A., Saja, C. Mechanistic and kinetic insights into the wet air oxidation of phenol with oxygen (CWAO) by homogeneous and heterogeneous transition-metal catalysts. Appl. Catal. B-Environ. 2010, 99, 321–328.

[133] Chen, H., Sayari, A., Adnot, A., Larachi, F. Composition-activity effects of Mn-Ce-O composites on phenol catalytic wet oxidation. Appl. Catal. B-Environ. 2001, 32, 195–204.

[134] Santiago, A. F. J., Sousa, J. F., Guedes, R. C., Jeronimo, C. E. M., Benachour, M. Kinetic and wet oxidation of phenol catalyzed by non-promoted and potassium-promoted manganese/cerium oxide. J. Hazard. Mater. 2006, 138, 325–330.

[135] Parvas, M., Haghighi, M., and Allahyari, S. Degradation of phenol via wet-air oxidation over CuO/CeO2-ZrO2 nanocatalyst synthesized employing ultrasound energy: physicochemical characterization and catalytic performance. Environ. Technol. 2014, 35, 1140–1149.

[136] Wang, S. Z., Yang, Q., Bai, Z. Y., Wang, S. D., Chen, H., Cao, Y. Catalytic wet air oxidation of wastewater of the herbicide fomesafen production with CeO2-TiO2 catalysts. Environ. Eng. Sci. 2015, 32, 389–396.

[137] Neto, R.C.R. and Schmal, M. Synthesis of CeO2 and CeZrO2 mixed oxide nanostructured catalysts for the iso-syntheses reaction. Appl. Catal. A-Gen. 2013, 450, 131–142.

[138] Balcaen, V., Roelant, R., Poelman, H., Poelman, D., Marin, G. B. TAP study on the active oxygen species in the total oxidation of propane over a CuO–CeO2/γ-Al2O3 catalyst. Catal. Today 2010, 157, 49–54.

[139] Guo, J. and Al-Dahhan, M. Catalytic wet air oxidation of phenol in concurrent downflow and upflow packed-bed reactors over pillared clay catalyst. Chem. Eng. Sci. 2005, 60, 735–746.

[140] Zhang, Z., Yang, R. Y., Gao, Y. S., Zhao, Y. F., Wang, J. Y., Huang, L., et al. Novel Na2Mo4O13/[alpha]-MoO3 hybrid material as highly efficient CWAO catalyst for dye degradation at ambient conditions. Scientific Reports (Nature Publisher Group) 2014, 4, 6797.

[141] Song, M. G., Wang, Y. S., Guo, Y., Wang, L., Zhan, W. C., Guo, Y. L., et al. Catalytic wet oxidation of aniline over Ru catalysts supported on a modified TiO2. Chin. J. Catal. 2017, 38, 1155–1165.

[142] Milone, C., Fazio, A., Pistone, A., Galvagno, S. Catalytic wet air oxidation of p-coumaric acid on CeO2, platinum and gold supported on CeO2 catalysts. Appl. Catal. B-Environ. 2006, 68, 28–37.

[143] Renard, B., Barbier-Jr, J., Duprez, D., Durecu, S. Catalytic wet air oxidation of stearic acid on cerium oxide supported noble metal catalysts. Appl. Catal. B-Environ. 2005, 55, 1–10.

[144] Masende, Z. P. G., Kuster, B. F. M., Ptasinski, K. J., Janssen, F. J. J. G., Katima, J. H. Y., Schouten, J. C. Support and dispersion effects on activity of platinum catalysts during wet oxidation of organic wastes. Top. Catal. 2005, 33, 87–99.

[145] Gomes, H. T., Serp, P., Kalck, P., Figueiredo, J. L., Faria, J. L. Carbon supported platinum catalysts for catalytic wet air oxidation of refractory carboxylic acids. Top. Catal. 2005, 33, 59–68.

[146] Pintar, A., Bercic, G., Besson, M., Gallezot, P. Catalytic wet-air oxidation of industrial effluents: total mineralization of organics and lumped kinetic modelling. Appl. Catal. B-Environ. 2004, 47, 143–152.

[147] Grosjean, N., Descorme, C., and Besson, M. Catalytic wet air oxidation of N,N-dimethylformamide aqueous solutions: deactivation of TiO2 and ZrO2-supported noble metal catalysts. Appl. Catal. B-Environ. 2010, 97, 276–283.

[148] Coughlin, R.W. Carbon as adsorbent and catalyst. Ind. Eng. Chem. Prod. Res. Dev. 1969, 8, 12–23.

[149] Aguilar, C., Garcia, R., Soto-Garrido, G., Arraigada, R. Catalytic oxidation of aqueous methyl and dimethylamines by activated carbon. Top. Catal. 2005, 33, 201–206.

[150] Podyacheva, O. Y., Ismagilov, Z. R., Boronin, A. I., Kibis, L. S., Slavinskaya, E. M., Noskov, A. S., et al. Platinum nanoparticles supported on nitrogen-containing carbon nanofibers. Catal. Today 2012, 186, 42–47.

[151] Zhou, Y. K., Neyerlin, K., Olson, T. S., Pylypenko, S., Bult, J., Dinh, H. N., et al. Enhancement of Pt and Pt-alloy fuel cell catalyst activity and durability via nitrogen-modified carbon supports. Energy Environ. Sci. 2010, 3, 1437–1446.

[152] Dobrynkin, N. M., Batygina, M. V., Noskov, A. S., Tsyrulnikov, P. G., Shlyapin, D. A., Schegolev, V. V., et al. Catalysts Ru–CeO2/Sibunit for catalytic wet air oxidation of aniline and phenol. Top. Catal. 2005, 33, 69–76.

[153] Masende, Z. P. G., Kuster, B. F. M., Ptasinski, K. J., Janssen, F. J. J. G., Katima, J. H. Y., Schouten, J. C. Platinum catalysed wet oxidation of phenol in a stirred slurry reactor – The role of oxygen and phenol loads on reaction pathways. Catal. Today 2003, 79, 357–370.

[154] Cybulski, A. and Trawczynski, J. Catalytic wet air oxidation of phenol over platinum and ruthenium catalysts. Appl. Catal. B-Environ. 2004, 47, 1–13.

[155] Pintar, A., Batista, J., and Tisler, T. Catalytic wet-air oxidation of aqueous solutions of formic acid, acetic acid and phenol in a continuous-flow trickle-bed reactor over Ru/TiO2 catalysts. Appl. Catal. B-Environ. 2008, 84, 30–41.

[156] Lee, D. K., Kim, D. S., Kim, T. H., Lee, Y. K., Jeong, S. E., Le, N. T., et al. Deactivation of Pt catalysts during wet oxidation of phenol. Catal. Today 2010, 154, 244–249.

[157] Nousir, S., Keav, S., Barbier-Jr, J., Bensitel, M., Brahmi, R., Duprez, D. Deactivation phenomena during catalytic wet air oxidation (CWAO) of phenol over platinum catalysts supported on ceria and ceria-zirconia mixed oxides. Appl. Catal. B-Environ. 2008, 84, 723–731.

[158] Keav, S., Martin, A., Barbier-Jr, J., Duprez, D. Nature of the deposit formed during catalytic wet air oxidation of phenol. C.R. Chim. 2010, 13, 508–514.

[159] Hamoudi, S., Sayari, A., Belkacemi, K., Bonneviot, L., Larachi, F. Catalytic wet oxidation of phenol over PtxAg1-xMnO2/CeO2 catalysts. Catal. Today 2000, 62, 379–388.

[160] Massa, P., Ivorra, F., Haure, P., Cabello, F. M., Fenoglio, R. Catalytic wet air oxidation of phenol aqueous solutions by 1% Ru/CeO2-Al2O3 catalysts prepared by different methods. Catal. Commun. 2007, 8, 424–428.

[161] Santos, A., Yustos, P., Quintanilla, A., Garcia-Ochoa, F. Influence of pH on the wet oxidation of phenol with copper catalyst. Top. Catal. 2005, 33, 181–192.

[162] Kouraichi, R., Delgado, J. J., Lopez-Castro, J. D., Stitou, M., Rodriguez-Izquierdo, J. M., Cauqui, M. A. Deactivation of Pt/MnOx-CeO2 catalysts for the catalytic wet oxidation of phenol: formation of carbonaceous deposits and leaching of manganese. Catal. Today 2010, 154, 195–201.

[163] Santos, A., Yustos, P., Durban, B., Garcia-Ochoa, F. Oxidation of phenol in aqueous solution with copper catalysts. Catal. Today 2001, 66, 511–517.

[164] Arena, F., Giovenco, R., Torre, T., Venuto, A., Parmaliana, A. Activity and resistance to leaching of Cu-based catalysts in the wet oxidation of phenol. Appl. Catal. B-Environ. 2003, 45, 51–62.

[165] Lumpkin, G.R. Crystal chemistry and durability of the spinel structure type in natural systems. Progress in Nuclear Energy 2001, 38, 447–454.

[166] Alejandre, A., Medina, F., Rodriguez, X., Salagre, P., Sueiras, J. E. Preparation and activity of Cu-Al mixed oxides via hydrotalcite-like precursors for the oxidation of phenol aqueous solutions. J. Catal. 1999, 188, 311–324.

[167] Xu, A. H., Yang, M., Du, H. Z., Sun, C. L. Influence of partial replacement of Cu by Fe on the CWO of phenol in the CU0.5-xFexZn0.5Al2O4 spinel catalysts. Catal. Commun. 2006, 7, 513–517.

[168] Xu, A. H., Yang, M., Qiao, R. P., Du, H. Z., Sun, C. L. Activity and leaching features of zinc-aluminum ferrites in catalytic wet oxidation of phenol. J. Hazard. Mater. 2007, 147, 449–456.

[169] Xu, A. H., Lu, X. Y., Yang, M., Du, H. Z., Sun, C. L. Activity and stability of ZnFe0.25Al1.75O4 catalyst in catalytic wet air oxidation of phenol. Chin. J. Catal. 2007, 28, 395–397.

Shuangde Li and Yunfa Chen

2 Catalytic oxidation of volatile organic compounds

2.1 Introduction

2.1.1 Definition of VOCs

Volatile organic compounds (VOCs) are organic chemicals with a high vapor pressure at ordinary room temperature. There exist diverse definitions of VOCs for different country legislation. The European Union defines VOCs as "any organic compound with an initial boiling point less than or equal to 250 °C under standard atmospheric pressure," which is commonly accepted. According to the Environmental Protection Agency of the United States, VOC means any compound of carbon that takes part in atmospheric photochemical reactions, excluding carbon monoxide, carbon dioxide, carbonic acid, metallic carbides or carbonates, and ammonium carbonate [1]. The main VOCs are alcohols, aldehydes, ketones, aromatic compounds, and others. VOCs are deemed to be the serious air pollutants similar to nitrogen oxides, sulfur oxides, and particulate matter [1].

2.1.2 Emission sources of VOCs

Generally, VOCs can be directly emitted from both natural sources, like plants from forests and grasslands, and anthropogenic sources, like industrial processes, road traffic, and other transport. The main industrial sources include petroleum refineries, papermaking, painting, printing, furniture, electronic, and solvents [2]. The different emission sources will normally generate varied VOC species. Petrochemicals, painting, and detergent will produce benzene, toluene, and ethyl-benzene. Alcohols and ketones are naturally occurring in cosmetics and personal care products, perfumes, and hair spray. Besides outdoor emission of VOCs, there are also many indoor emission of VOCs [3] originating from cooking [4–8] or indoor decoration, like furniture and painting [9–11].

Shuangde Li and Yunfa Chen, State Key Laboratory of Multi-phase Complex Systems, Institute of Process Engineering, Chinese Academy of Sciences, Beijing, China

https://doi.org/10.1515/9783110544183-002

2.1.3 Environmental and health impact of VOCs

VOCs are not only classified indirectly as ozone/smog precursors, but also as contributors of major toxic substances to the environment. The large amount of VOCs in the air is also hazardous to human health. Some of the VOCs can give rise to eye, nose, and throat irritation, and some of the VOCs are highly toxic [12–15]. Getting in touch with a low level of aldehyde may increase throat irritation, while exposure to a high level of aldehyde may cause chronic toxicity to human health. Benzene series compounds are not only toxic but also carcinogenic or genotoxic. Unconsciousness or dizziness is reported upon inhalation of high concentration of aromatic compounds [2].

2.1.4 Purification of VOCs

VOC emission control usually includes the following methods based on absorption, adsorption, membrane separation, condensation, plasma, biodegradation, thermal incineration, photocatalytic, thermal catalytic oxidation, and so on, which are normally nominated by either recovery or destruction methods [2]. Absorption is usually used as a recovery-based method, which is suitable for the treatment of noncomplicated emission mixture under an economic model, while catalytic abatement is suitable for complicated VOC discharge condition with varied VOC species and wide concentration range. One of the effective methods for the purification of the diluted VOCs in a flue gas stream is physical adsorption with activated carbon or zeolite or polymeric adsorbents. However, the frequent replacement and regeneration of the adsorbent after full adsorption will limit its practical application. Thermal incineration is quite suitable for high-concentration VOC degradation under high temperatures (>1,000 °C) with additional fuel requirements, and will produce undesirable byproducts such as nitrogen oxides [2]. Catalytic oxidation is one of the best technologies to remove VOCs (roughly 10^2–10^3 ppm) to CO_2 and H_2O under relative low temperature (250–500 °C) and atmospheric pressure with proper catalytic materials, leading to many advantages such as energy savings, low cost, and operation safety.

Catalytic oxidation of VOCs

Compared with the thermal incineration, the catalytic oxidation technique is quite environmentally friendly with the formation of less noxious products for their low-temperature operation. Regeneration catalytic oxidation is a more efficient method, equipped with a heat exchanger after a catalytic oxidation chamber. Catalytic oxidation is suitable for treating exhaust gas with varying concentrations of VOCs and flow rates with high efficiency. The commonly used catalysts for oxidation of VOCs are the precious metals and transition metal oxides. The monolithic noble metals (platinum, palladium, etc.) supported on oxides are reported to exhibit higher activity at relatively low temperature with the exception of high cost [16]. Transition metal oxides (CuO, Co_3O_4, NiO, and MnO_2) as potential alternatives to precious metals are highly concerned because of their superior properties such as low cost, high resistance to poisoning, and good reducibility [17–18]. The qualities of catalyst (mainly focused on activity, selectivity, and durability) differ largely in laboratory and industrial applications. So for meeting the high practical emission reduction requirements, the required catalysts may contain a variety of different noble metals, metal oxides, and their mixtures. The structure of catalytic materials, such as morphology, size, and crystallinity, quite affects their catalytic activity. How to select the proper catalysts for abatement of VOCs from the large number of available catalysts is one of the main challenges.

Catalytic oxidation mechanism

The catalytic oxidation of VOCs has been widely studied for many years, and three different reaction mechanisms are proposed for the oxidation of VOCs: the Langmuir–Hinshelwood (L–H) mechanism, the Eley–Rideal (E–R) mechanism, and the Mars–van Krevelen (MvK) mechanism. However, it is still difficult to elucidate a clear reaction mechanism because of the different pollutants, catalysts, and reaction conditions. The L–H mechanism assumes that the VOCs and oxygen molecule are separately adsorbed on the surface of the catalyst first, and then the reaction takes place between the adsorbed VOCs and the adsorbed oxygen. The VOCs and oxygen may adsorb on similar type of active sites (single-site L–H model) or two different types of active sites (dual-site L–H model) [2]. The controlling step of the L–H mechanism involves the surface reaction between two adsorbed molecules. The E–R mechanism considers that oxygen molecules are adsorbed first on the surface of the catalyst. Then the adsorbed oxygen species will react with reactant molecules in the gas phase, leading the controlling step to the reaction between the adsorbed molecule and the gas phase molecule [2]. The MvK mechanism assumes that the reaction occurs between the adsorbed VOCs and the adsorbed or lattice oxygen of the catalyst rather than the oxygen in the gas phase. It takes two steps for the oxidation of VOCs according to the MvK mechanism. First, the adsorbed VOCs react with oxygen in the catalyst, resulting in the reduction of the metal oxide. Oxygen vacancies are left behind when the oxidation products desorb from the oxide's surface. Then, the reduced metal oxide is reoxidized by the gas phase oxygen present in the feed, which will annihilate the oxygen vacancies [2, 19]. The MvK mechanism is also recognized as an oxidation-reductive mechanism, because of reduction first and then reoxidation processes over the catalyst. The controlling step of the MvK mechanism is the reaction between the adsorbed VOC molecule and oxygen on different redox sites of the catalyst. MvK mechanism has been widely used for the total oxidation of VOCs [20].

2.2 Synthetic methods for catalyst

2.2.1 Synthetic methods for supported noble catalyst

Impregnation

Impregnation method is widely used for preparing supported noble metal nanoparticles (NPs) for catalytic removal of VOCs. Impregnation means that exposing the supported oxide to precursor solution containing noble salt, following thermal decomposition of noble metal salts at elevated temperature. Impregnation normally includes dry (incipient wetness) and wet impregnation, which need equal and excess volume precursor solution with the saturated adsorption capacity of the support separately. Metal nitrates are favorably used instead of metal chlorides as impregnation precursor for their easily decomposition. Abbasi et al. [21] reported 1% Pt/Al_2O_3–CeO_2 nanocatalysts prepared via wet impregnation method for catalytic oxidation of benzene series (BTX). The alumina support is added to the cerium nitrate hexahydrate solution, then dried and calcined to obtain ceria on alumina. Then Al_2O_3–CeO_2 support is added to hexachloroplatinic acid solution via wet impregnation, then calcination at 500 °C for 4 h under air flow to obtain Pt/Al_2O_3–CeO_2 nanocatalysts. Transmission electron microscope (TEM) images show that platinum particles with an average size of 5–20 nm are well dispersed on Al_2O_3–CeO_2.

Deposition–precipitation

Deposition–precipitation (DP) is another usually used method similar to impregnation for preparing supported noble catalysts. The choice of precipitant, like urea or NaOH, and its concentration would affect the reaction rate, thus affecting the final size and morphology of noble metal particles. Furthermore, the surface area and species of the oxides support will also have an obvious influence on the oxidation states of noble metal particles. Given two examples elucidate the effect of precipitant and surface area of support. For the first example, Chen et al. [22] prepared two kinds of Au/CeO_2 catalysts by DP method using urea or NaOH. An aqueous solution of $HAuCl_4$ is first mixed with CeO_2 and then a designated amount of urea or NaOH is added as the precipitating agent. The two kinds of solid products are filtered first from the liquid, then following the calcination with proper temperature in air. The Au/CeO_2 catalyst through urea method shows higher activity than the NaOH-derived catalyst, with

100% conversion of HCHO into CO_2 and H_2O at room temperature. This is because urea-derived catalyst will produce more amounts of active surface oxygen species resulting from the strong Au–CeO_2 interaction. For the second example, Li et al. [23] synthesized gold catalysts supported on high and low surface area CeO_2 and tested for low-temperature formaldehyde oxidation. High surface area CeO_2 with 270 m^2 g^{-1} is synthesized using a surfactant-templated method, while conventional CeO_2 with surface area of 37 m^2 g^{-1} is synthesized by a citrate sol–gel (SG) method. Then 2.0 g of CeO_2 is added to $HAuCl_4$ (100 mL, 1 mg$_{Au}$ mL^{-1}) aqueous solution with pH adjusted to 7 by 1 M NaOH solution under vigorous stirring at 70 °C for 1 h. The solid is then isolated by centrifugation and dried followed by calcination at 300 °C for 4 h. The reactivity is found to be greatly enhanced, which could be attributed to the fact that Au species mainly in high oxidation states are formed on the increasing high surface area of CeO_2.

Reduction of metal salts

Metallic nanoparticles (NPs) are usually synthesized by the reduction of metal salts under organic solvent or aqueous media with proper reducing agents. The choice of reducing agents has a significant effect on controlling the reaction rate as well as the particle size. Li et al. [24] prepared a series of highly active TiO_2-supported Pt NPs by impregnation methods via different reduction processes and used for benzene oxidation. During the preparation of Pt–TiO_2 catalysts, three different reducing agents, sodium borohydride ($NaBH_4$), sodium citrate, both $NaBH_4$ and sodium citrate are introduced to get different Pt NP sizes supported on TiO_2, which exhibit different catalytic activity for the decomposition of benzene. The Pt/TiO_2 catalyst obtained by the sodium citrate reduction with smallest Pt particle size of 1.9 nm exhibits best total benzene oxidation at approximately 160 °C. Their high benzene catalytic activities are probably attributed to both the metallic Pt NPs with strong oxygen activation capacity, and the rich chemisorbed oxygen with negative charges on the surface of smaller metallic Pt NPs.

2.2.2 Synthetic methods for metal oxide catalysts

Solid-state synthesis

Two or more kinds of molecular compounds (salts, oxides, etc.) are mechanically intermixed as solid particulates followed by being heated at high temperature in oxygen, forming a homogeneous phase. This method is suitable for the synthesis of the metal oxides with thermodynamic stability. de Rivas et al. [25] report the preparation of the Co_3O_4 nanocrystalline for chlorinated VOC decomposition by solid-state reaction with two different procedures. One catalyst is obtained by grinding a mixture of cobalt(II) nitrate and ammonium hydrogen carbonate in an agate mortar for 30 min. The solids are washed thoroughly with distilled water and collected by filtration. The other catalyst with surface area of 80 m^2 g^{-1} is prepared by grinding a mixture of Co(II) basic carbonate and citric acid, which is first premixed by hand grinding for 5 min, then ball milled at a speed of 600 rpm for 6 h.

Co-precipitation method

A solution of the mixed salts with needed molar ratio is treated with a precipitant, such as NaOH, urea, and ammonia, then all the cation ions simultaneously precipitate. The precipitate is washed, dried, and calcined at high temperature to get the targeted metal oxides. Morales et al. [26] prepared Mn–Cu mixed oxide catalysts with co-precipitation (CP) method and used them for ethanol and propane decomposition. The preparation is carried out by mixing aqueous solutions of $Cu(NO_3)_2$, $Mn(NO_3)_2$, and Na_2CO_3 at different aging time. The aging time will have influence on the crystal phases. The activity and the selectivity to CO_2 are enhanced with the increasing aging time. Tang et al. [27] synthesized mesoporous manganese oxides in large quantities with high surface area (355 m^2 g^{-1}) and well-defined mesopores using a novel template-free oxalate route. A clear $(NH_4)_2C_2O_4$ solution is quickly added to $Mn(NO_3)_2$ solution under vigorous stirring at room temperature. Then the precipitation happens in several seconds. After stirring for 40 min, the precipitate is obtained by filtration directly with microtube structure. Compared with manganese oxides prepared by NaOH or NH_4HCO_3 as a precipitant, manganese oxides prepared by oxalate route exhibit much good completely catalytic decomposition activities for benzene, toluene, and o-xylene. The oxalate route is good for keeping the appropriate distribution of manganese oxidation states, which is important for increasing the activity on MnO_x.

Sol–gel synthesis

In this method, the precursors are normally through hydrolysis or citric complex approach to form uniform cation sol, then gradually turn into gel after evaporation of the solvent. The catalysts will be obtained through the calcination of gel at proper temperature. One of the advantages of SG method is inducing the sample maintaining high specific area, controllable pore system, and good dispersion of active phase compared to the traditional impregnation method. Yu et al. [28] prepared a mesoporous MnO_x–CeO_2/TiO_2 by SG method through tetrabutyl titanate hydrolysis for low-temperature toluene decomposition. A transparent sol is first prepared through the mixture of tetrabutyl titanate, acetic acid, ethanol and water with certain mole ratio under vigorous stirring at room temperature. Then certain amount of manganese nitrate and cerium (III) nitrate hexahydrate solution are added to the above sol. After aging for 14 days, the sol transforms to gel, then dry followed by calcined at 500 °C for 4 h. The loaded MnO_x and ceria could disperse well because of the strong interactions with TiO_2, which is the main reason for the enhanced toluene oxidation activity. A series of $La_{(1-x)}Sr_xCo_{(1-y)}Fe_yO_3$ samples are prepared by SG method using EDTA and citric acid as complexing agents [29]. The stoichiometric amounts of $Co(NO_3)_2·6H_2O$, $Fe(NO_3)_3·9H_2O$, $Sr(NO_3)_2$, and $La(NO_3)_3·6H_2O$ are dissolved in EDTA–citric acid solution under heating and stirring with the mole ratio of EDTA:citric acid: total metal ions 1:1.5:1. The solution pH is adjusted to 9 by $NH_3·H_2O$. The prepared perovskites are pure phase with good structural homogeneity and high specific surface area.

Layered double hydroxide precursors transformation

Layered double hydroxides (LDHs) are a class of anion clay materials generally expressed by the formula $[M^{2+}_{1-x}M^{3+}_x (OH)_2](A^{n-})_{x/n} mH_2O$, in which M^{II} and M^{III} cations disperse in an ordered and uniform manner in brucite-like layers, and A^{n-} is a charge-compensating anion such as NO_3^-, $·CO_3^{2-}$, or other ligands [30]. The arrangement of cations in the LDH layers is uniform dispersion, especially fully ordered for Mg:Al ratios of 2:1, which could enhance their functional optimization in oxidation catalysis [31]. The cation composition of LDH-related mixed oxide catalysts can be simply adjusted during the synthetic process of LDH precursors. Therefore, M^{II} and M^{III} metals of the mixed oxide catalysts obtained through the calcination of LDHs will be finely dispersed from a topotactic transformation of LDH [32]. Furthermore, the abundant interfaces over LDH-derived catalysts due to the existing edge and corner sites with less coordinative unsaturation active sites will have great potential to increase catalytic activity for enhanced synergistic effect, leading to low-temperature reducibility and more lattice oxygen [33–34]. In the past few years, there have been some important reviews for the significant development related to the synthesis and application of LDHs. In 2012, Wang and O'Hare [35] first reviewed recent advances in the synthesis and application of LDH nanosheets, then Fan et al. [36] reviewed recent advances in the applications of LDHs with smart design in heterogeneous catalysis, both as directly prepared or after thermal treatment and/or reduction. Furthermore, supported catalysts based on LDHs for catalytic oxidation and hydrogenation are reviewed by Feng et al. [37] together with general functionality and promising application prospects. Li et al. reported the fabrication of porous CoNiAlO [38] or CoCuAlO [39] composite oxides derived from LDH precursors, which exhibit excellent catalytic activity toward complete oxidation of benzene, in comparison with single transition metal Co/Ni/CuAlO catalysts. The reason is ascribed to the low-temperature reducibility and more surface oxygen species arousing from the coordination effect.

Flame spray pyrolysis

Flame spray pyrolysis (FSP) is a gas phase synthesis method, which is suitable for preparation of composite metal oxides [40], like $NiO–Co_3O_4$, $NiO–MoO_3$ [41], and $V_2O_5–TiO_2$ [42]. High-temperature and oxygen-rich condition in the flame may facilitate crystal growth for metal precursors. Then the rapid quench of the prepared metal oxides will maintain vacancies and metastable structure, which will in turn influence catalytic activity [43]. Liu et al. [43] synthesized Ce–Mn oxides using cerium acetate and manganese acetate as precursors in one-step FSP method for catalytic oxidation of benzene, which show SSAs of 20–50 m^2 g^{-1}. Crystalline Ce–Mn oxides with small particles size <40 nm show strong interaction between cerium and manganese oxides within the catalyst particles, which increases the existence of multiple chemical states of Mn ions. The well mixing of Ce and Mn species for flame-made synthesis induces that 12.5% Ce–Mn oxide exhibits excellent benzene catalytic activity.

2.3 VOC purification by supported noble metal nanostructures

Noble metal-based catalysts such as supported platinum and palladium show good reactivity for complete oxidation of VOCs at low temperatures. Table 2.1 lists examples of the most recent catalytic systems reported in the literature for supported noble metal catalysts. Liotta et al. [20] reviewed the catalytic oxidation of VOCs on supported noble metals before the year 2010. The particle size, morphology, and structure dependence caused by preparation method will produce varied oxidation performance of VOCs. Noble metal catalysts supported on conventional substrates such as Al_2O_3, SiO_2, and TiO_2 are mostly studied, and show good oxidation activities and stability [44–45]. While for noble metal catalysts supported on transition metal oxides, especially those with variable valences, the role of the support and redox properties have an important influence for noble metal activities through changing the electronic structure or chemical state. Noble metal catalysts are usually inclined to sintering at high temperature, which will induce the decreased catalytic oxidation activity. Hence, how to improve the stability at elevated reaction temperature is paying much attention for many leading research groups.

2.3.1 Platinum (Pt)-based nanostructures

Platinum-based catalysts are widely used for oxidation of VOCs mainly because of their high activity and stability. It is quite necessary for desiring the catalyst with

Table 2.1: Summary of publications in recent years on noble metal-based nanostructures for catalytic oxidation of VOCs.

Catalyst	Method	VOCs	Space velocity	Temp (°C)	Conversion	Year	References
0.25wt% Pt /CeO$_2$–Al$_2$O$_3$	Sol–gel	1,000 ppm n-butanol	60,000 h^{-1}	165	90%	2014	[46]
1wt% Pt/10wt% CeO$_2$-activated carbon	Impregnation	1,000 ppm ethanol	4,000 h^{-1}	160	100%	2015	[47]
1wt% Pt/10wt% CeO$_2$-activated carbon	Impregnation	1,000 ppm toluene	4,000 h^{-1}	180	100%	2015	[47]
Pt-1.9nm/ZSM-5	Mixture of the two kinds of particles	1,000 ppm toluene	60,000 mL·g^{-1}·h^{-1}	156	98%	2015	[48]
Pt/SBA-15	Wet impregnation and reduction	200 ppm hexane	80 h^{-1}	170	90%	2017	[49]
0.2% Pt/6%Nd/MCM-41	Impregnating with equal volumes	1,000 ppm benzene	20,000 h^{-1}	220	100%	2017	[50]
0.27 wt% Pt/K–Al–SiO$_2$	Ethylene glycol reduction method and colloid impregnation method	800 ppm methyl-ethyl-ketone	42,600 mL·g^{-1}·h^{-1}	163	90%	2017	[51]
Pt/CeO$_2$	Pt nanocrystals dissolving CeO$_2$ solution	1,000 ppm toluene	48,000 mL·g^{-1}·h^{-1}	143	90%	2017	[52]
Pd/Co$_3$O$_4$(3D)	Nanocasting route	150 ppm o-xylene	60,000 mL·g^{-1}·h^{-1}	204	90%	2013	[53]
0.5 wt% Pd/OMS-2	Deposition precipitation	2,000 ppm toluene	240,000 mL·g^{-1}·h^{-1}	285	90%	2017	[54]
0.5 wt% Pd/OMS-2	Deposition precipitation	2,000 ppm ethyl acetate	240,000 mL·g^{-1}·h^{-1}	200	90%	2017	[54]

(continued)

Table 2.1 (continued)

Catalyst	Method	VOCs	Space velocity	Temp (°C)	Conversion	Year	References
0.5 wt% Pd(shell)–Au (core)/TiO$_2$	Impregnation	1,000 ppm toluene	60,000 mL·g^{-1}·h^{-1}	240	100%	2012	[55]
0.5 wt% Pd(shell)–Au (core)/TiO$_2$	Impregnation	3,000 ppm propene	60,000 mL·g^{-1}·h^{-1}	210	100%	2012	[55]
1 wt% AuPd/3DOM Co$_3$O$_4$	PVA-protected reduction route	1,000 ppm toluene	40,000 mL·g^{-1}·h^{-1}	168	90%	2015	[56]
Pt–Pd/MCM-41	One-step synthesis method	500 ppm toluene	10,000 h^{-1}	180	100	2016	[57]

both low content and high activity due to the high cost of noble metals. An effective way for increasing the catalytic activities of unit mass noble metals is to improve the dispersion of active elements through an appropriate depositing method [50]. Support will also deeply affect the catalytic performance for noble metal-loaded catalysts.

Chen et al. [48] observed that the particle size of platinum over Pt/ZSM-5 catalysts plays a vital role in complete oxidation of toluene. Pt-x/ZSM-5 with size-controlled Pt NPs (x, mean diameter) ranging from 1.3 to 2.3 nm are successfully fabricated. Pt-1.9/ZSM-5 exhibits the best toluene oxidation performance (Figure 2.1), which may be due to a balance of Pt dispersion and Pt^0 proportion. Peng et al. [52] also proved the size effect of Pt NPs from 1.3 to 2.5 nm is an important factor and can affect the performance of catalytic oxidation over Pt/CeO$_2$ catalysts prepared by the adsorption method. Pt/CeO$_2$-1.8 sample has the highest activity for toluene oxidation due to the balance of both Pt dispersion and oxygen vacancy concentration of ceria.

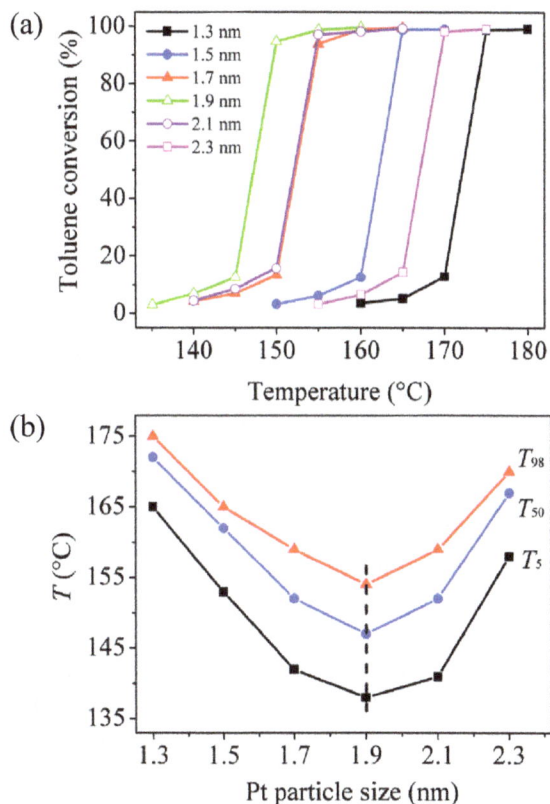

Figure 2.1: (a) Dependence of the catalytic activity on reaction temperature and (b) the dependence of T_5, T_{50}, and T_{98} of toluene on Pt particle size in the complete oxidation of toluene over the Pt-x/ZSM-5 catalysts. Reproduced from Ref. [48] with permission from the Royal Society of Chemistry.

Li et al. [24] studied the effect of reduction treatment on structural properties of Pt–TiO$_2$ catalysts prepared by impregnation methods via different reduction processes and their catalytic activity for benzene oxidation. There exist a number of apparent differences between the oxidized and reduced Pt–TiO$_2$ catalysts, such as particle size, chemical state, and electronic property of Pt NPs, and surface oxygen. Pt–TiO$_2$ catalysts obtained by the sodium citrate reduction show lowest Pt particles size of 1.9 nm, while Pt–TiO$_2$ obtained by H$_2$ or NaBH$_4$ reduction has 2.5 nm. The smallest Pt NPs show best benzene oxidation activity at approximate 160 °C, which are probably responsible for their strong capacity for oxygen activation.

Zuo et al. [50] evaluated the effects of rare earth element-doped MCM-41 on benzene abatement over 0.2% Pt/Nd/MCM-41 samples. The results show that 6 wt% Nd doping sample is highly active and durable for benzene combustion, which is due to optimized Nd content, smaller size Pt particles, and improved Pt dispersion on MCM-41. Abdelouahab-Reddam et al. [47] observed the effect of the dispersion of support for oxidation of VOCs by comparing the two prepared samples, platinum supported on highly dispersed ceria on activated carbon and supported on bulk CeO$_2$. The former catalyst has better performances due to an optimum synergistic interaction between highly dispersed CeO$_2$ and Pt particles.

2.3.2 Palladium (Pd)-based nanostructures

Palladium catalysts have a relatively higher thermal and hydrothermal resistance compared to other noble metal catalysts. Metal–support interaction for improved activity in the catalytic combustion of VOCs is quite investigated. Okumura et al. [45] studied the metal–support interaction using Pd supported on metal oxides with different acid–base properties. The acid–base character of metal oxide over MgO, Al$_2$O$_3$, SiO$_2$, SnO$_2$, Nb$_2$O$_5$, and WO$_3$ influences the affinity for oxygen on Pd surface [45]. So the experimental results prove that Pd loading on strong acidic or basic oxide support samples exhibits poor activity for decomposition of VOCs, compared with samples with Pd loading on weak acidic or basic metal oxides.

The nature of the support plays a key role in total oxidation of VOCs. Tidahy et al. [58] synthesized high surface area macro-mesoporous ZrO$_2$, TiO$_2$, and ZrO$_2$–TiO$_2$ used as catalytic supports for Pd loading via impregnation method, which exhibit powerful catalytic activity for total oxidation of VOCs. Pd/TiO$_2$ owns the highest toluene oxidation performance shown in Figure 2.2, which is attributed to the lowest toluene adsorption enthalpy, and the easiest reducibility for PdO particles. The oxidation mechanism over these Pd-impregnated catalysts should undergo first Pd^0 oxidation by O$_2$ to form active [Pd^{2+}O^{2-}] species, which oxidize VOCs, then the Pd^{2+} cation is simultaneously reduced to Pd^0.

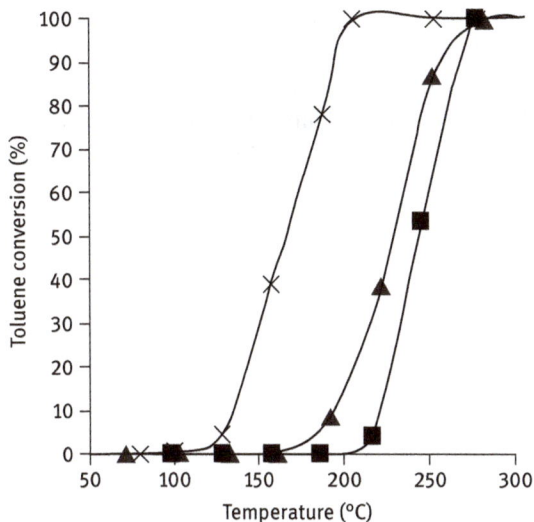

Figure 2.2: Toluene total conversion versus the temperature for the Pd-supported catalysts: (■) Pd/ZrO$_2$, (×) Pd/TiO$_2$, (▲) Pd/ZrO$_2$–TiO$_2$. Reproduced from Ref. [58] with permission from Elsevier B.V. All rights reserved.

A more ordered mesostructure and well-dispersed PdO species for Pd-supported Co$_3$O$_4$ are proved as the main factor for o-xylene oxidation reported by Wang et al. [53] Pd/Co$_3$O$_4$(3D) catalyst prepared by in situ nanocasting method displays higher activity, achieving 100% o-xylene decomposition at around 200 °C than Pd/Co$_3$O$_4$ (3DL) prepared by post-impregnation methods shown in Figure 2.3.

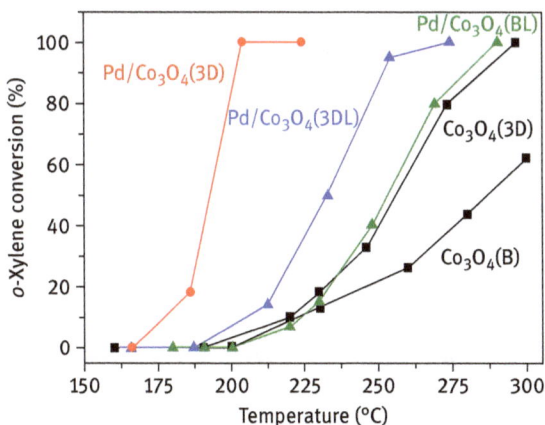

Figure 2.3: o-Xylene conversion as a function of reaction temperature over the catalysts under the condition of o-xylene concentration = 150 ppm in air, the total flow is 100 mL min^{-1}. (3D = 3 dimension, B = bulk). Reproduced from Ref. [53] with permission from Elsevier B.V. All rights reserved.

Effect of preparation method on the surface characteristics and activity of the Pd/
OMS-2 catalysts synthesized by the deposition-precipitation (DP), pre-incorporation
(PI), and ion-exchanging (EX) methods, for the oxidation of VOCs is further studied
by Liu et al. [54] The 0.5 wt% Pd/OMS-2-DP sample exhibits the best catalytic activity (Figure 2.4), with the $T_{50\%}$ and $T_{90\%}$ being 25 and 55 °C for CO oxidation, 240

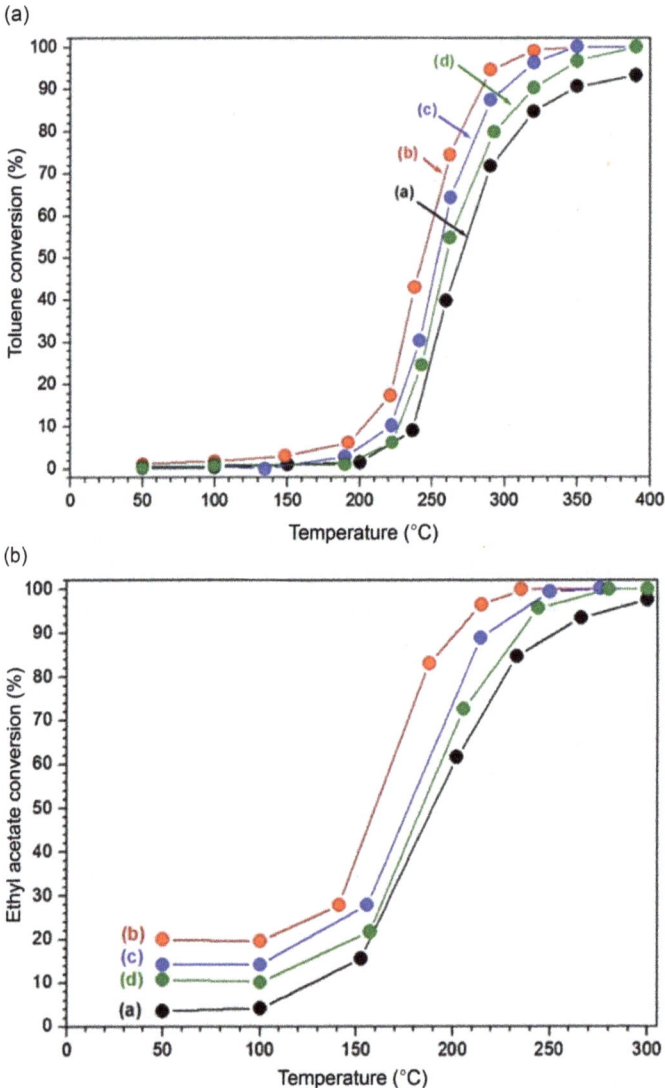

Figure 2.4: Catalytic activity of the OMS-2 and 0.5 wt% Pd/OMS-2 samples synthesized by different
methods for the oxidation of (a) toluene at SV = 240,000 mL/(g h), (b) ethyl acetate at SV =
240,000 mL/(g h). (a) OMS-2, (b) 0.5 wt% Pd/OMS-2-DP, (c) 0.5 wt% Pd/OMS-2-PI, and (d) 0.5 wt%
Pd/OMS-2-EX. Reproduced from Ref. [54] with permission from Elsevier B.V. All rights reserved.

and 285 °C for toluene oxidation, and 160 and 200 °C for ethyl acetate oxidation, respectively. The excellent catalytic activity is related to 0.5 wt% Pd/OMS-2-DP sample having the highest surface Pd concentration, $(Mn^{2+} + Mn^{3+})/Mn^{4+}$ and O_{ads}/O_{latt} ratios, and acid sites.

2.3.3 Mixed noble metal catalysts

In order to further improve the activity of the supported noble metal catalyst, the bimetallic catalyst is widely studied, since there is a synergic interaction between the particles. Influence of Pt–Pd ratio on the complete benzene oxidation over Pt–Pd bimetal catalyst supported on γ-alumina is reported by Kim et al. [59] Pt–Pd bimetallic catalysts with the exposed Pt and Pd active sites show superior catalytic activity than monometallic Pd or Pt catalysts. Figure 2.5 shows that 03Pt2Pd bimetallic catalyst with the optimum ratio of Pt and Pd possess the highest activity in benzene combustion. However, the addition of an excess amount of Pt-modified Pt–Pd bimetal catalysts will decrease the benzene oxidation activity due to overlap or blockage of active sites. Moreover, long-term stability test in Figure 2.6 indicates that the addition of Pt to Pd/γ-Al_2O_3 catalyst is effective to prevent the deactivation of catalyts.

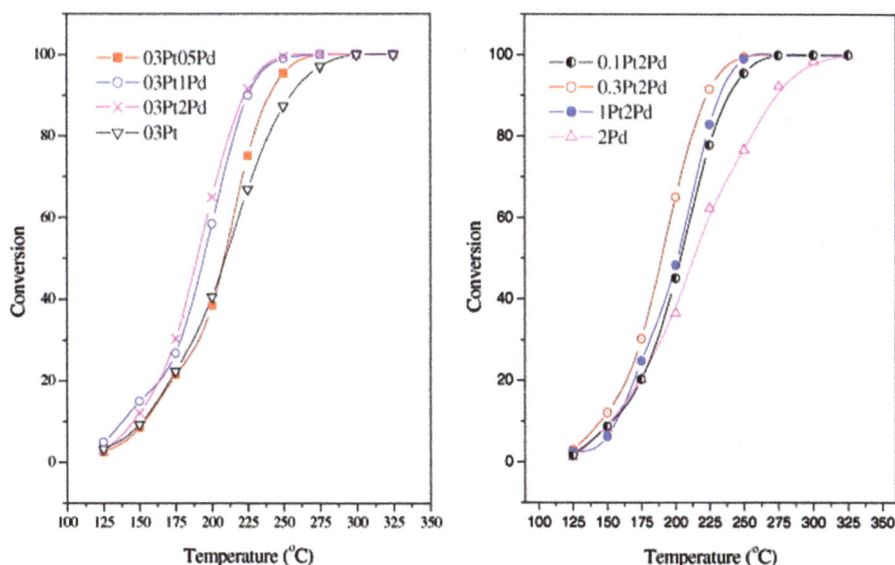

Figure 2.5: Light-off curves for benzene combustion over mPt2Pd and 0.3PtnPd. (m: 0.3, 1, 0; n: 0.5, 1, 2). Reproduced from Ref. [59] with permission from Elsevier B.V. All rights reserved.

Pd–Pt/SiO_2–OA catalysts are successfully prepared to estimate the synergistic effect of Pd and Pt for toluene oxidation by improved incipient wetness impregnation

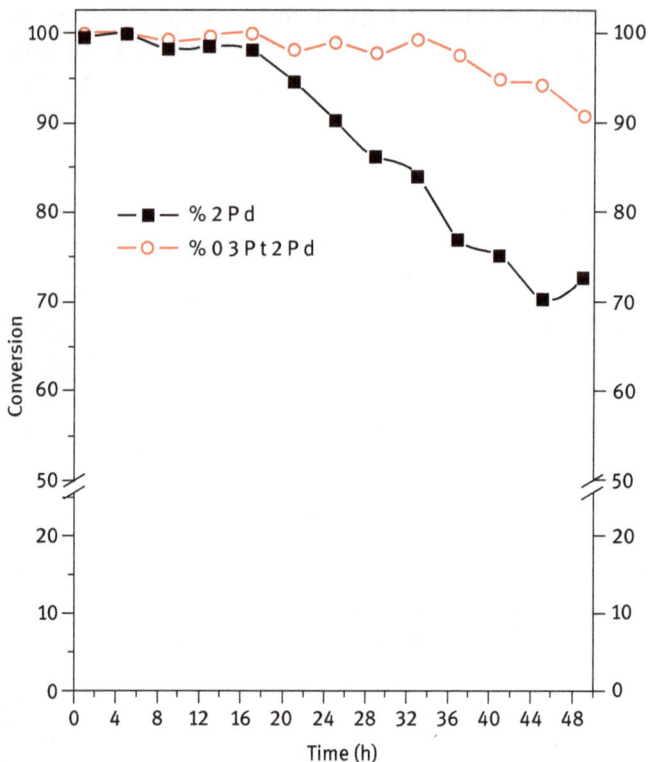

Figure 2.6: Stability test of 2Pd and 03Pt2Pd catalysts for 48 h. Reproduced from Ref. [59] with permission from Elsevier B.V. All rights reserved.

with the addition of oleic acid (OA) by Wang et al. [60] As shown in Figure 2.7, the bimetallic catalyst 0.25% Pd–0.25% Pt/SiO$_2$–OA with the same metal loading displays better catalytic activity than the other two catalysts with monometallic loading of 0.5% Pd/SiO$_2$–OA or 0.5% Pt/SiO$_2$–OA. The result shows the bimetallic loading sample having high dispersion of Pd and Pt particles, which is responsible for the effective and promoting oxidation activities and anticoking performance.

Hosseini et al. [55] prepared three kinds of Pd and Au deposited on mesoporous TiO$_2$ catalysts by different metal deposition order Pd(shell)–Au(core)/TiO$_2$, Pd–Au (alloy)/TiO$_2$, and Au(shell)–Pd(core)/TiO$_2$ shown in Figure 2.8, and evaluated the effect of metal deposition on total oxidation of VOC. As shown in Figure 2.9, Pd(shell)–Au(core)/TiO$_2$ shows significantly higher toluene destruction catalytic activity than the other two catalysts, due to the unique core–shell morphology. The oxygen and VOC molecules are adsorbed in competition on the surface of catalyst following L–H mechanism. Therefore, Pd-shell and Au-core morphology have strong ability of the preferential adsorption and activation of oxygen molecules, while Au-shell is correlated with its lower ability for oxygen affinity and polarization.

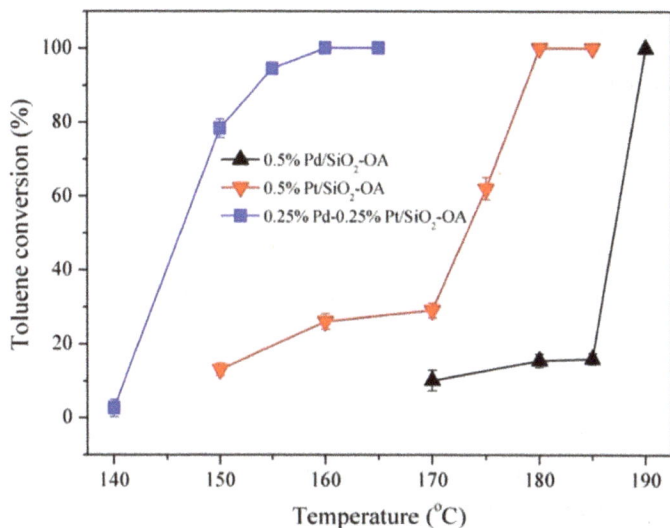

Figure 2.7: Catalytic combustion of toluene over the 0.5% Pd/SiO₂–OA, 0.5% Pd/SiO₂–OA, and 0.25% Pd–0.25% Pt/SiO₂–OA catalysts ($n = 3$). Reproduced from Ref. [60] with permission from Elsevier B.V. All rights reserved.

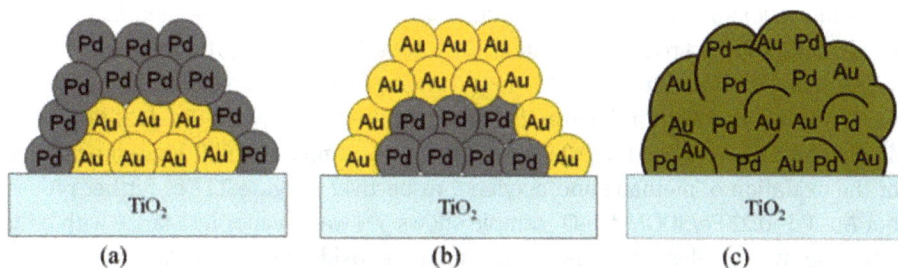

Figure 2.8: Models proposed for: (a) Pd(shell)–Au(core)/TiO₂, (b) Au(shell)–Pd(core)/TiO₂, and (c) Pd–Au(alloy)/TiO₂. Reproduced from Ref. [55] with permission from Elsevier B.V. All rights reserved.

Alloying of noble metals is another effective method for promoting oxidation of VOCs, which has been demonstrated in many works. Pd and Au alloy catalyst shows high activity and stability for toluene oxidation due to better oxygen activation ability and stronger noble metal and support oxide interaction [56]. To further decrease the amount of the noble metal for practical application, doping a certain amount of transition metals has been widely reported and accepted as a facile way to reduce the high cost of these noble metal-based materials. The transition metal-doped catalysts usually perform well in many VOC catalytic reactions. Doping 1.86–1.97 wt% transition metal, M (M = Mn, Cr, Fe, and Co) to Au–Pd NPs supported on

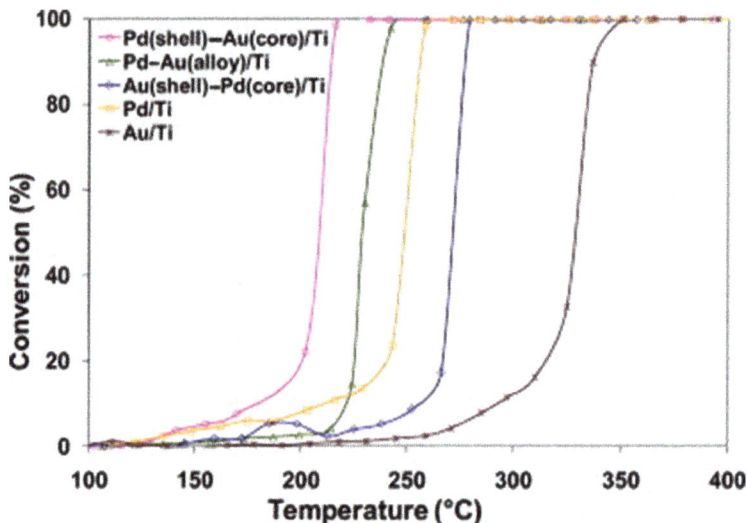

Figure 2.9: Conversion of toluene on Pd(shell)–Au(core)/TiO$_2$, Au(shell)–Pd(core)/TiO$_2$, Pd–Au (alloy)/TiO$_2$, Pd/TiO$_2$ and Au/TiO$_2$. Reproduced from Ref. [55] with permission from Elsevier B.V. All rights reserved.

three-dimensionally ordered macroporous (3DOM) Mn$_2$O$_3$ using the modified polyvi-nyl alcohol-protected reduction method is reported by Xie et al. [61] Figure 2.10 shows the Au–Pd–xM NPs with a size of 3.6–4.4 nm are highly dispersed on the surface of 3DOM Mn$_2$O$_3$. As shown in Figures 2.11 and 2.12, the 1.94 wt% Au–Pd–0.21 Co/3DOM Mn$_2$O$_3$ and 1.94 wt% Au–Pd–0.22 Fe/3DOM Mn$_2$O$_3$ samples exhibit highest activities for the oxidation of methane and o-xylene, respectively. Figure 2.12(b) further proves that Au–Pd–0.22 Fe/3DOM Mn$_2$O$_3$ sample shows 5% water vapor resistance with only 10% drop over o-xylene conversion. The methane oxidation rate of Au–Pd–0.21 Co/ 3DOM Mn$_2$O$_3$ at 340 °C is three times higher compared with the corresponding Co ab-sent sample, and the o-xylene reaction rate of Au–Pd–0.22Fe/3DOM Mn$_2$O$_3$ at 140 °C is two times higher than that of corresponding Fe absent sample. Doping a certain amount of the transition metal to Au–Pd/3DOM Mn$_2$O$_3$ will improve the affinity for methane adsorption and oxygen activation, which could be correlated with the modi-fication of the microstructure of the alloy NPs.

2.4 Mixed metal oxide catalysts

There are many reports on catalytic oxidation of VOCs on transition metal oxides [1–2], [19, 62–65]. Nonnoble metal oxide catalysts such as single oxides, mixed ox-ides, and perovskite oxides have been considered as alternatives to low cost to

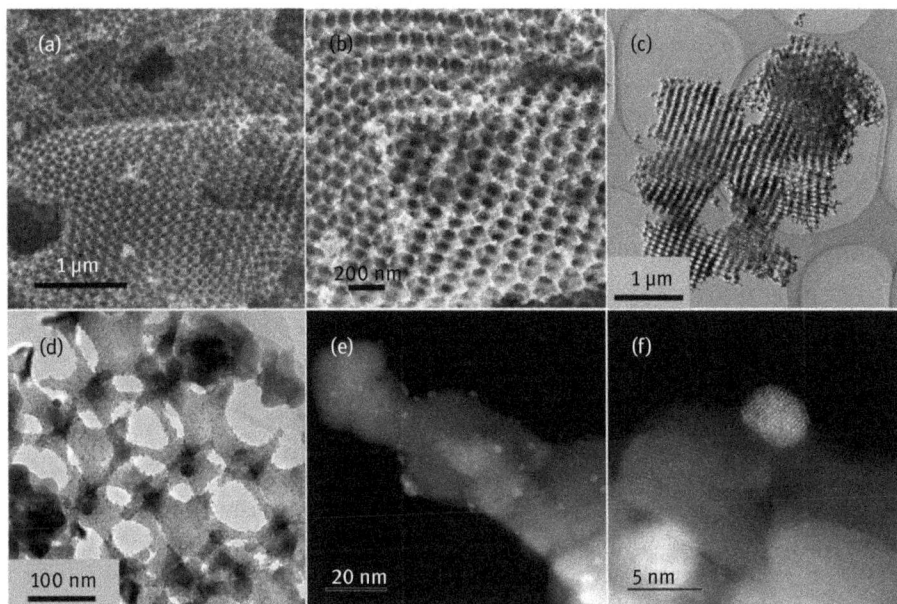

Figure 2.10: (a, b) SEM, (c, d) TEM, and (e, f) HAADF-STEM images of the Au–Pd–0.47Mn/3DOM Mn$_2$O$_3$ sample. Reproduced from Ref. [61] with permission from Elsevier B.V. All rights reserved.

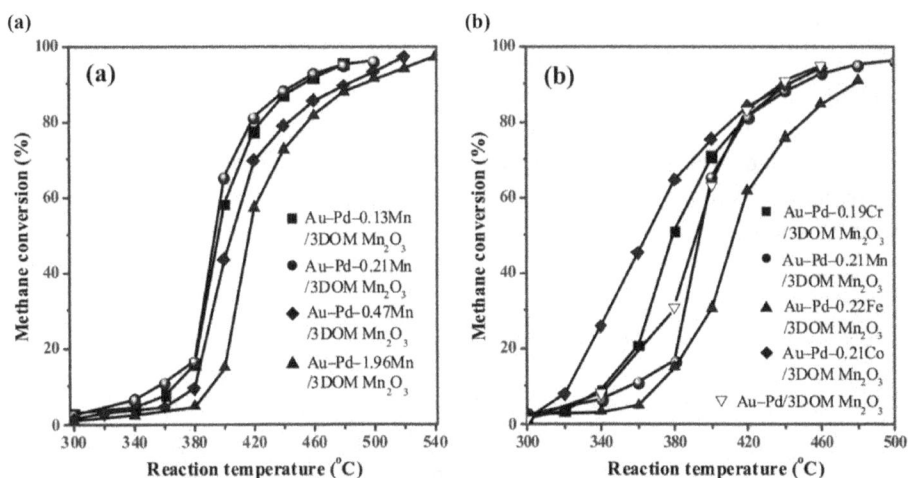

Figure 2.11: Methane conversion as a function of reaction temperature over (a) Au–Pd–xMn/3DOM Mn$_2$O$_3$ (x = 0.13–1.96) and (b) Au–Pd–xM/3DOM Mn$_2$O$_3$ (x = 0.19–0.22; M = Mn, Cr, Fe, and Co) at SV = 40,000 mL/(g h). Reproduced from Ref. [61] with permission from Elsevier B.V. All rights reserved.

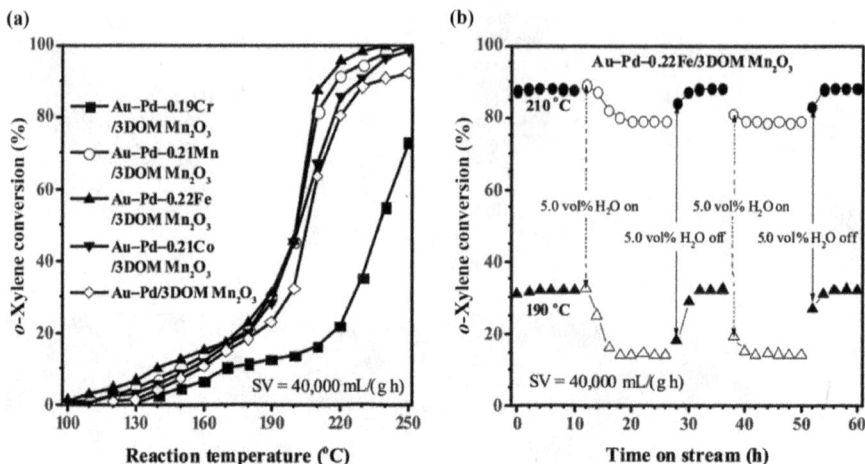

Figure 2.12: (a) *o*-Xylene conversion versus reaction temperature over Au–Pd–xM/3DOM Mn$_2$O$_3$ (M = Mn, Cr, Fe, and Co) and (a) *o*-xylene conversion as a function of on-stream reaction time in the presence or absence of 5.0 vol% water vapor at different temperatures over Au–Pd–0.22 Fe/3DOM Mn$_2$O$_3$. Reproduced from Ref. [61] with permission from Elsevier B.V. All rights reserved.

platinum- and palladium-based catalysts, which are partially limited by high cost in the industrial application. Researchers have been seeking to improve the oxidation activity, resistance to water vapor and to poisoning of metal oxide catalysts. McFarland and Metiu [19] reviewed catalysis by doped oxides through a novel perspective for the improvement of the catalytic activity of an oxide by substituting a small fraction of the cations of a "host oxide" with a different cation, which is called substitutional doping or doping. Li et al. [64] reviewed nanostructured transition metal oxide materials for the catalytic removal of VOCs. Table 2.2 lists some of the typical transition metal oxide catalysts as a representative of the possible innovative directions.

2.4.1 Single metal oxide catalysts

Many active single transition metal oxides, Mn, Co, Ni, and Ce are cheaper and easier to be obtained as catalysts for oxidation of VOCs. Among the active oxide catalysts, manganese oxides such as Mn$_3$O$_4$, Mn$_2$O$_3$, and MnO$_2$ with varied morphology and controlled size through different preparation method are reported to display good activity and also considered as environmentally friendly materials. Lahousse et al. [74] prepared a metal oxide γ-MnO$_2$ and a noble metal Pt/TiO$_2$ catalyst to evaluate their performance for oxidation of VOCs, based on the activity, sensitivity to competition effects between VOCs, resistance to water vapor, and the stability.

Table 2.2: Summary of publications in recent years on transition metal-based nanostructures for catalytic oxidation of VOCs.

Catalyst	Method	VOCs	Space velocity	Temp (°C)	Conversion	Year	References
Mn_3O_4	Purchase	1,000 ppm toluene	15,000 mL·g^{-1}·h^{-1}	270	100%	2010	[66]
0.5 wt% K/Mn_3O_4	Impregnation-vaporization	1,000 ppm toluene	15,000 mL·g^{-1}·h^{-1}	260	100%	2010	[66]
MnO_x	Precipitation with oxalate route	500 ppm benzene	60,000 mL·g^{-1}·h^{-1}	209	90%	2014	[27]
Mn_2O_3	Rapid preparation method	1,000 ppm toluene	60,000 mL·g^{-1}·h^{-1}	274	90%	2017	[67]
Co_3O_4	Dispersion–precipitation synthesis	1,000 ppm propane	12,000 h^{-1}	225	90%	2017	[68]
Mesoporous $Cu_{0.3}Ce_{0.7}O_x$	Self-precipitation protocol	1,000 ppm toluene	36,000 h^{-1}	212	90%	2014	[69]
CuCo/halloysite	Wet impregnation	600 ppm toluene	60,000 mL·g^{-1}·h^{-1}	301	90%	2015	[70]
$CuCe_{0.75}Zr_{0.25}O/ZSM$-5	Impregnation	ethyl acetate unknown concentration	24,000 h^{-1}	270	100%	2016	[71]
$LaMnO_3$	Citrate sol–gel	1,000 ppm toluene	15,000 mL·g^{-1}·h^{-1}	213	90%	2014	[72]
$LaMnO_3$	Glycine combustion method	1,000 ppm toluene	15,000 mL·g^{-1}·h^{-1}	260	90%	2014	[72]
$LaMnO_3$	Co-precipitation	1,000 ppm toluene	15,000 mL·g^{-1}·h^{-1}	250	90%	2014	[72]
Co_2NiAlO	LDH-derived method	100 ppm toluene	60,000 mL·g^{-1}·h^{-1}	227	90%	2016	[38]
$Cu_{0.5}Co_{2.5}Al$–MMO	LDH-derived method	1,000 ppm ethyl acetate	60,000 mL·g^{-1}·h^{-1}	290	90%	2015	[39]
Co/La–CeO_2	Incipient wetness impregnation	1,000 mg cm^{-3} toluene and ethyl acetate	12,0000 mL·g^{-1}·h^{-1}	240	100%	2014	[73]

The γ-MnO$_2$ catalyst is proved to be more active and less affected by interferences from VOC mixture (benzene, ethyl acetate, and n-hexane) than those of Pt/TiO$_2$ catalyst. Conversely, the activity of the Pt/TiO$_2$ appears slightly more stable under water vapor. Kim and Shim [66] investigated the catalytic performance of a series of manganese oxide and obtained activities in the following sequence Mn$_3$O$_4$ > Mn$_2$O$_3$ > MnO$_2$, which is in accordance with the ability of oxygen mobility of the catalysts. Addition of potassium, calcium, and magnesium, acting as promoters to Mn$_3$O$_4$ catalyst will further boost the catalytic activity of Mn$_3$O$_4$, which could be ascribed to the defect-oxide or a hydroxyl-like group. A novel method is applied to synthesize mesoporous manganese oxides with high surface area reported by Tang et al. [27], which is employing a template-free oxalate route to fabricate manganese oxides nanorod with surface area of 355 m^2 g^{-1}. The sample exhibits best VOC degradation activities for benzene, toluene, and o-xylene, among other prepared manganese oxides through NaOH, NH$_4$HCO$_3$, and nanocasting method. For example, the temperature for 90% benzene conversion achieves at 209 °C, while it is 341 °C for the catalyst prepared by NaOH route. The prompt performance of manganese oxides by oxalate route is ascribed to their high surface area, low-temperature reducibility, and distribution of surface species. Li et al. [75] reported the performance for benzene removal of the following three types of hierarchical MnO$_2$ microspheres with the order of S3 > S2 > S1 (Figure 2.13). S1 represents β-MnO$_2$ assembled by uniform nanorods, S2 represents hollow double-walled β/α-MnO$_2$ constructed by two-categorical nanorods, and S3 represents hollow α-MnO$_2$ consisting of nanorods and nanowires.

Russo and coworkers [76] evaluated three different valence mesoporous manganese oxide catalysts (Mn$_2$O$_3$, Mn$_3$O$_4$, and Mn$_x$O$_y$) prepared by solution combustion synthesis, using complete oxidation of ethylene, propylene, toluene, and their mixture as probe molecules. The results shown in Figure 2.14 display that Mn$_3$O$_4$ catalyst owns the best oxidation of VOCs, which is correlated with the highest amount of electrophilic surface oxygen.

Besides Mn-based catalysts, ceria-based catalysts for their unique oxygen storage capacity exhibit high VOC oxidation activity based on the literature survey [77–79]. The VOC oxidation mechanism over ceria is generally accepted by the MvK type with the rate control steps related to oxygen supply and reoxidation capacity over the reducible oxide. Torrente-Murciano et al. [80] reported shape-dependent naphthalene oxidation activity over nanostructured CeO$_2$. The base concentration and temperature are the two key factors to depend on the final ceria morphology with NPs, nanorods, and nanocubes during the hydrothermal synthesis, and the obtained morphological phase diagram is shown in Figure 2.15. Catalytic activity trend normalized by the unit of surface area is nanorods < nanocubes < NPs. The trend is quite in accordance with the surface oxygen vacancies concentration depending on the preferential exposure of the (110) and (100) planes.

López et al. [78] further revealed that surface oxygen defects of nanostructured ceria with varied morphologies play a very importance role in the oxidation activity

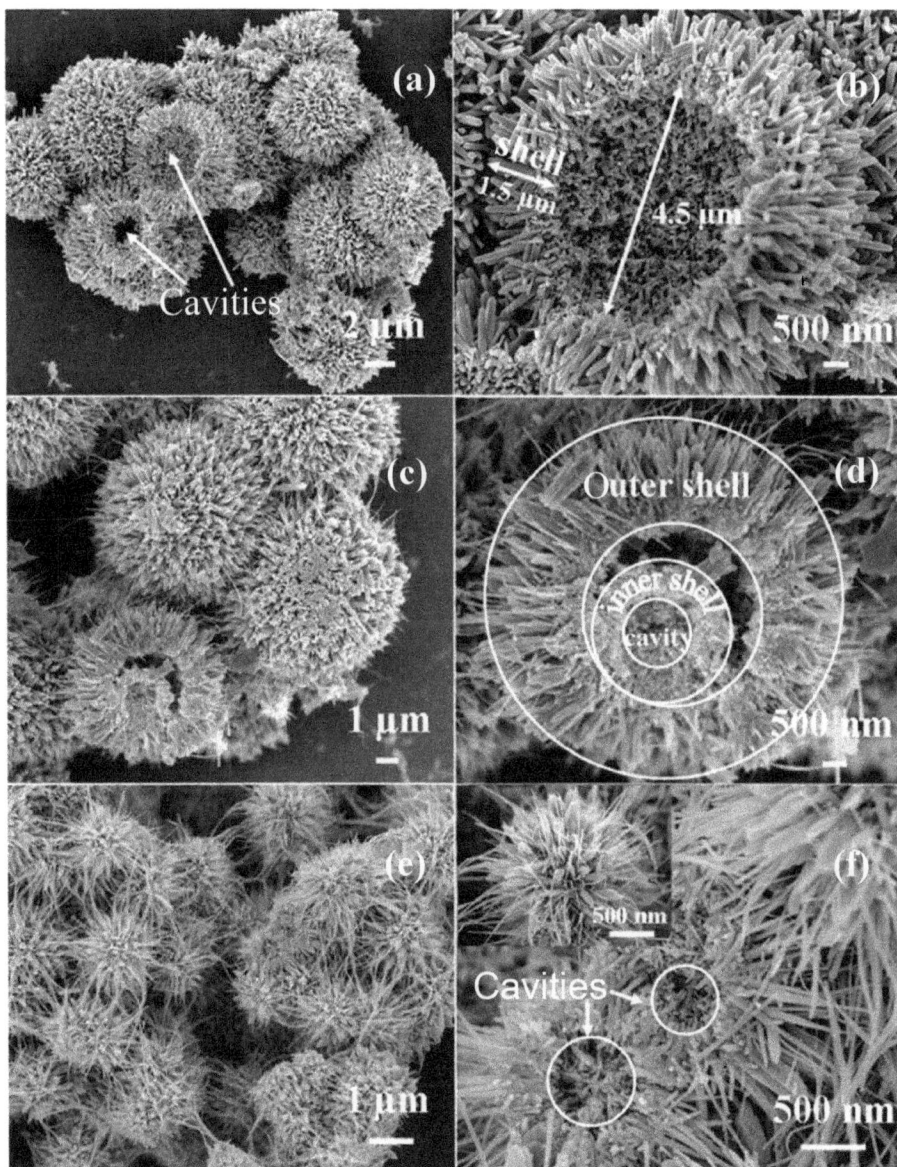

Figure 2.13: SEM images of sample S1 (a and b), S2 (c and d), and S3 (e and f) obtained by a hydrothermal method. Reproduced from Ref. [75] with permission from The Royal Society of Chemistry.

of VOCs. Ceria nanorods present higher oxidation activities than nanocubes, because nanorods with higher surface area show higher concentration of bulk and surface defects. Figure 2.16 shows the influence of the oxygen defects on the

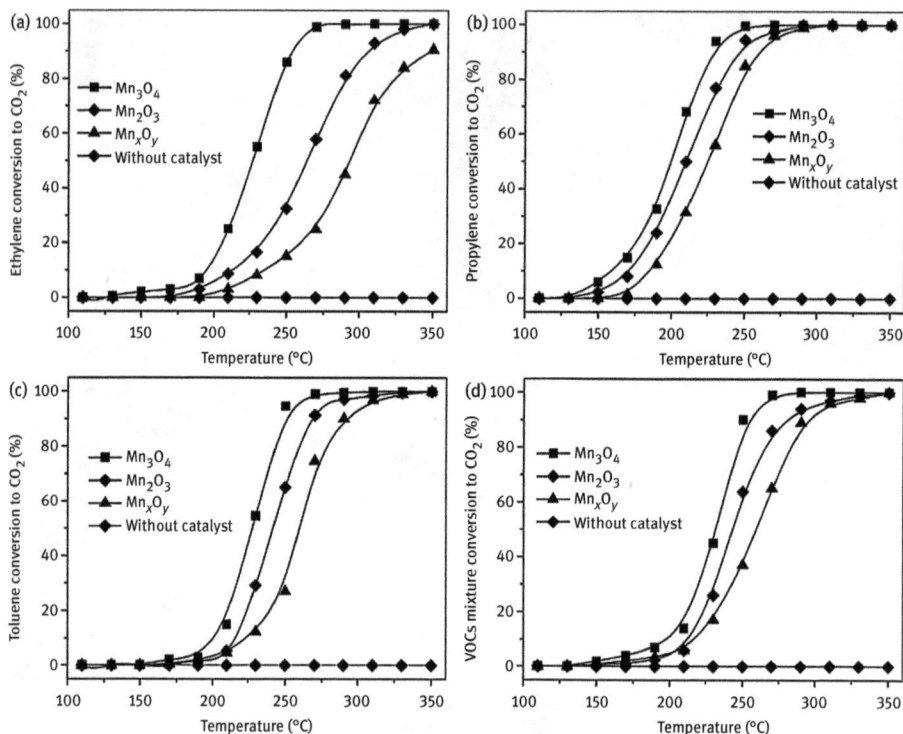

Figure 2.14: Catalytic results of powder catalysts for the total oxidation of (a) ethylene, (b) propylene, (c) toluene, and (d) VOC mixture (ethylene, propylene, and toluene) as a function of the reaction temperature. Reproduced from Ref. [76] with permission from Elsevier B.V. All rights reserved.

catalytic performance of ceria. The ratio of $Ce^{3+}/(Ce^{3+} +Ce^{4+})$ (Figure 2.16A) is related with the surface amount of the presence of oxygen vacancies determined by X-ray photoelectron spectroscopy (XPS), which indicates that the activity of toluene abatement is quite linearly dependent with the concentration of surface Ce^{3+}. The intensity ratio of I600/I460 determined by Raman spectra represents the oxygen vacancy concentration, and the high full width at half maximum (FWHM460) is associated with a high amount of oxygen vacancies in the CeO_2 structure (Figure 2.16B). The results tell us that the ceria nanorods are directly related to the concentration of surface oxygen defects.

There is a growing potential to promote the removal performance of VOCs by preparing the nanostructured Co-based metal oxides [81–82]. Spinel Co_3O_4 is a p-type semiconductor, where O^{2-} is cubic close packed, Co^{2+} and Co^{3+} are in tetrahedral and octahedral coordination, respectively. Co_3O_4 has a lower Co–O bond strength and a higher capability of the activation of oxygen molecule, which is reported to play crucial roles in catalytic oxidation of VOCs [81]. Yan et al. [81] fabricated Co_3O_4

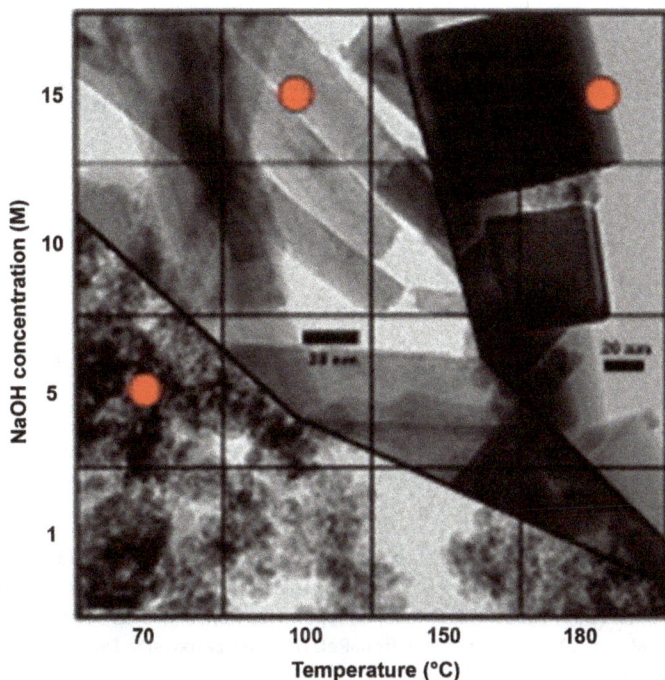

Figure 2.15: Morphological phase diagram of CeO$_2$ after 10 h of hydrothermal treatment. Phase boundaries shown do not imply sharp transitions to pure phases. Circles show the conditions of the representative samples of each morphology taken for the catalytic study. Reproduced from Ref. [80] with permission from Elsevier B.V. All rights reserved.

nanoflower clusters by self-assembly under low-temperature hydrothermal process shown in Figure 2.17. The Co$_3$O$_4$ sample with good crystallinity and porous structure shows much higher toluene catalytic oxidation activity and stability than the Co$_3$O$_4$ blocks under the same reaction conditions. This is correlated with the Co$_3$O$_4$ nano-flowers exposed to abundant Co^{3+} cations on the surface, which proves that Co^{3+} cation may play a key role in toluene adsorption.

2.4.2 Mixed metal oxide catalysts

Mixed metal oxide catalysts such as Cu–Mn–O, Mn–Ce–O, and Mn–Co–O have been reported to exhibit higher catalytic oxidation activity of VOCs than single metal oxides because of the widely present synergistic effects [43, 69–71, 83–92]. However, there are still many challenges to maximize the synergistic effects for achieving high catalytic activity for mixed metal oxide catalysts. Recently, many de-signs have been created to promote the synergistic effect such as the fabrication of spherical shaped structures [90], finely dispersed structures [89], mesoporous

(a)

(b)

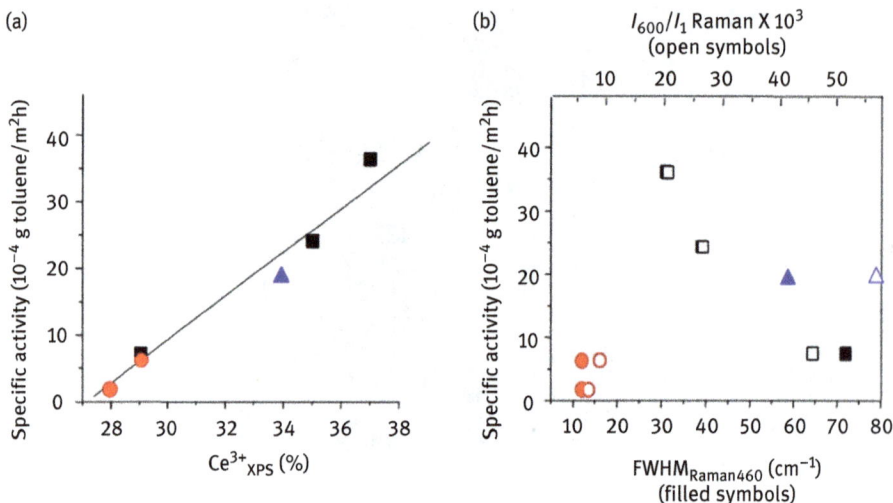

Figure 2.16: Influence of the oxygen defects on the catalytic performance of ceria nanorods. (a) Specific rate versus $Ce^{3+}/(Ce^{3+} +Ce^{4+})$ determined by XPS and (b) specific rate versus FWHM460 Raman and I600/I460. Symbols: (●,○) cubes, (■,□) rods, nanoparticles from sigma (▲). Reaction conditions in text. Reaction rates at 225 °C. Reproduced from Ref. [78] with permission from Elsevier B.V. All rights reserved.

sturctures [69], and solid solution [83]. The literature aimed at creating a highly active interface between different species, or making different metal ions within the same crystal structure, which is expected to increase the catalytic activities.

Tang et al. [83] prepared porous Mn–Co mixed oxide nanorod by an oxalate route, which shows an increased ethyl acetate and n-hexane degradation with T_{90} at 194 and 210 °C, respectively, in comparison with their single oxides. The obtained Mn–Co oxide with higher surface area is due to the formation of solid solution with spinel structure, which inhibits the growth of NPs. The lower temperature reducibility is ascribed to the strong synergistic effect of Mn–Co solid solution. Tang et al. [84] furthermore synthesized a series of mesoporous Cu–Mn oxides with different molar ratio of Cu:Mn owning high surface area (~221 $m^2 g^{-1}$) through mesoporous SBA-15 as hard template shown in Figure 2.18. The T_{90} for benzene decomposition over Cu0.6Mn oxides achieves at 234 °C, which is 131 °C lower than that of Cu–Mn oxide with low surface area prepared by NaOH CP method (Figure 2.19). SBA-15 plays a key role as a limited nanoreactor for the formation of Cu–Mn oxide. The Cu–Mn oxide possesses better catalytic activity owing to small particle size, high surface area, and rich surface-adsorbed oxygen species emerging from the promoting interaction of Cu–Mn species.

Liu et al. [43] synthesized crystalline Ce–Mn oxides with varied Ce and Mn ratio in one step for benzene decomposition utilizing FSP method, with particles of size

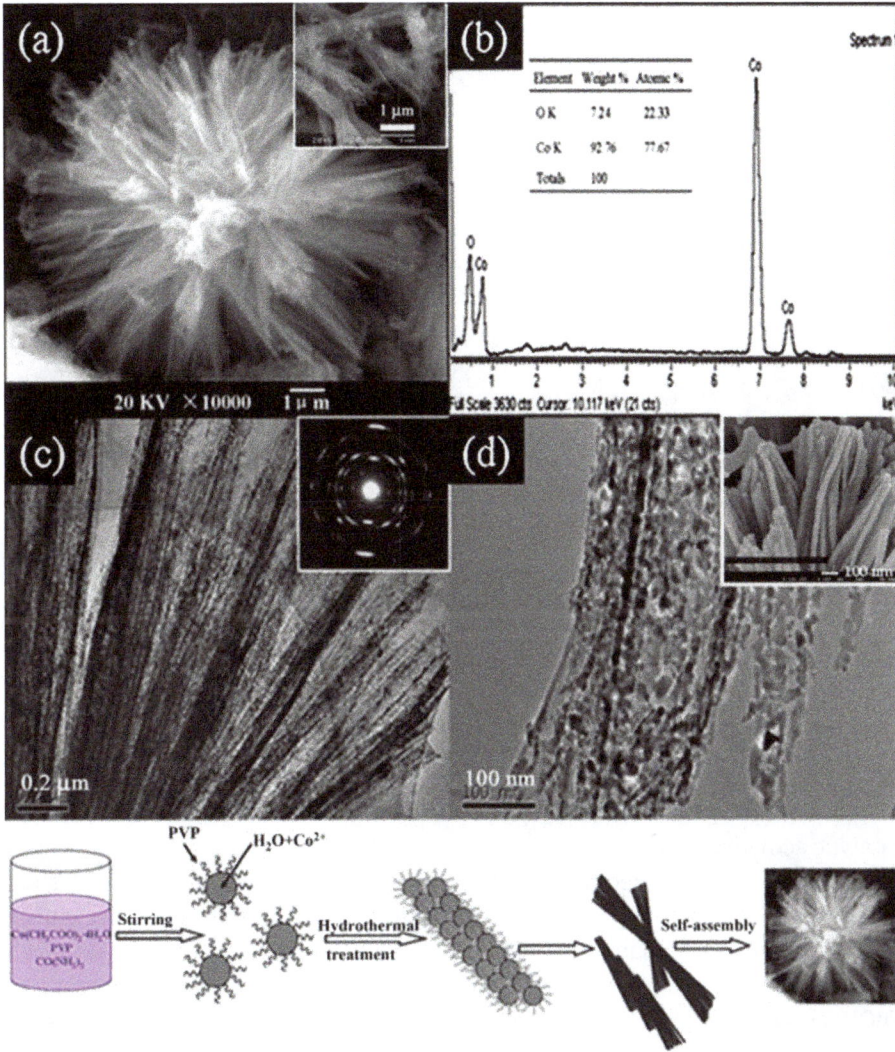

Element	Weight %	Atomic %
O K	7.24	22.33
Co K	92.76	77.67
Totals	100	

Figure 2.17: (a) SEM image, (b) EDX spectrum, (c and d) TEM, and SAED images of the as-prepared Co_3O_4 sample. Reproduced from Ref. [81] with permission from Elsevier B.V. All rights reserved.

<40 nm and specific surface area of 20–50 m^2 g^{-1}. About 12.5% Ce–Mn oxide exhibits the best catalytic activity with T_{95} about 260 °C. All the catalysts show good performance because of Mn ions evidenced in multiple chemical states and strong interaction between Ce and Mn brought about during the high-temperature flame process. Hierarchical layer-stacked and mesoporous Ce–Mn composite oxides shown in Figure 2.20 are prepared by an oxalate acid precipitation following decomposition procedure [93]. They are characterized to exhibit higher manganese

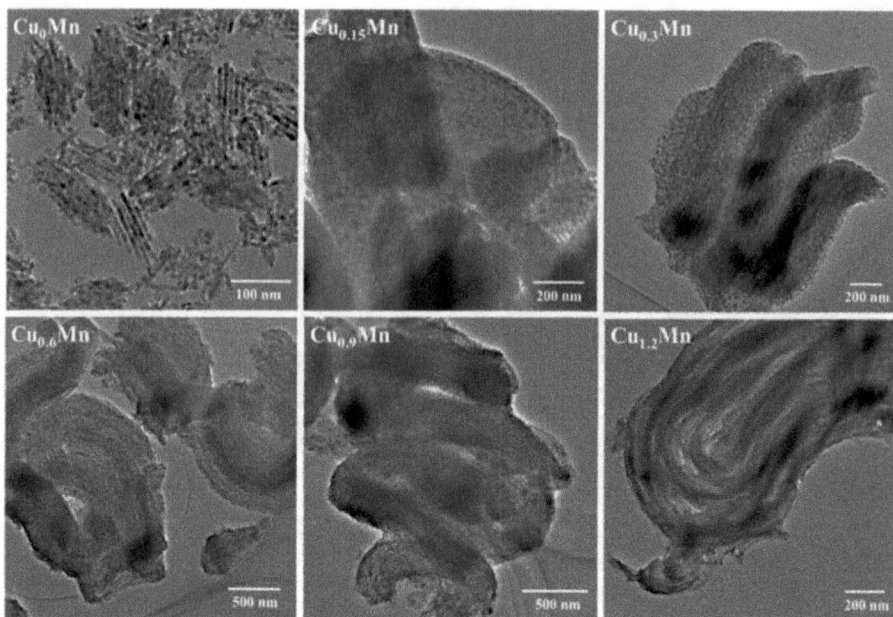

Figure 2.18: TEM images of the Cu–Mn composite oxides prepared by nanocasting strategy. Reproduced from Ref. [84] with permission from Elsevier B.V. All rights reserved.

oxidation state and richer adsorbed surface oxygen species, compared with Mn–Ce composite oxide prepared by Na_2CO_3 precipitation, which leads to prompt toluene catalytic activities.

He et al. [69] evaluated mesoporous $CuCeO_x$ catalysts for removal of VOCs obtained via a simple self-precipitation procedure with the formation mechanism shown in Figure 2.21. Many Cu^{2+} ions enter into CeO_2 lattice to form $Cu_xCe_{1-x}O_{2-\delta}$ solid solution, bringing about large amounts of oxygen vacancies in the interface of CuO_x and CeO_2 oxides. Meanwhile, the formed $Cu^{2+}-O^{2-}-Ce^{4+}$ bridges in the solid solution will promote their reducibility for the convenience of oxygen transfer between Cu and Ce. Increasing the Cu content gradually, the activity enhances to the maximum, then starts to decrease. $Cu_{0.3}Ce_{0.7}O_x$ sample exhibits the highest toluene and propanal conversion efficiency, with T_{90} around 212 and 192 °C, respectively, at GHSV of 36,000 h^{-1}. However, propanal oxidation can be remarkably suppressed under the coexistence of the toluene, because $CuCeO_x$ catalysts exhibit stronger surface affinity for toluene molecules (Figure 2.22). The results demonstrate that the superior activity over $CuCeO_x$ catalysts are ascribed to the higher surface oxygen ad-species concentration and better low-temperature reducibility.

Ahn et al. [90] compared the effect of amorphous and crystalline spherical-shaped Cu–Mn oxides for benzene abatement performance obtained by a wet granulation process under different calcination temperatures. The low-temperature

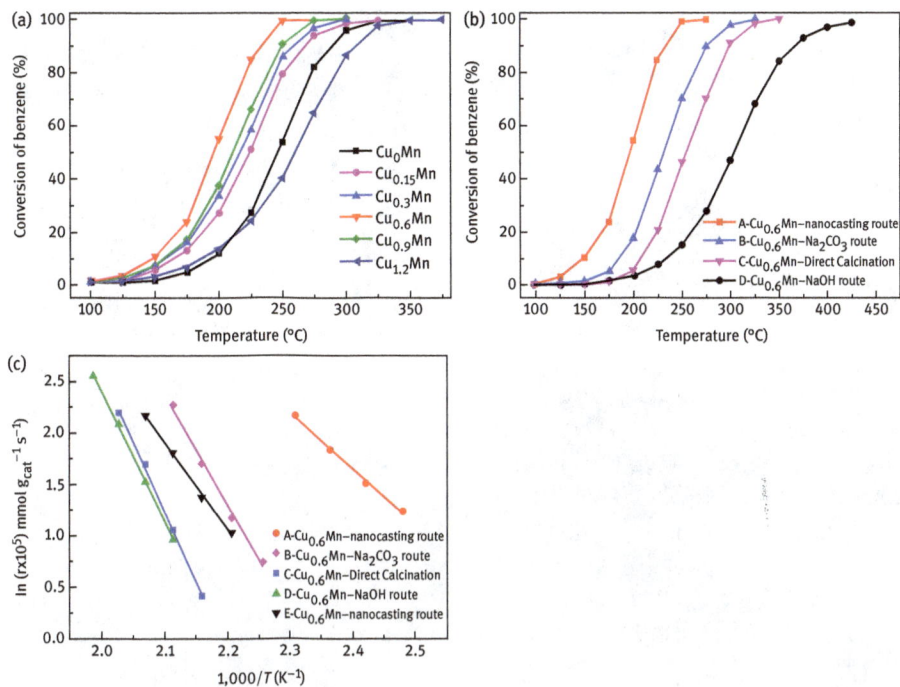

Figure 2.19: (a) and (b) Benzene conversion as a function of reaction temperature over Cu–Mn composite oxides prepared by nanocasting strategy and other methods, (c) Arrhenius plots for the oxidation of benzene over the Cu–Mn composite oxides synthesized by different methods. Reproduced from Ref. [84] with permission from Elsevier B.V. All rights reserved.

(400 °C) calcination leads to an amorphous phase, which shows higher activity for benzene abatement than the high-temperature calcined catalyst with crystalline phase from the amorphous transition. The amorphous phase owns a larger BET SSA and lower temperature reducibility than the crystalline phase catalyst, which is responsible for their catalytic discrepancy. Moreover, the higher concentrations of Cu^+ and Mn^{4+} species during crystallization are known to be deactivated species, which will restrain their catalytic performance.

It is known that depositing metal oxide in an adequate support like molecular sieve or active carbon will advance more of the catalytic performance than the corresponding bulk catalyst, through enhancing the dispersion of the active phase [71]. The halloysite nanotube (HNT), $Si_2Al_2O_5(OH)_4 \cdot 2H_2O$ is considered to be the support candidate with exposed Si–O and Si–OH groups in the outer surface, which are potential sites for anchoring metal oxide particles for oxidation of VOCs [70]. Carrillo and Carriazo prepared CuO and Co_3O_4 oxides supported on HNTs with an elevated activity for toluene oxidation [70]. The halloysite support shows minor loss of surface area and structural stability throughout the preparation process of supported

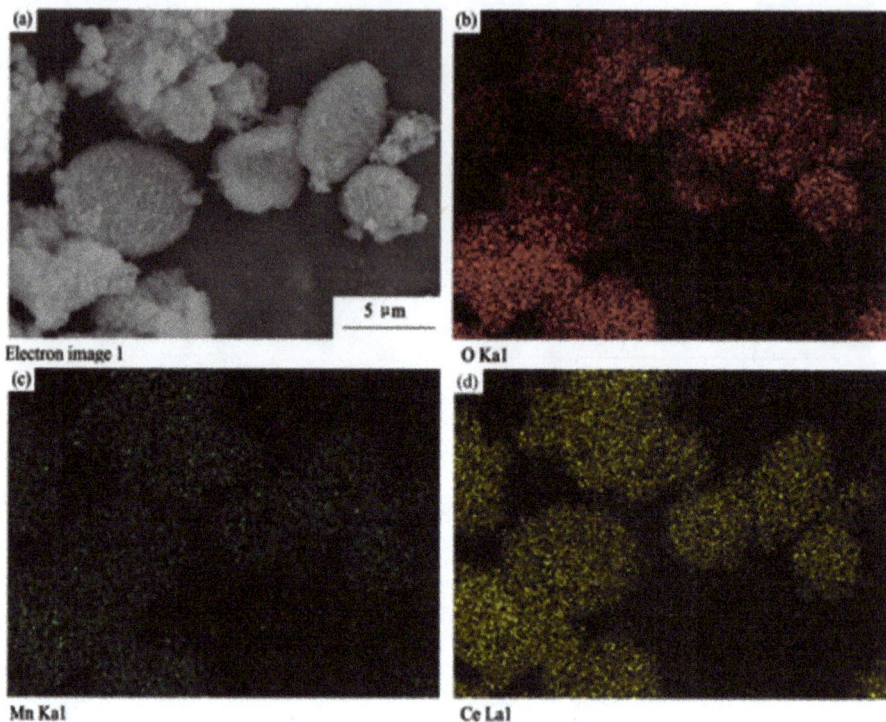

Figure 2.20: (a) SEM image and (b–d) element distribution of hierarchical layer-stacking Mn–Ce composite oxide. Reproduced from Ref. [93] with permission from Elsevier B.V. All rights reserved.

catalysts [70]. Li et al. [71] designed copper/ZSM-5-supported Ce/Zr catalysts with the superiority of CuO, CeO_2, and ZrO_2 species with high dispersion on the surfaces of ZSM-5 support, which exhibits promotion effect in highly efficient catalytic removal of ethyl acetate. Cu^{2+} and Cu^+ states are observed to coexist in all the $CuCe_xZr_{1-x}$/ZSM-5 catalysts as shown in Figure 2.23. Higher Cu^{2+}/Cu^+ ratio in $CuCe_{0.75}Zr_{0.25}$/ZSM-5 catalyst results in the optimum oxidation performance with complete ethyl acetate conversion at 270 °C (Figure 2.24). The synergistic effect of Cu–Ce–Zr gives rise to the increased amount of the active oxygen, and acidity of strong acid sites, which play key roles in the $CuCe_xZr_{1-x}$/ZSM-5 catalytic performance.

2.4.3 Perovskite catalysts

Perovskites, represented by the general formula ABO_3 (Figure 2.25), have been reported for decades as important VOC oxidation catalysts or supports, because of their high structural stability, low cost, great diversity, high thermal stability, and mechanical strength [94]. Significant efforts have been made to substitute the A- and/or B-site

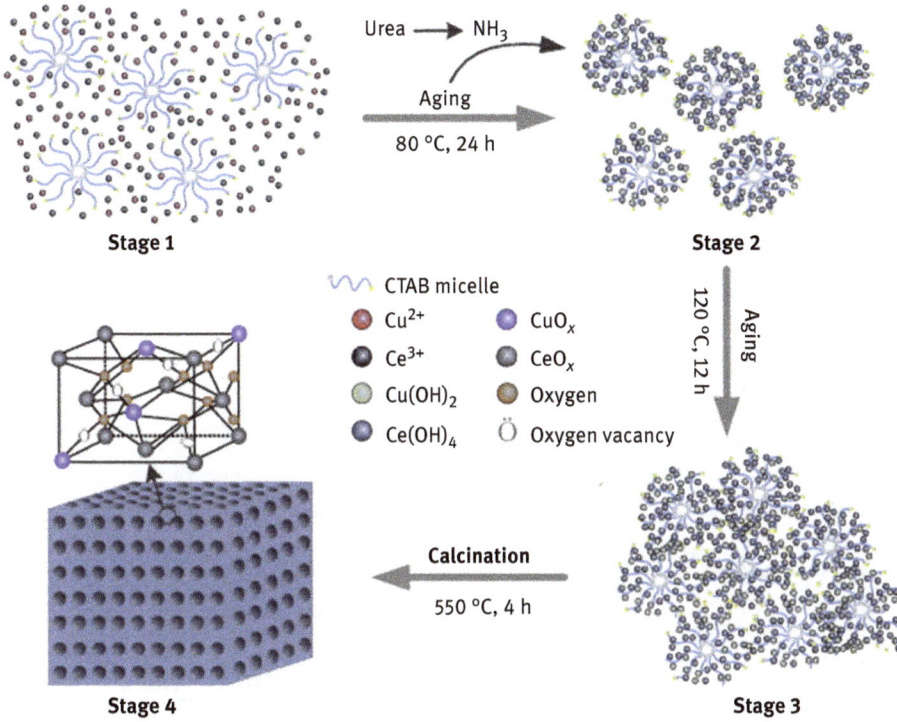

Figure 2.21: Schematic illustration of the formation mechanism for the mesoporous CuCeO$_x$ oxide. Reproduced from Ref. [69] with permission from Elsevier B.V. All rights reserved.

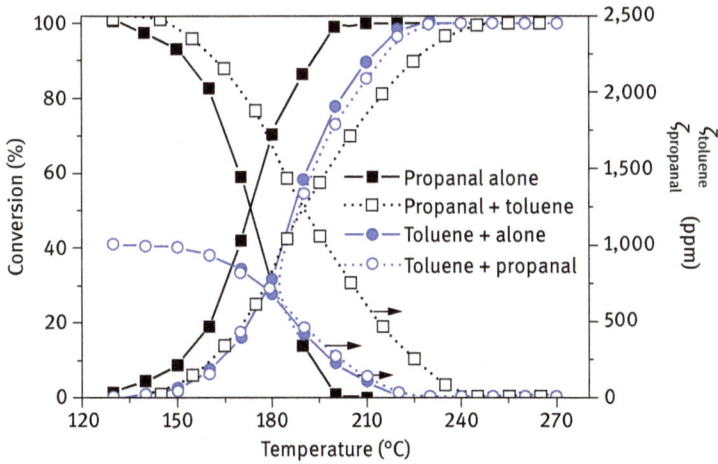

Figure 2.22: The co-combustion of toluene and propanal over Cu$_{0.3}$Ce$_{0.7}$O$_x$ catalyst (300 mg catalyst, 1,000 ppm toluene, 2,500 ppm propanal, 21% O$_2$, N$_2$ balance, GHSV = 36,000 h^{-1}). Reproduced from Ref. [69] with permission from Elsevier B.V. All rights reserved.

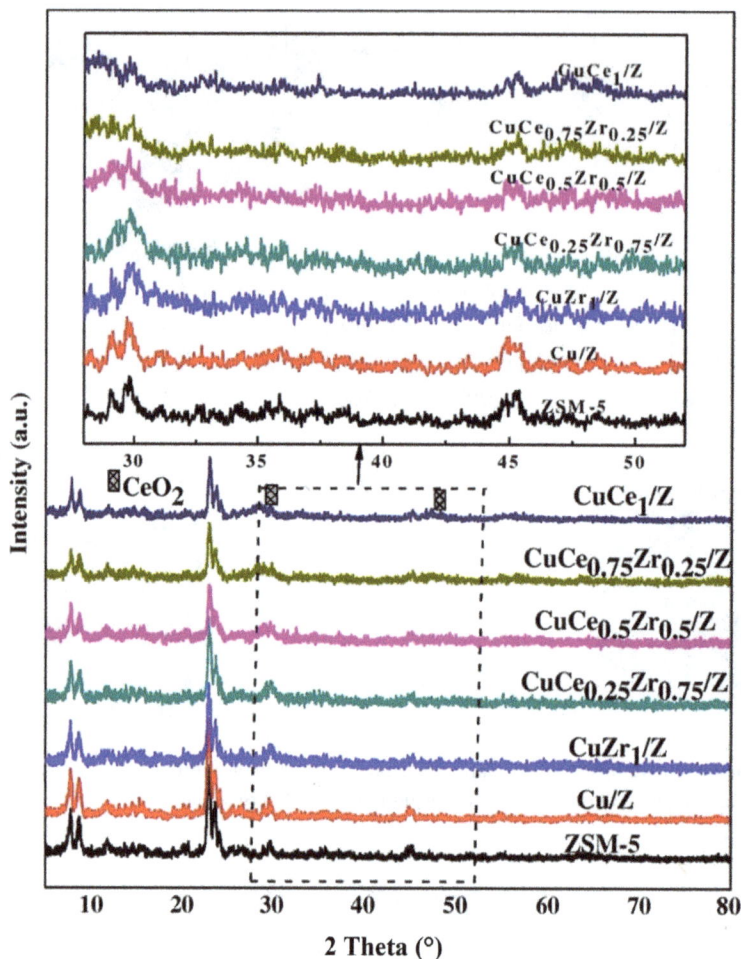

Figure 2.23: XRD spectra of CuCe$_x$Zr$_{1-x}$O$_y$/ZSM-5 catalysts. Reproduced from Ref. [71] with permission from Elsevier B.V. All rights reserved.

cation by other cations with different oxidation state or radius, leading to the ABO$_3$ molecule possessing varied oxidation states of B-site cation and oxygen vacancy. Many literatures have reported the structural modification of the A and/or B sites to develop advanced perovskite with more defective structures and better catalytic performances. Zhu et al. [94] reviewed the recent applications of lanthanum-based perovskite oxides in energy conversion and traditional heterogeneous catalytic reactions. Zhu et al. [95] summarized systematically perovskite oxides in catalysis based on the synthesis, characterizations, and applications.

Zhang et al. [72] evaluated the activity for toluene oxidation over three kinds of LaMnO$_3$ (LMO) catalysts synthesized by citrate SG, glycine combustion (GC), and CP

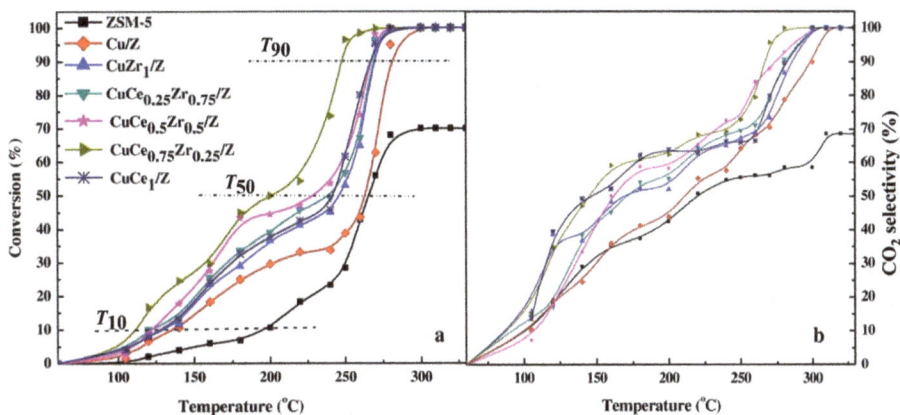

Figure 2.24: Ethyl acetate conversion (a) and CO_2 selectivity (b) as a function of reaction temperature for $CuCe_xZr_{1-x}O_y$/ZSM-5 catalysts. Reproduced from Ref. [71] with permission from Elsevier B.V. All rights reserved.

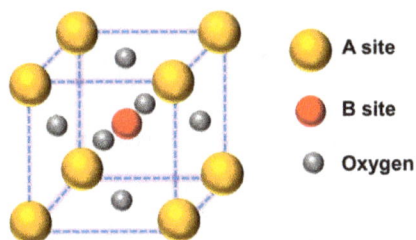

- A site
- B site
- Oxygen

Figure 2.25: Schematic illustration of an ideal ABO_3 perovskite unit cell. Reproduced from Ref. [94] with permission from American Chemical Society.

methods, respectively. LMO-SG and LMO-GC samples show well-formed perovskite structures. Catalytic activity follows the order: LMO-SG > LMO-CP > LMO-GC, which conforms to their SSA, low-temperature reducibility, and concentration of surface-ad-sorbed oxygen species. O_2-TPD is a good technique for investigating oxygen species of the perovskite oxides, which is closely correlated with their oxidation capability of VOCs as depicted in Figure 2.26. O_α represents the weakest physisorbed and/or chem-isorbed surface oxygen molecules, and can be easily removed at relatively low tem-perature. O_β is ascribed to overstoichiometric oxygen, and the produced excess oxygen charge will be compensated by cation vacancies. O_γ represents the migrated lattice oxygen in the bulk perovskite structure. LMO-SG exhibits the highest capacity for surface-adsorbed oxygen, LMO-CP owns medium O_α, and LMO-GC has poor O_α, and the ranking of O_α has positive correlation with LMO oxidation activity.

For the investigation of the effect of element doping on perovskite, a series of $La_{1-x}Sr_xM_{1-y}Fe_yO_3$ (M = Mn, Co; x = 0, 0.4; y = 0.1, 1.0) catalysts have been ob-tained through citric acid complexation followed by hydrothermal treatment

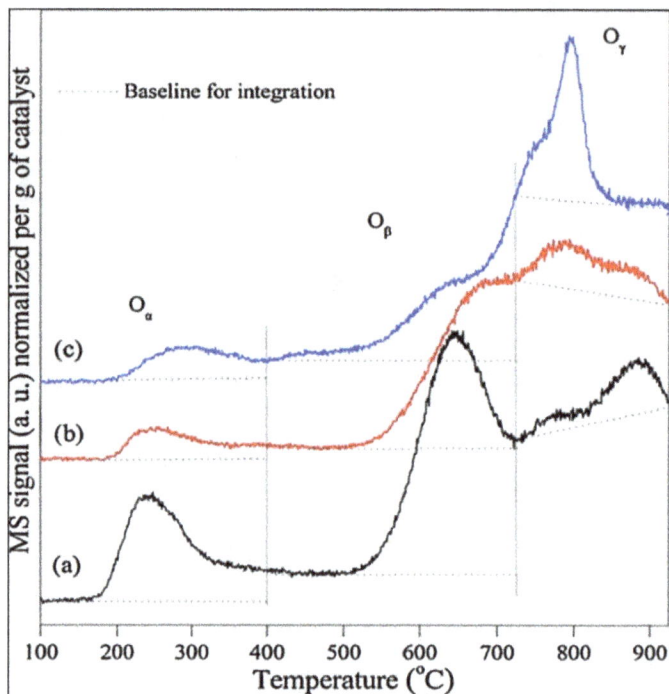

Figure 2.26: O_2-TPD profiles of samples (a) LMO-SG, (b) LMO-GC and (c) LMO-CP. Reproduced from Ref. [72] with permission from Elsevier B.V. All rights reserved.

by Deng et al. [96] The partial substitution of Mn and Co with Fe in $La_{0.6}Sr_{0.4}MnO_3$ and $La_{0.6}Sr_{0.4}CoO_3$ samples will show prompt catalytic performance in the order: $La_{0.6}Sr_{0.4}Co_{0.9}Fe_{0.1}O_3 > La_{0.6}Sr_{0.4}FeO_3 > La_{0.6}Sr_{0.4}Mn_{0.9}Fe_{0.1}O_3 > LaFeO_3 > La_{0.6}Sr_{0.4}CoO_3 > La_{0.6}Sr_{0.4}MnO_3$. The temperatures for 100% toluene oxidation over $La_{0.6}Sr_{0.4}Mn_{0.9}Fe_{0.1}O_3$ and $La_{0.6}Sr_{0.4}Co_{0.9}Fe_{0.1}O_3$ are 50 and 85 °C lower than their corresponding Fe-free perovskite oxides, respectively. It is well known that pure Fe perovskite-type material exhibits localized electrons, whereas pure Co perovskite-type material has delocalized electronic conduction. Hence, the partial substitution of cobalt with iron will induce the change of electronic structure, which is responsible for the different catalytic activity.

Besides the transition metal modification, alkaline-earth metals substitution will also yield defective structure of perovskite, which will result in high catalytic activities [97]. Zhang et al. [97] studied the synergistic effect of Ca^{2+} and Mg^{2+} co-substitution on $LaCoO_3$ perovskite phase formation and toluene catalytic activity. The crystal growths of $La(OH)_3$ and $Co(OH)_2$ are characterized to be hindered by doping Ca^{2+} and Mg^{2+} ions in A and B sites. Only A-site doping by Ca^{2+} in $LaCoO_3$ lattice will evoke trace amounts of oxygen vacancies, which depresses the low-temperature reducibility of the perovskite. Only A-site doping by Mg^{2+} will induce the transition of partial Co^{3+}

to Co^{4+}, increasing the weakly chemisorbed oxygen and lattice oxygen of the perovskite. Doping both A and B sites in $LaCoO_3$ will promote superficial oxidation process due to the synergistic effect, which is beneficial for the decrease in the apparent activation energy (E_a) to only 34 kJ mol^{-1} in this catalyst system.

2.4.4 LDH-derived mixed oxide catalysts

As we know, among many advantages through obtaining mixed oxides from the transition of LDH decomposition, there are three key ones: first, active species can be effectively immobilized due to LDH precursor as the integration supports; second, active species such as M^{2+} and M^{3+} cations in the layers acting as excellent solid base-like catalysts will disperse uniformly because of the ordered arrangement of cations embedding in LDH layers; third, hierarchical structure, porous, high surface area, and synergistic effect for the obtained mixed oxide catalysts will be beneficial for promoting catalytic activity and stability. Consequently, there are great interests in rationally designing and controllably preparing mixed metal oxide catalysts derived from LDH precursors to evoke the catalytic activities of VOCs [38, 98–104].

Mesoporous CoMnAl mixed metal oxide catalysts with various Co/Mn atomic ratios by the NH_4OH coprecipitation–hydrothermal method followed by calcination at a high temperature are prepared, which show outstanding low-temperature benzene oxidation performance reported by Mo et al. [98] The $Co_{3-x}Mn_xAlO$ catalysts exhibit enhanced catalytic activities in comparison with single Co_3AlO and Mn_3AlO catalysts. Among them, $CoMn_2AlO$-450 shows the optimal activity with T_{90}, reaction rate, and activation energy for benzene conversion at 238 °C, 0.24 mmol g_{cat}^{-1} h^{-1}, and 65.77 kJ mol^{-1}, respectively (Figure 2.27(a) and (c)). The effect of calcination temperature on benzene decomposition further investigates through evaluating the samples under 350, 450 and 550 °C calcination. The $CoMn_2AlO$-550 catalyst exhibits higher catalytic activity, stability, and water vapor resistance because of the highest low-temperature reducibility, the most abundant surface Co^{3+}, Mn^{4+}, and O_{ads} species (Figure 2.27(b) and (d)).

Li et al. [39] investigated the effect of Cu:Co molar ratio on benzene oxidation over porous $Cu_xCo_{1-x}AlO$ catalysts derived from LDHs by a CP method. Increasing Cu content gradually, the XRD diffraction peak intensities of the Co_3O_4 peaks decrease, and CuO phase develops with increasing intensities (Figure 2.28). The TEM images of the $Cu_xCo_{1-x}AlO$ catalysts shown in Figure 2.29 maintain the plate-like morphology of the original LDH precursor. TEM also proves the coexistence of Co_3O_4 spinel and CuO phases identified by their corresponding crystal fringes, which is expected to alter the physicochemical properties of the catalysts and advance the catalytic activities. Among them, the $Cu_{0.5}Co_{2.5}AlO$ sample displays remarkably more enhanced reaction activity with the value 2.41 mmol g^{-1} under 290 °C, achieving 90%

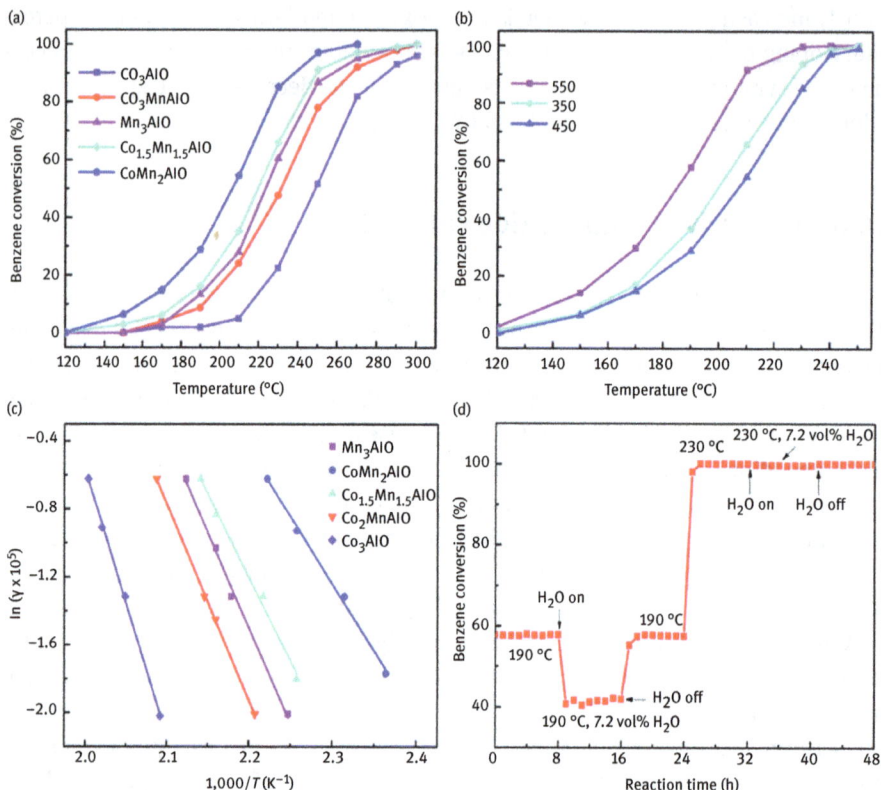

Figure 2.27: (a) Benzene conversion as a function of reaction temperature over $Co_{3-x}Mn_xAlO$ catalysts calcined at 450 °C; (b) benzene conversion over $CoMn_2AlO$ catalysts calcined at different temperatures (350, 450 and 550 °C); (c) Arrhenius plots for the oxidation of benzene over the $Co_{3-x}Mn_xAlO$ catalysts; (d) effect of water vapor on the activities of $CoMn_2AlO$-550 catalysts at 190 °C and 230 °C. Benzene concentration = 100 ppm, water concentration = 7.2 vol% and SV = 60 000 mL g^{-1} h^{-1}. Reproduced from Ref [98] with permission from the Royal Society of Chemistry.

benzene conversion than that of only 8% conversion for Cu_3Al–MMO sample at space velocity of 60,000 mL g^{-1} h^{-1}.

There is an interesting phenomenon for increased oxidation performance of VOCs by adjusting Co valence derived from $Co^{II}Co^{III}LDH$ reported by Mo et al. [105] $Co^{II}_{2.8}Co^{III}_1LDH$ is fabricated through a facile topochemical transformation route under pumping dynamic oxygen. Self-assembled coral-like CoAlLDH nanostructures are also constructed via homogeneous precipitation method under different ammonia-releasing reagents and solvents for comparison. Catalytic activities for $Co^{II}Co^{III}O$ and CoAlO oxides derived from their corresponding LDH precursors are evaluated for benzene and toluene oxidation. $Co^{II}Co^{III}O$ catalyst in which Co^{3+} ions substituted for Al^{3+} ions in the bruited-like layer has a higher catalytic activity with

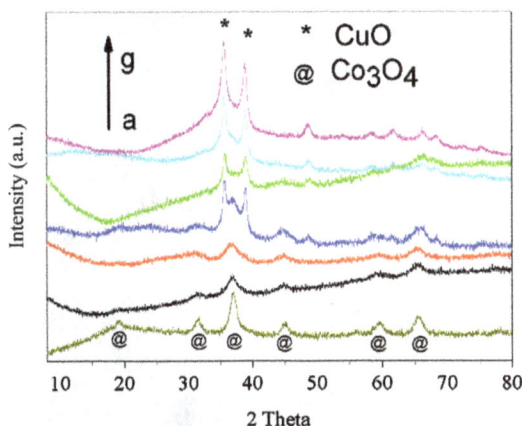

Figure 2.28: (a–g) XRD patterns of Co$_3$Al–MMO, Cu$_{0.5}$Co$_{2.5}$Al–MMO, Cu$_1$Co$_2$Al–MMO, Cu$_{1.5}$Co$_{1.5}$Al–MMO, Cu$_2$Co$_1$Al–MMO, Cu$_{2.5}$Co$_{0.5}$Al–MMO, Cu$_3$Al–MMO samples. Reproduced from Ref. [39] with permission from Elsevier B.V. All rights reserved.

T_{99} benzene and toluene at 210 °C and 220 °C, respectively. Raman scattering is very sensitive for studying lattice defects. The five Raman-active modes ($A_{1g} + E_g + 3F_{2g}$) of Co$_3$O$_4$ from Co^{II}Co^{III}O sample exhibit a red shift, which demonstrates that there are more abundant Co^{3+} ions in Co^{II}Co^{III}O than CoAlO samples (Figure 2.30). XPS, a powerful tool for analyzing surface elemental compositions and valence state, further evidences richer surface Co^{3+} species in Co^{II}Co^{III}O than CoAlO samples, which are detected by a slight shift of the main peak of the Co 2p$_{3/2}$ and Co 2p$_{1/2}$ peaks for Co^{II}Co^{III}O sample to lower binding energies (Figure 2.31(a)). Furthermore, the binding energy of O$_{1s}$ for Co^{II}Co^{III}O sample shifts to lower binding energies with 0.6 eV (Figure 2.31(b)), because of the varied chemical state of the Co species in the near-surface region, in which the Co–O bond strength can impact with oxygen desorption. The quantitative result from the O$_{1s}$ spectrum reveals that the surface O$_{ads}$/O$_{latt}$ molar ratio (0.611) for Co^{II}Co^{III}O sample is higher than those of other samples. The higher surface oxygen vacancy density and higher surface O$_{ads}$/O$_{latt}$ molar ratio, induced by the existence of more surface Co^{3+} species, are favorable for the oxidation process, prompting the catalyst activity.

2.5 Monolithic catalysts and application

2.5.1 Monolithic catalysts

The monolithic catalysts are generally divided into noble metals and metal oxide catalysts for laboratory scale or practical catalytic combustion of VOCs. Monolithic

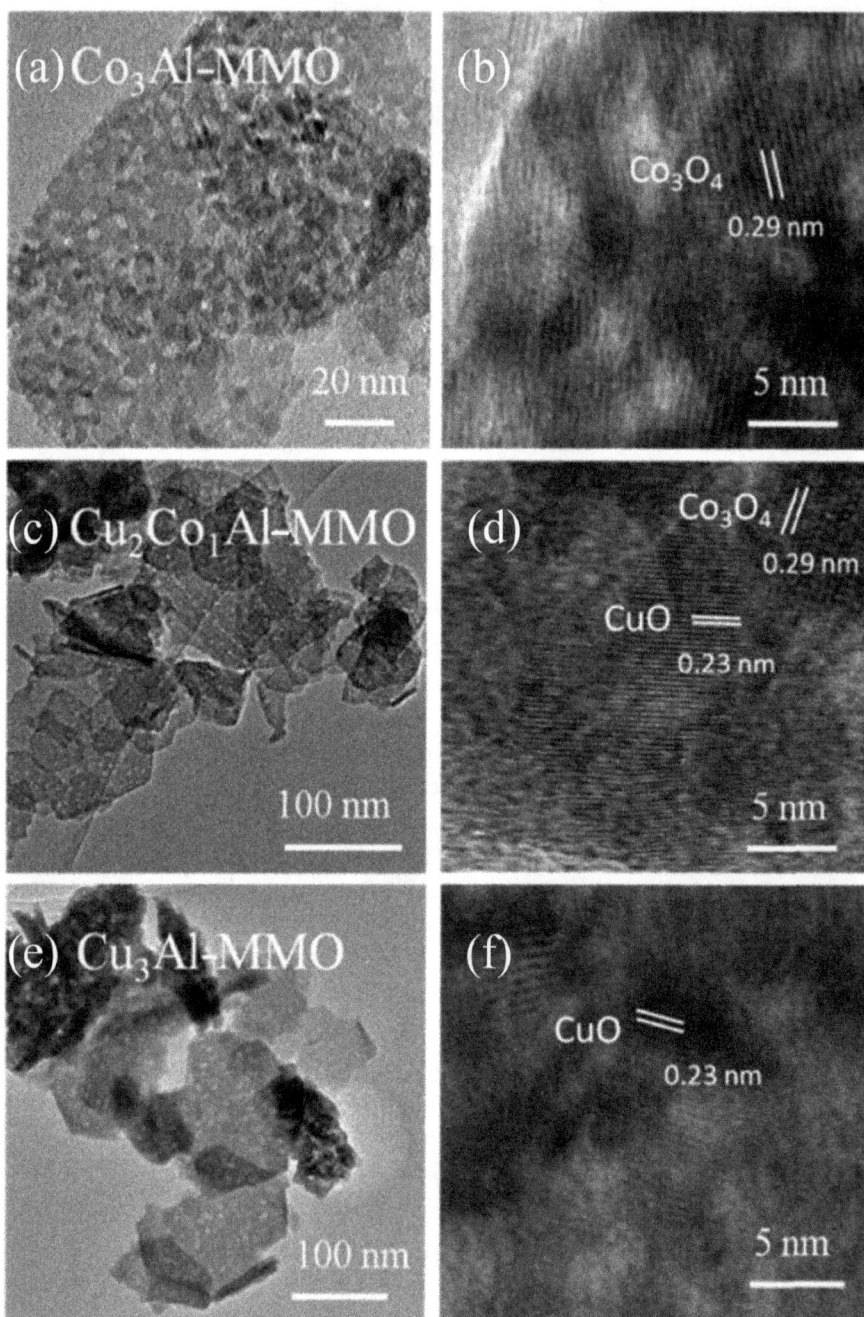

Figure 2.29: (a, c, e) TEM images of the catalysts, Co_3Al–MMO, Cu_2Co_1Al–MMO, and Cu_3Al–MMO. (b, d, f) The corresponding HRTEM images in (a), (c), and (e), respectively. Reproduced from Ref. [39] with permission from Elsevier B.V. All rights reserved.

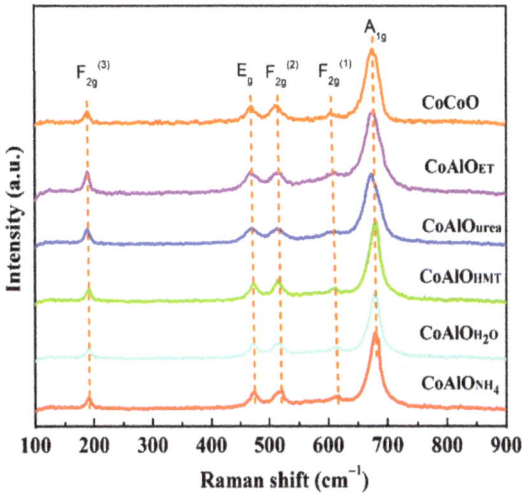

Figure 2.30: Raman spectra profiles of all catalysts. All rights reserved. CoAlO$_x$ represents the preparation in different ammonia-releasing reagents (NH$_4$OH, HMT, and urea dissolved in methanol) and reaction solvents (H$_2$O and ethanol with urea as precipitant). Reproduced from Ref. [105] with permission from Elsevier B.V.

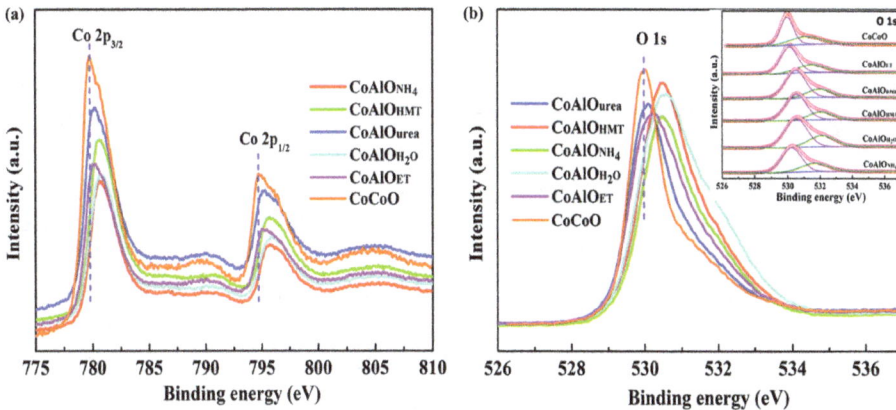

Figure 2.31: The Co$_{2p}$ and O$_{1s}$ XPS spectra of all catalysts. Reproduced from Ref. [105] with permission from Elsevier B.V. All rights reserved.

catalysts are commonly composed of carrier, washcoating, and active components. The component of washcoating is inclined to choose oxides such as Al$_2$O$_3$, SiO$_2$, and CeO$_2$ with high surface area, cohesive and porous structure, which is coated on carriers like cordierite or metallic honeycomb substrate. Then the active component such as noble or metal oxides together with additives is co-adsorbed on the surface of washcoating [106]. Noble metals (Pt, Pd, or Rh) supported on high surface area

γ-Al$_2$O$_3$, then washcoated on monolithic substrate are commonly used for their higher efficiency than metal oxide catalysts for abatement of VOCs. Such monolithic structure offers the major advantage of low-pressure drop compared with more traditional pellet-shaped or powder catalysts [20, 107].

Jiang et al. [106] prepared monolithic Pt–Pd bimetallic catalysts supported on γ-Al$_2$O$_3$, then deposited cordierite honeycomb ceramics. The optimal preparation factors with superior catalytic activities for benzene and other aromatic hydrocarbons are as follows, active phases 0.1% Pt–Pd, the molar ratio of additives Ce:Zr with 3:1, and the calcination temperature lower than 500 °C.

Besides noble metal monolithic catalysts, metal oxide monolithic catalysts are also extensively researched for oxidation of VOCs, due to some advantages, like low cost and potentially more resistant to deactivation than precious metal catalysts. However, their lower oxidation activities than precious metal normally limit their wide application, which impel researchers and companies to seek high activity metal oxide catalysts [73, 108–109].

Two kinds of Mn-, Ce-, and Zr-coated cordierite monolithic catalysts are synthesized to evaluate their n-butanol oxidation performance [108]. The "one-pot synthesis" method through Ce, Mn, and Zr salt precursors mixed together leads to a homogeneous distribution of Mn species, while the "two-step loading" method through first coating Ce–Zr washcoat on cordierite, then impregnation of Mn over Ce–Zr phases produces inhomogeneous distribution of Mn species. The mixed oxide active phases prepared by the two methods are well anchored on the cordierite. The introduction of Mn species will increase specific reaction rate, which represents Mn is crucial for n-butanol catalytic oxidation. The specific reaction rate (Figure 2.32) of CeZrMn(0.48)/monolithic catalyst with "one-pot synthesis" is much higher than Mn/CeZr/M using two-step loading method. This is ascribed to the high specific surface area, good dispersion, and reducibility of the mixed oxides, caused by a thin layer coating on cordierite through "one-pot synthesis."

Ren et al. [109] provided a feasible approach to design a series of large-scale $M_xCo_{3-x}O_4$ (M = Co, Ni, Zn) nanoarray catalysts integrated onto cordierite monolithic substrates with scalable fabrication, cost-effectiveness, and rational manipulation for tunable low-temperature catalytic performance of VOCs. Cordierite substrates first in situ grow basic carbonate nanowire arrays through urea hydrolysis process under continuous mechanical agitation, which will obtain a uniform deposition of nanowire arrays. Then the spinel cobaltite $M_xCo_{3-x}O_4$ nanoarrays (M = Co, Zn, Ni; $x = 0.5$) are obtained after annealing at 300 °C under air condition.

2.5.2 Monolithic catalyst application

There are quite discrepancy for conditions of laboratory and industrial pollution of VOCs, for example, flow rate, concentration, and components. Therefore, there are

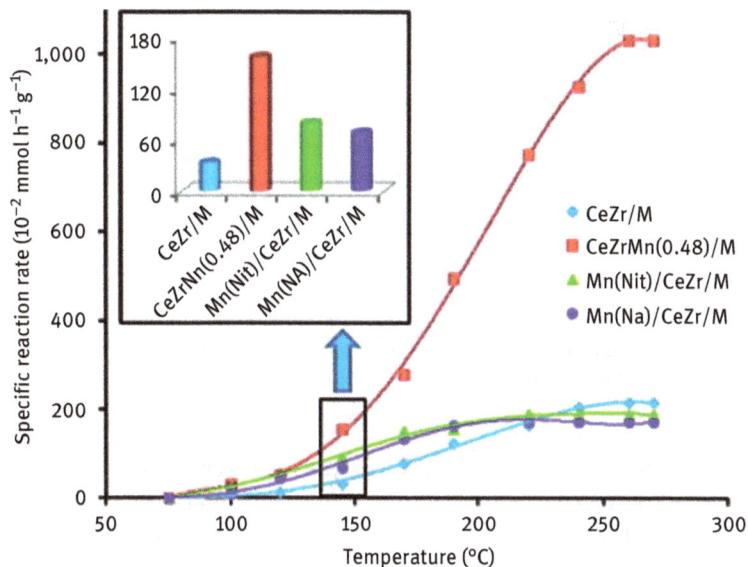

Figure 2.32: Specific reaction rates of *n*-butanol oxidation as a function of the reaction temperature and the monolithic catalyst. Reproduced from Ref. [108] with permission from Elsevier B.V. All rights reserved.

more essential factors for achieving successful industrial purification of VOCs, for example, the significant low-temperature activity, selectivity and efficiency, treatment principle, technology, and equipment.

Zhang et al. [65] summarized the three major principles that are necessarily considered. Safety is always at the top of the list. The purification process should emit VOCs to maximum limit effectively, but not emit any toxic product or by-product throughout the treatment process. Efficiency of the equipment is also very important. The selected technology should be highly efficient to remove VOCs continuously, and satisfy the strict environmental regulations. Economy should further be considered to avoid increasing excessive burden like running expense for manufacturing company.

Regenerative catalytic oxidation (RCO) is widely accepted for industrial purification of VOCs like painting and printing, because RCO is considered as one of the most energy-saving techniques for VOCs removal running under 300–450 °C among the market of pollution control of VOCs. A similar technology about regenerative thermal oxidation needs running at high temperature above 800 °C under the absence of catalysts. RCO is normally composed of two or more separate beds containing honeycomb ceramics with high specific heat materials packing for heat transfer media. There are three necessary parts for a typical and simple two-bed RCO as shown in Figure 2.33, ceramics layer, catalyst layer, and natural gas burner or

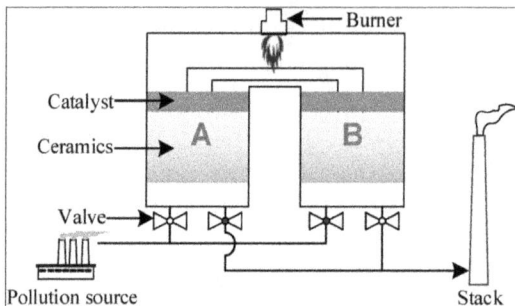

Figure 2.33: The schematic diagram of regenerative catalytic oxidizer (RCO). Reproduced from Ref. [65] with permission from Elsevier B.V. All rights reserved.

electrical heater, with their functions as heat storage, reaction media, and heat supply, respectively [65]. The whole cabinet is first heated by gas or electric vehicles to proper temperature. VOC flow is then switched to pass through the ceramic layer in cabinet A, which will be heated by ceramic layer with its temperature increased by about 200–300 °C. Second, the preheated VOC flow contacts with catalyst layer, and VOCs are decomposed effectively under proper catalytic oxidation temperature heated by gas burner. The catalyst layer temperature is normally set up slightly higher than the light-off temperature of catalyst. Meanwhile, oxidation process of VOCs will release some heat for increasing the temperature of the chamber. Third, the VOCs flow after decomposition goes down through the ceramic layer in cabinet B, and most of the heat can be reserved by the ceramics. The heated ceramic layer in cabinet B will be ready to preheat the inlet VOCs flow in the next cycle (from cabinet B to cabinet A). The switch of VOCs flow is controlled by four combined valves through programmable logic controller, with every 60–150 s interval finishing on cycle from cabinet A to B or vice versa [65].

References

[1]   Ojala S., Pitkaaho S., Laitinen T., Koivikko N.N., Brahmi R., Gaalova J., Matejova L., Kucherov A., Paivarinta S., Hirschmann C., Nevanpera T., Riihimaki M., Pirila M., Keiski R.L., Catalysis in VOC abatement. Top. Catal. 2011, 54, 1224–1256.
[2]   Kamal M.S., Razzak S.A., Hossain M.M., Catalytic oxidation of volatile organic compounds (VOCs) – A review. Atmos. Environ. 2016, 140, 117–134.
[3]   Mølhave L., Clausen G., Berglund B., De Ceaurriz J., Kettrup A., Lindvall T., Maroni M., Pickering A.C., Risse U., Rothweiler H., Seifert B., Younes, M., Total volatile organic compounds (TVOC) in indoor air quality investigations. Indoor Air 1997, 7, 225–240.
[4]   Schauer J.J., Kleeman M.J., Cass G.R., Simoneit B.R.T., Measurement of emissions from air pollution sources. 1. C-1 through C-29 organic compounds from meat charbroiling. Environ. Sci. Technol. 1999, 33, 1566–1577.

[5]   Schauer J.J., Kleeman M.J., Cass G.R., Simoneit B.R.T., Measurement of emissions from air
       pollution sources. 4. C-1-C-27 organic compounds from cooking with seed oils. Environ. Sci.
       Technol. 2002, 36, 567–575.
[6]   Endo Y., Hayashi C., Yamanaka T., Takayose K., Yamaoka M., Tsuno T., Nakajima S., Linolenic
       acid as the main source of acrolein formed during heating of vegetable oils. J. Am. Oil Chem.
       Soc. 2013, 90, 959–964.
[7]   Ontanón I., Culleré L., Zapata J., Villanueva B., Ferreira V., Escudero A., Application of a new
       sampling device for determination of volatile compounds released during heating olive and
       sunflower oil: Sensory evaluation of those identified compounds. Eur. Food Res. Technol.
       2013, 236, 1031–1040.
[8]   Yu K.-P., Yang K.R., Chen Y.C., Gong J.Y., Chen Y.P., Shih H.-C., Lung S.-C.C., Indoor air
       pollution from gas cooking in five Taiwanese families. Build. Environ. 2015, 93, 258–266.
[9]   Lee S.C., Guo H., Li W.M., Chan L.Y., Inter-comparison of air pollutant concentrations in
       different indoor environments in Hong Kong. Atmos. Environ. 2002, 36, 1929–1940.
[10]   Duan H., Liu X., Yan M., Wu Y., Liu Z., Characteristics of carbonyls and volatile organic
       compounds (VOCs) in residences in Beijing, China. Front. Env. Sci Eng. 2016, 10, 73–84.
[11]   Chang T., Ren D., Shen Z., Huang Y., Sun J., Cao J., Zhou J., Liu H., Xu H., Zheng C., Pan H.,
       He C., Indoor air pollution levels in decorated residences and public places over Xi'an, China.
       Aerosol Air Qual. Res. 2017, 17, 2197–2205.
[12]   Jones A.P., Indoor air quality and health. Atmos. Environ. 1999, 33, 4535–4564.
[13]   Ahmed F.E., Toxicology and human health effects following exposure to oxygenated or
       reformulated gasoline. Toxicol. Lett. 2001, 123, 89–113.
[14]   Cachot J., Geffard O., Augagneur S., Lacroix S., Le Menach K., Peluhet L., Couteau J., Denier
       X., Devier M. H., Pottier D., Budzinski H., Evidence of genotoxicity related to high PAH
       content of sediments in the upper part of the Seine estuary (Normandy, France). Aquat.
       Toxicol. 2006, 79, 257–267.
[15]   Haritash A. K., Kaushik C. P., Biodegradation aspects of polycyclic aromatic hydrocarbons
       (PAHs): A review. J. Hazard. Mater. 2009, 169, 1–15.
[16]   Zhao S., Li K., Jiang S., Li J., Pd–Co based spinel oxides derived from Pd nanoparticles
       immobilized on layered double hydroxides for toluene combustion. Appl. Catal. B: Environ.
       2016, 181, 236–248.
[17]   Garcia T., Agouram S., Sanchez-Royo J.F., Murillo R., Maria Mastral A., Aranda A.,
       Vazquez I., Dejoz A., Solsona B., Deep oxidation of volatile organic compounds using
       ordered cobalt oxides prepared by a nanocasting route. Appl. Catal. A: Gen. 2010, 386,
       16–27.
[18]   Bai G., Dai H., Deng J., Liu Y., Wang F., Zhao Z., Qiu W., Au C.T., Porous Co_3O_4 nanowires and
       nanorods: Highly active catalysts for the combustion of toluene. Appl. Catal. A: Gen. 2013,
       450, 42–49.
[19]   McFarland E.W., Metiu H., Catalysis by doped oxides. Chem. Rev. 2013, 113, 4391–4427.
[20]   Liotta L.F., Catalytic oxidation of volatile organic compounds on supported noble metals.
       Appl. Catal. B: Environ. 2010, 100, 403–412.
[21]   Abbasi Z., Haghighi M., Fatehifar E., Saedy S., Synthesis and physicochemical
       characterizations of nanostructured Pt/Al2O3-CeO2 catalysts for total oxidation of VOCs.
       J. Hazard Mater. 2011, 186, 1445–54.
[22]   Chen B.-B., Shi C., Crocker M., Wang Y., Zhu A.-M., Catalytic removal of formaldehyde at
       room temperature over supported gold catalysts. Appl. Catal. B: Environ. 2013, 132–133,
       245–255.
[23]   Li H.-F., Zhang N., Chen P., Luo M.-F., Lu J.-Q., High surface area Au/CeO_2 catalysts for low
       temperature formaldehyde oxidation. Appl. Catal. B: Environ. 2011, 110, 279–285.

[24] Li Z., Yang K., Liu G., Deng G., Li J., Li G., Yue R., Yang J., Chen Y., Effect of reduction treatment on structural properties of TiO_2 supported pt nanoparticles and their catalytic activity for benzene oxidation. Catal. Lett. 2014, 144, 1080–1087.

[25] de Rivas B., López-Fonseca R., Jiménez-González C., Gutiérrez-Ortiz J. I., Synthesis, characterisation and catalytic performance of nanocrystalline Co_3O_4 for gas-phase chlorinated VOC abatement. J. Catal. 2011, 281, 88–97.

[26] Morales M., Barbero B., Cadus L., Total oxidation of ethanol and propane over Mn-Cu mixed oxide catalysts. Appl. Catal. B: Environ. 2006, 67, 229–236.

[27] Tang W., Wu X., Li D., Wang Z., Liu G., Liu H., Chen Y., Oxalate route for promoting activity of manganese oxide catalysts in total VOCs' oxidation: effect of calcination temperature and preparation method. J. Mater. Chem. A 2014, 2, 2544–2554.

[28] Yu D., Liu Y., Wu Z., Low-temperature catalytic oxidation of toluene over mesoporous $MnO_x–CeO_2/TiO_2$ prepared by sol–gel method. Catal. Commun. 2010, 11, 788–791.

[29] Rousseau S., Loridant S., Delichere P., Boreave A., Deloume J. P., Vernoux P., $La_{(1-x)}Sr_xCo_{1-y}Fe_yO_3$ perovskites prepared by sol–gel method: Characterization and relationships with catalytic properties for total oxidation of toluene. Appl. Catal. B: Environ. 2009, 88, 438–447.

[30] Zhao Y., Li F., Zhang R., Evans D.G., Duan X., Preparation of layered double-hydroxide nanomaterials with a uniform crystallite size using a new method involving separate nucleation and aging steps. Chem. Mater. 2002, 14, 4286–4291.

[31] Sideris P.J., Nielsen U.G., Gan Z., Grey C.P., Mg/Al ordering in layered double hydroxides revealed by multinuclear NMR spectroscopy. Science 2008, 321, 113–117.

[32] Lamonier J.-F., Boutoundou A.-B., Gennequin C., Pérez-Zurita M.J., Siffert S., Aboukais A., Catalytic removal of toluene in air over Co−Mn−Al nano-oxides synthesized by hydrotalcite route. Catal. Lett. 2007, 118, 165–172.

[33] Xu Y., Wang Z., Tan L., Zhao Y., Duan, H. Song Y.-F., Fine tuning the heterostructured interfaces by topological transformation of layered double hydroxide nanosheets. Ind. Eng. Chem. Res. 2018, 57, 10411–10420.

[34] Xu Y., Wang Z., Tan L., Yan H., Zhao Y., Duan H., Song Y.-F., Interface engineering of high-energy faceted Co_3O_4/ZnO heterostructured catalysts derived from layered double hydroxide nanosheets. Ind. Eng. Chem. Res. 2018, 57, 5259–5267.

[35] Wang Q., O'Hare D., Recent advances in the synthesis and application of layered double hydroxide (LDH) nanosheets. Chem. Rev. 2012, 112, 4124–4155.

[36] Fan G., Li F., Evans D. G., Duan X., Catalytic applications of layered double hydroxides: Recent advances and perspectives. Chem. Soc. Rev. 2014, 43, 7040–7066.

[37] Feng J., He Y., Liu Y., Du Y., Li D., Supported catalysts based on layered double hydroxides for catalytic oxidation and hydrogenation: General functionality and promising application prospects. Chem. Soc. Rev. 2015, 44, 5291–5319.

[38] Li S., Mo S., Li J., Liu H., Chen Y., Promoted VOC oxidation over homogeneous porous CoxNiAlO composite oxides derived from hydrotalcites: Effect of preparation method and doping. RSC Adv. 2016, 6, S6874–S6884.

[39] Li S., Wang H., Li W., Wu X., Tang W., Chen Y., Effect of Cu substitution on promoted benzene oxidation over porous CuCo-based catalysts derived from layered double hydroxide with resistance of water vapor. Appl. Catal. B: Environ. 2015, 166–167, 260–269.

[40] Li C., Hu Y., Yuan W., Nanomaterials synthesized by gas combustion flames: Morphology and structure. Particuology 2010, 8, 556–562.

[41] Azurdia J.A., McCrum A., Laine R.M., Systematic synthesis of mixed- metal oxides in NiO–Co_3O_4, NiO–MoO_3, and NiO–CuO systems via liquid-feed flame spray pyrolysis. J. Mater. Chem. A 2010, 18, 3249–3258.

[42] Schimmoeller B., Jiang Y., Pratsinis S.E., Baiker A., Structure of flame made vanadia/silica and catalytic behavior in the oxidative dehydrogenation of propane. J. Catal. 2010, 274, 64–75.
[43] Liu G., Yue R., Jia Y., Ni Y., Yang J., Liu H., Wang Z., Wu X., Chen Y., Catalytic oxidation of benzene over Ce–Mn oxides synthesized by flame spray pyrolysis. Particuology 2013, 11, 454–459.
[44] Kwak J.H., Hu J., Mei D., Yi C.W., Kim D.H., Peden C.H.F., Allard L.F., Szanyi J., Coordinatively unsaturated Al^{3+} centers as binding sites for active catalyst phases of platinum on g-Al$_2$O$_3$. Science 2009, 325, 1670–1673.
[45] Okumura K., Kobayashi T., Tanaka H., Niwa M., Toluene combustion over palladium supported on various metal oxide supports. Appl. Catal. B: Environ. 2003, 44, 325–331.
[46] Sedjame H.-J., Fontaine C., Lafaye G., Barbier Jr J. On the promoting effect of the addition of ceria to platinum based alumina catalysts for VOCs oxidation. Appl. Catal. B: Environ. 2014, 144, 233–242.
[47] Abdelouahab-Reddam Z., Mail R.E., Coloma F., Sepúlveda-Escribano A., Platinum supported on highly-dispersed ceria on activated carbon for the total oxidation of VOCs. Appl. Catal. A: Gen. 2015, 494, 87–94.
[48] Chen C., Chen F., Zhang L., Pan S., Bian C., Zheng X., Meng X., Xiao F. S., Importance of platinum particle size for complete oxidation of toluene over Pt/ZSM-5 catalysts. Chem Commun (Camb) 2015, 51(27), 5936–5938.
[49] Uson L., Hueso J. L., Sebastian V., Arenal R., Florea I., Irusta S., Arruebo M., Santamaria J., In-situ preparation of ultra-small Pt nanoparticles within rod-shaped mesoporous silica particles: 3-D tomography and catalytic oxidation of n -hexane. Catal. Commun. 2017, 100, 93–97.
[50] Zuo S., Wang X., Yang P., Qi C., Preparation and high performance of rare earth modified Pt/MCM-41 for benzene catalytic combustion. Catal. Commun. 2017, 94, 52–55.
[51] Jiang Z., He C., Dummer N.F., Shi J., Tian M., Ma C., Hao Z., Taylor S.H., Ma M., Shen Z., Insight into the efficient oxidation of methyl-ethyl-ketone over hierarchically micro-mesostructured Pt/K-(Al)SiO$_2$ nanorod catalysts: Structure-activity relationships and mechanism. Appl. Catal. B: Environ. 2018, 226, 220–233.
[52] Peng R., Li S., Sun X., Ren Q., Chen L., Fu M., Wu J., Ye D., Size effect of Pt nanoparticles on the catalytic oxidation of toluene over Pt/CeO$_2$ catalysts. Appl. Catal. B: Environ. 2018, 220, 462–470.
[53] Wang Y., Zhang C., Liu F., He H., Well-dispersed palladium supported on ordered mesoporous Co$_3$O$_4$ for catalytic oxidation of o-xylene. Appl. Catal. B: Environ. 2013, 142–143, 72–79.
[54] Liu L., Song Y., Fu Z., Ye Q., Cheng S., Kang T., Dai H., Effect of preparation method on the surface characteristics and activity of the Pd/OMS-2 catalysts for the oxidation of carbon monoxide, toluene, and ethyl acetate. Appl. Surf. Sci. 2017, 396, 599–608.
[55] Hosseini M., Barakat T., Cousin R., Aboukaïs A., Su B. L., De Weireld G., Siffert S., Catalytic performance of core–shell and alloy Pd–Au nanoparticles for total oxidation of VOC: The effect of metal deposition. Appl. Catal. B: Environ. 2012, 111–112, 218–224.
[56] Xie S., Deng J., Zang S., Yang H., Guo G., Arandiyan H., Dai H., Au–Pd/3DOM Co$_3$O$_4$: Highly active and stable nanocatalysts for toluene oxidation. J. Catal. 2015, 322, 38–48.
[57] Fu X., Liu Y., Yao W., Wu Z., One-step synthesis of bimetallic Pt-Pd/MCM-41 mesoporous materials with superior catalytic performance for toluene oxidation. Catal. Commun. 2016, 83, 22–26.
[58] Tidahy H.L., Siffert S., Lamonier J.F., Zhilinskaya E.A., Aboukais A., Yuan Z.Y., Vantomme A., Su B.L., Canet X., De Weireld G., Frere M., N'Guyen T.B., Giraudon J.M., Leclercq G., New Pd/

hierarchical macro-mesoporous ZrO_2, TiO_2 and ZrO_2-TiO_2 catalysts for VOCs total oxidation. Appl. Catal. A: Gen. 2006, 310, 61–69.

[59] Kim H.S., Kim T.W., Koh H.L., Lee S.H., Min B.R., Complete benzene oxidation over Pt-Pd bimetal catalyst supported on γ-alumina: Influence of Pt-Pd ratio on the catalytic activity. Appl. Catal. A: Gen. 2005, 280, 125–131.

[60] Wang H., Yang W., Tian P., Zhou J., Tang R., Wu S., A highly active and anti-coking Pd-Pt/SiO 2 catalyst for catalytic combustion of toluene at low temperature. Appl. Catal. A: Gen. 2017, 529, 60–67.

[61] Xie S., Liu Y., Deng J., Zhao X., Yang J., Zhang K., Han Z., Arandiyan H., Dai H., Effect of transition metal doping on the catalytic performance of Au–Pd/3DOM Mn_2O_3 for the oxidation of methane and o-xylene. Appl. Catal. B: Environ. 2017, 206, 221–232.

[62] Li W.B., Wang J.X., Gong H., Catalytic combustion of VOCs on non-noble metal catalysts. Catal. Today 2009, 148, 81–87.

[63] Scire S., Liotta L.F., Supported gold catalysts for the total oxidation of volatile organic compounds. Appl. Catal. B: Environ. 2012, 125, 222–246.

[64] Li J., Liu H., Deng Y., Liu G., Chen Y., Yang J., Emerging nanostructured materials for the catalytic removal of volatile organic compounds. Nanotechnol. Rev. 2016, 5, 147–181.

[65] Zhang Z., Jiang Z., Shangguan W., Low-temperature catalysis for VOCs removal in technology and application: A state-of-the-art review. Catal. Today 2016, 264, 270–278.

[66] Kim S.C., Shim W.G., Catalytic combustion of VOCs over a series of manganese oxide catalysts. Appl. Catal. B: Environ. 2010, 98, 180–185.

[67] Sihaib Z., Puleo F., Garcia-Vargas J.M., Retailleau L., Descorme C., Liotta L.F., Valverde J.L., Gil S., Giroir-Fendler A., Manganese oxide-based catalysts for toluene oxidation. Appl. Catal. B: Environ. 2017, 209, 689–700.

[68] Zhang W., Wu F., Li J., You Z., Dispersion–precipitation synthesis of highly active nanosized Co_3O_4 for catalytic oxidation of carbon monoxide and propane. Appl. Surf. Sci. 2017, 411, 136–143.

[69] He C., Yu Y., Yue L., Qiao N., Li J., Shen Q., Yu W., Chen J., Hao Z., Low-temperature removal of toluene and propanal over highly active mesoporous CuCeOx catalysts synthesized via a simple self-precipitation protocol. Appl. Catal. B: Environ. 2014, 147, 156–166.

[70] Carrillo A.M., Carriazo J.G., Cu and Co oxides supported on halloysite for the total oxidation of toluene. Appl. Catal. B: Environ. 2015, 164, 443–452.

[71] Li S., Hao Q., Zhao R., Liu D., Duan H., Dou B., Highly efficient catalytic removal of ethyl acetate over Ce/Zr promoted copper/ZSM-5 catalysts. Chem. Eng. J. 2016, 285, 536–543.

[72] Zhang C., Guo Y., Guo Y., Lu G., Boreave A., Retailleau L., Baylet A., Giroir-Fendler, A., $LaMnO_3$ perovskite oxides prepared by different methods for catalytic oxidation of toluene. Appl. Catal. B: Environ. 2014, 148, 490–498.

[73] Gómez D.M., Gatica J.M., Hernández-Garrido J.C., Cifredo G.A., Montes M., Sanz O., Rebled J. M., Vidal H., A novel CoOx/La-modified-CeO_2 formulation for powdered and washcoated onto cordierite honeycomb catalysts with application in VOCs oxidation. Appl. Catal. B: Environ. 2014, 144, 425–434.

[74] Lahousse C. Grange A.B.P., Delmon B., Papaefthimiou P., Ioannides T., and Verykios X., Evaluation of γ-MnO_2 as a VOC removal catalyst: Comparison with a noble metal catalyst. J. Catal. 1998, 178, 214–225.

[75] Li D., Yang J., Tang W., Wu X., Wei L., Chen Y., Controlled synthesis of hierarchical MnO_2 microspheres with hollow interiors for the removal of benzene. RSC Adv. 2014, 4, 26796–26803.

[76] Piumetti M., Fino D., Russo N., Mesoporous manganese oxides prepared by solution combustion synthesis as catalysts for the total oxidation of VOCs. Appl. Catal. B: Environ. 2015, 163, 277–287.

[77] Zhou G., Gui B., Xie H., Yang F., Chen Y., Chen S., Zheng X., Influence of CeO_2 morphology on the catalytic oxidation of ethanol in air. J. Ind. Eng. Chem. 2014, 20, 160–165.

[78] López J. M., Gilbank A. L., García T., Solsona B., Agouram S., Torrente-Murciano L., The prevalence of surface oxygen vacancies over the mobility of bulk oxygen in nanostructured ceria for the total toluene oxidation. Appl. Catal. B: Environ. 2015, 174–175, 403–412.

[79] Montini T., Melchionna M., Monai M., Fornasiero P., Fundamentals and catalytic applications of CeO_2-based materials. Chem. Rev. 2016, 116, 5987–6041.

[80] Torrente-Murciano L., Gilbank A., Puertolas B., Garcia T., Solsona B., Chadwick D., Shape-dependency activity of nanostructured CeO_2 in the total oxidation of polycyclic aromatic hydrocarbons. Appl. Catal. B: Environ. 2013, 132–133, 116–122.

[81] Yan Q., Li X., Zhao Q., Chen G., Shape-controlled fabrication of the porous Co_3O_4 nanoflower clusters for efficient catalytic oxidation of gaseous toluene. J. Hazard Mater. 2012, 209–210, 385–91.

[82] Chen Z., Wang S., Liu W., Gao X., Gao D., Wang M., Wang S., Morphology-dependent performance of Co_3O_4 via facile and controllable synthesis for methane combustion. Appl. Catal. A: Gen. 2016, 525, 94–102.

[83] Tang W., Wu X., Li S., Li W., Chen Y., Porous Mn-Co mixed oxide nanorod as a novel catalyst with enhanced catalytic activity for removal of VOCs. Catal. Commun. 2014, 56, 134–138.

[84] Tang W., Wu X., Li S., Shan X., Liu G., Chen Y., Co-nanocasting synthesis of mesoporous Cu–Mn composite oxides and their promoted catalytic activities for gaseous benzene removal. Appl. Catal. B: Environ. 2015, 162, 110–121.

[85] Delimaris D., Ioannides T., VOC oxidation over MnO_x–CeO_2 catalysts prepared by a combustion method. Appl. Catal. B: Environ. 2008, 84, 303–312.

[86] LiottaL.F., Ousmane M., Di Carlo G., Pantaleo G., Deganello G., Marci G., Retailleau L., Giroir-Fendler A., Total oxidation of propene at low temperature over Co_3O_4–CeO_2 mixed oxides: Role of surface oxygen vacancies and bulk oxygen mobility in the catalytic activity. Appl. Catal. A: Gen. 2008, 347, 81–88.

[87] Delimaris D., Ioannides T., VOC oxidation over CuO–CeO_2 catalysts prepared by a combustion method. Appl. Catal. B: Environ. 2009, 89, 295–302.

[88] Saqer S. M., Kondarides D. I., Verykios X. E., Catalytic oxidation of toluene over binary mixtures of copper, manganese and cerium oxides supported on γ-Al_2O_3. Appl. Catal. B: Environ. 2011, 103, 275–286.

[89] Carabineiro S.A.C., Che, X., Konsolakis M., Psarras A.C., Tavares P.B., Órfão J.J.M., Pereira M. F.R., Figueiredo J.L., Catalytic oxidation of toluene on Ce–Co and La–Co mixed oxides synthesized by exotemplating and evaporation methods. Catal. Today 2015, 244, 161–171.

[90] Ahn C.-W., You Y.-W., Heo I., Hong J. S., Jeon J.-K., Ko Y.-D., Kim Y., Park H., Suh, J.-K., Catalytic combustion of volatile organic compound over spherical-shaped copper–manganese oxide. J. Ind. Eng. Chem. 2017, 47, 439–445.

[91] Tang W.X., Liu G., Li D.Y., Liu, H.D., Wu, X.F., Han, N., Chen, Y.F., Design and synthesis of porous non-noble metal oxides for catalytic removal of VOCs. Sci. China Chem. 2015, 58, 1359–1366.

[92] Tang W.X., Wu X.F., Chen Y.F., Catalytic removal of gaseous benzene over Pt/SBA-15 catalyst: the effect of the preparation method. React. Kinet. Mech. Cat. 2015, 114, 711–723.

[93] Tang W., Wu X., Liu G., Li S., Li D., Li W., Chen Y., Preparation of hierarchical layer-stacking Mn-Ce composite oxide for catalytic total oxidation of VOCs. J. Rare Earths 2015, 33, 62–69.

[94] Zhu H., Zhang P., Dai S., Recent advances of lanthanum-Based perovskite oxides for catalysis. ACS Catal. 2015, 5, 6370–6385.

[95] Zhu J., Li H., Zhong L., Xiao P., Xu X., Yang X., Zhao Z., Li J., Perovskite oxides: preparation, characterizations, and applications in heterogeneous catalysis. ACS Catal. 2014, 4, 2917–2940.

[96] Deng J., Dai H., Jiang H., Zhang L., Wang G., He H., Au C.T., Hydrothermal fabrication and catalytic properties of $La_{1-x}Sr_xM_{1-y}Fe_yO_3$ (M = Mn, Co) that are highly active for the removal of toluene. Environ. Sci. Technol. 2010, 44, 2618–2623.

[97] Zhang J., Tan D., Meng Q., Weng X., Wu Z., Structural modification of $LaCoO_3$ perovskite for oxidation reactions: The synergistic effect of Ca^{2+} and Mg^{2+} co-substitution on phase formation and catalytic performance. Appl. Catal. B: Environ. 2015, 172, 18–26.

[98] Mo S., Li S., Li W., Li J., Chen J., Chen Y., Excellent low temperature performance for total benzene oxidation over mesoporous CoMnAl composited oxides from hydrotalcites. J. Mater. Chem. A 2016, 4, 8113–8122.

[99] Zhang F., Li M., Yang L., Ye S., Huang L., Ni-Mg-Mn-Fe-O catalyst derived from layered double hydroxide for hydrogen production by auto-thermal reforming of ethanol. Catal. Commun. 2014, 43, 6–10.

[100] Cao A., Liu G., Yue, Y. Zhang, L. Liu, Y. Nanoparticles of Cu-Co alloy derived from layered double hydroxides and their catalytic performance for higher alcohol synthesis from syngas. Rsc Advances 2015, 5, 58804–58812.

[101] Li D., Ding Y., Wei X., Xiao Y., Jiang L., Cobalt-aluminum mixed oxides prepared from layered double hydroxides for the total oxidation of benzene. Appl. Catal. A: Gen. 2015, 507, 130–138.

[102] Shao Y., Li J., Chang H., Peng Y., Deng Y., The outstanding performance of LDH-derived mixed oxide Mn/CoAlOx for HgO oxidation. Catal. Sci. Technol. 2015, 5, 3536–3544.

[103] Castaño M. H., Molina R., Moreno S., Cooperative effect of the Co–Mn mixed oxides for the catalytic oxidation of VOCs: Influence of the synthesis method. Appl. Catal. A: Gen. 2015, 492, 48–59.

[104] Mo S., Li S., Li J., Peng S., Chen J., Chen Y., Promotional effects of Ce on the activity of Mn-Al oxide catalysts derived from hydrotalcites for low temperature benzene oxidation. Catal. Commun. 2016, 87, 102–105.

[105] Mo S., Li S., Li J., Deng Y., Peng S., Chen J., Chen Y., Rich surface Co(III) ions-enhanced Co nanocatalyst benzene/toluene oxidation performance derived from Co(II)Co(III) layered double hydroxide. Nanoscale 2016, 8, 15763–15773.

[106] Jiang L., Yang N., Zhu J., Song C., Preparation of monolithic Pt–Pd bimetallic catalyst and its performance in catalytic combustion of benzene series. Catal. Today 2013, 216, 71–75.

[107] Liotta L. F., Longo A., Pantaleo G., Di Carlo G., Martorana A., Cimino S., Russo G., Deganello G., Alumina supported Pt(1%)/$Ce_{0.6}Zr_{0.4}O_2$ monolith: Remarkable stabilization of ceria-zirconia solution towards $CeAlO_3$ formation operated by Pt under redox conditions. Appl. Catal. B: Environ. 2009, 90, 470–477.

[108] Azalim S., Brahmi R., Agunaou M., Beaurain A., Giraudon J.M., Lamonier J.F., Washcoating of cordierite honeycomb with Ce–Zr–Mn mixed oxides for VOC catalytic oxidation. Chem. Eng. J. 2013, 223, 536–546.

[109] Ren Z., Botu V., Wang S., Meng Y., Song W., Guo Y., Ramprasad R., Suib S.L., Gao P.X., Monolithically integrated spinel $M_{(x)}Co_{(3-x)}O_{(4)}$ (M = Co, Ni, Zn) nanoarray catalysts: scalable synthesis and cation manipulation for tunable low-temperature $CH_{(4)}$ and CO oxidation. Angew. Chem. Int. Ed. Engl. 2014, 53, 7223–7227.

Ubong Jerome Etim, Ziyi Zhong, Zifeng Yan and Peng Bai

3 Functional catalysts for catalytic removal of formaldehyde from air

3.1 Introduction

Air pollution is the emission of substances into the air that have negative environmental and health implications. The pollutants may be a single substance or a combination of various substances formed through chemical reactions. Generally, air pollution is of serious concern because of its high prevalence in many parts of the world. The air quality degradation by the emission of various kinds of air pollutants such as ozone (O_3), particulate matter ($PM_{2.5}$ and PM_{10}), sulfur dioxide (SO_2), carbon monoxide (CO), carbon dioxide (CO_2), nitrogen dioxide (NO_2), and volatile organic compounds (VOCs) inflicts harms to the health of humans. Recent data from the World Health Organization (WHO) indicate that 9 out of 10 persons inhale highly polluted air, and about 7 million people die yearly as a result of both outdoor and indoor pollution [1]. Among the list, VOCs comprising indoor air pollutants deserve particular attention.

VOCs are a class of air pollutants that have high vapor pressure and are easily vaporized from ambient temperature up to 250 °C at atmospheric pressure [2, 3]. They are among the most common air pollutants emitted from chemical, petrochemical, and allied industries, buildings, and furniture as well as transportation activities [4]. Ubiquitously present and persisting in the air with high toxicity, VOCs have widespread environmental implications. For example, their reactions with NO_x and other airborne chemicals in the presence of sunlight form ground-level ozone, which primarily contributes to smog and respiratory complications, as well as the enhancement of the global greenhouse effect, which leads to depletion of the stratospheric and tropospheric ozone layers [5–11]. The regulation of both indoor and outdoor VOCs emissions is necessary to prevent the formation of ground-level ozone. In recent times, indoor emission of VOCs has become a global subject due to the adverse impact of VOCs on the health of exposed human beings. Indeed, poor air quality or high exposure to VOCs has been linked with reduced productivity and poor learning in schools, and with certain diseases such as cancer, liver damage, leukemia, and

Ubong Jerome Etim, College of Engineering, Guangdong Technion Israel Institute of Technology (GTIIT), Shantou, Guangdong, China
Ziyi Zhong, College of Engineering, Guangdong Technion Israel Institute of Technology (GTIIT), Shantou, Guangdong, China; Technion–Israel Institute of Technology (IIT), Haifa, Israel
Zifeng Yan, Peng Bai, State Key Laboratory of Heavy Oil Processing, Key Laboratory of Catalysis CPNC, College of Chemical Engineering, China University of Petroleum (East China), Qingdao, China

https://doi.org/10.1515/9783110544183-003

damage to the central nervous system [12]. In this regard, the reduction of VOCs emissions to meet the international ambient air quality standard (0.1 mg^{-1} m^3) is mandated under the Clean Air Act of many countries [4].

VOCs include a variety of compounds listed in Table 3.1. The rising problems of these compounds in recent years are directly related to the rapid industrial development around the globe. In China, for example, the total amount of VOCs emitted from industries ranged from 31 to 57 Tgyr^{-1} between 2005 and 2012, and a forecast of VOCs increase has aroused the concern of the Chinese government to mark VOCs pollution control as a major focus of its air pollution prevention and control measures, and set VOCs emission control as an indicator in its Thirteenth

Table 3.1: Physiochemical and thermodynamic properties of common VOCs.

Compound	Formula	BP (°C)	C_p (J/mol °C) @25 °C	G_f (KJ/mol)	H_f (KJ/mol)	H_c (KJ/mol) @25 °C
Acetylene	C_2H_2	−84.0	44.1	209.2	226.7	−1299.6
Ethylene	C_2H_4	−103.7	42.9	68.4	52.5	−1411.1
Propane	C_3H_8	−42.0	73.8	−23.4	−103.8	−2220.0
Formaldehyde	CH_2O	−19.0	35.4	−109.9	−115.9	−563.4
Acetaldehyde	C_2H_4O	20.5	53.7	−133.2	−166.4	−1192.3
Dichloromethane	CH_2Cl_2	39.6	50.8	−68.9	−95.5	−583.8
Acetone	C_3H_6O	56.0	125.0	−155.3	−248.1	−1789.9
Trichloromethane	$CHCl_3$	61.2	65.8	−68.5	−101.3	−435.2
n-Hexane	C_6H_{14}	68.0	195.0	−4.0	−198.8	−4163.1
Tetrachloromethane	CCl_4	76.72	133.9	−62.5	−132.8	−260.7
Benzene	C_6H_6	80.1	136.1	124.5	49.0	−3267.6
Trichloroethylene	C_2HCl_3	87.2	80.02	6.7	−19.1	−910.8
Toluene	C_7H_8	110.6	166.0	114.09	12.0	−3909.8
Acetic acid	$C_2H_4O_2$	115.9	123.4	−389.2	−484.1	−875.2
Tetrachloroethylene	C_2Cl_4	121.1	95.6	20.6	−14.2	−772.8
Ethylbenzene	C_8H_{10}	136.0	185.9	120.0	−12.5	−4564.7
p-xylene	C_8H_{10}	138.4	181.7	110.0	−24.4	−4552
m-xylene	C_8H_{10}	139.1	184.5	107.7	−25.4	−4549
o-xylene	C_8H_{10}	144.4	188.8	110.8	−24.4	−4552.8

Data adapted from Ref. [3] with permission from Taylor and Francis.

Five-Year Plan (2016–2020) [13–16]. In 2015, the Air Pollution Prevention and Control Action Plan was put to effect in China, with significant improvements in policy implementations, especially in the comprehensive control of VOCs. These control policies are pushing companies to adopt and develop new environmental-friendly technologies [17].

Majority of VOCs are emitted from fossil fuel combustions, which constitute the major source of green energy production (e.g., hydrogen). Other sources include petroleum refineries, chemical industries, pharmaceutical plants, automobile industries, packing materials, printing presses, coating materials, painting, and cleaning [18, 19]. The occurrence of VOCs is a serious source of concern to humans as some members of VOCs are both carcinogenic and mutagenic. Growing environmental awareness has prescribed stringent regulations for the control of VOCs emissions. Consequently, it becomes mandatory for all VOCs-emitting industries or facilities to control VOCs because they are known to create health hazards in the indoor environment. Attention to indoor air quality has been continuously increasing since the early 1950s when it became known the correlation between indoor air pollution and allergies and some chronic ailments [20, 21]. The most important VOC that is toxic and highly reactive at ambient temperature is formaldehyde whose molecular structure is presented in Figure 3.1.

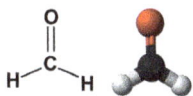

Figure 3.1: Molecular structure of formaldehyde.

3.2 Formaldehyde

3.2.1 Physical properties, chemistry, and applications

Formaldehyde (HCHO or CH_2O) is one of the two aldehydes belonging to VOCs found mostly in the indoor air, although it also exists in the outdoor air [22]. Outdoor formaldehyde pollution is less serious since ambient levels are generally low, typically around 1–4 μg^{-1} m^3. Formaldehyde exists as a colorless and flammable gas that is highly reactive at room temperature. At an elevated level, it can also be recognized by a characteristic pungent and irritating smell toxic to humans. It has a boiling point of −19 °C, a melting point of −92 °C, a density of 815 kg^{-1} m^3, and dissolves in water, ethanol, ether, and acetone. Formaldehyde solution in water (30–50% by weight) is commercially sold as *Formalin*. It is formed as a major intermediate gas in the photochemical oxidation of methane and many other hydrocarbons [23]. Due to its high reactivity, formaldehyde can undergo hydration and form hemiacetals with alcohols or thio-hemiacetals with thiols [24]. It also reacts with the hydroxyl radicals to give formic acid with an estimated half-life of about 1 h depending on the environmental conditions. The reaction of formaldehyde with amines forms Schiff bases and cross-links proteins by forming methylene bridges between amino groups [25, 26]. Conventional atmospheric chemistry considers formaldehyde as one of many target compounds in the complex reaction schemata of atmospheric components. It is an important source of the OH radical in the atmosphere and contributes to the formation of organic aerosols and interacts with other dissolved species in the cloud. The intensive solar radiation combined with high concentrations of reactive organic compounds such as alkenes leads to photosmog, which contributes to the prevalence of formaldehyde in highly industrialized and metropolitan areas, such as Beijing in China and New York, Los Angeles, and Chicago in the United States. Cities with high photochemical air pollution typically have an average formaldehyde concentration of between 20 ppb and 30 ppb, with peaks of 40 ppb to 50 ppb [22].

Formaldehyde is an important precursor for many chemical compounds. It sales as one of the most important industrial organic base chemicals, and the worldwide production capacity is over 30 million tons per year [27]. Formaldehyde has wide applications in the production of formaldehyde resins, plastics, and other chemical intermediates; the synthesis of urea-, phenol-, and melamine-formaldehyde resins (UF, PF, and MF resins) alone account for nearly 70% of the world consumption of formaldehyde in 2017 [28]. These resins are commonly used in building materials and everyday products. By 2022, the global formaldehyde production revenue is estimated to reach 103.81 million dollars [29].

3.2.2 Associated effects of formaldehyde on human health

Formaldehyde causes adverse health effects associated with irritation of the eyes and respiratory tract complications to humans and animals. The ingestion of 30 mL of a solution containing 37% formaldehyde by an adult human reportedly caused death according to a report by the Agency for Toxic Substances & Disease Registry [30]. Generally, acute exposure to formaldehyde is linked with many health problems including respiratory irritation and nasal tumors, eye and skin irritation, nasopharyngeal cancer, and multiple neuropsychological abnormalities. Studies have also shown that formaldehyde exposure can cause multiple problems to the nervous, immune, and reproductive systems, with characteristic developmental, genotoxic, and carcinogenic effects [31]. Moreover, the association of formaldehyde inhalation exposure with tissue damage, increase in cell proliferation, DNA damage, inflammation, and changes in microRNA expression patterns is documented [31]. As a consequence of the serious carcinogenic effect, a guideline maximum level of 0.1 ppm in the air was proposed by the German Federal Agency of Health to limit human exposure in dwelling places [21].

3.2.3 Historical facts

Since the first description of formaldehyde in 1855 by Alexander Butlerov, its first synthesis by dehydration of methanol was achieved by the German chemist August Wilhelm von Hofmann in 1867 [20]. Formaldehyde-based resins were already considered as an important adhesive for wood and wood composites between 1900 and 1930, the development that led to the publication of the first study titled "Formaldehyde" by Walker in 1944 [32]. Formaldehyde exposure-related health issues from prefabricated houses were first discovered in the mid-1960s. These health concerns including irritation of the eyes and respiratory tracts were traced to wooden materials bonded with UF resins, and formaldehyde became known as an indoor air pollutant in 1962 following a publication by Wittmann in which he described the postmanufacture release of formaldehyde from particleboard [33]. Consequently, the German Federal Agency of Health advocated for limited human exposure to formaldehyde in dwellings places in 1977 set a limit of 0.1 ppm ($0.125 \ mg^{-1} \ m^3$ under standard atmospheric conditions) [21]. In 1981, the criteria for the limitation and regulation of formaldehyde emissions from wood-based materials were established in Germany and Denmark, and the United States set their first formaldehyde exposure limit around 1985 following the report of potential carcinogenic properties of formaldehyde in 1980 after publication of long-term inhalation exposure results of rats and mice [20, 34, 35]. The WHO began to treat formaldehyde as an indoor pollutant in 1983 [36] and in 1987 it published an indoor guideline level of $0.1 \ mg^{-1} \ m^3$ (0.08 ppm) [37]. With those developments, formaldehyde has been regarded as one of the most dangerous indoor air pollutants up till now. Over the years efforts made by various government agencies have led to the decrease of the indoor formaldehyde, especially in developed countries, whereas concentrations in ambient air are on the increase, triggered by the rapid industrialization or urbanization in fast-developing countries such as China. Although the focus on formaldehyde as important indoor pollution reduced in the 1990s, its classification as a type 1 carcinogenic substance for humans in 2004 renewed interest and brought back its research to prominence [20]. As a result, various authorities and institutions proposed new indoor air guidelines, giving levels that are nearly ubiquitous. In 2006, the German Ad hoc Working Group approved the German indoor guideline level of 0.1 ppm as safe regarding the carcinogenic effect of formaldehyde in humans. In 2010, the WHO also set its guideline threshold for formaldehyde as $0.1 \ mg^{-1} \ m^3$ and also marked it as group 1 carcinogen for humans [38, 39]. These guidelines intended to prevent acute and chronic sensory irritation infections in the respiratory tracts and all types of cancer [40].

In response, various countries and institutions have set different guidelines for the occupational indoor formaldehyde exposure. For example, the American National Institute for Occupational Safety and Health (NIOSH) recommended exposure limits as an 8- h time-weighted average (TWA) (0.016 ppm) and a 15-min short-term exposure limit (STEL) (0.1 ppm), which are significantly lower than the workplace exposure limits decided by the UK's Health and Safety Executive (2 ppm for both the TWA and a 10-min STEL) [29]. In contrast, China, New Zealand, Finland, Israel, and Canada all designate formaldehyde occupational exposure limits in terms of a ceiling (C). According to this index, the American Conference of Governmental Industrial Hygienists for many years adopted a threshold limit value ceiling (TLV-C) (0.3 ppm), and the European Scientific Committee on Occupational Exposure Limits recently proposed a formaldehyde-related TWA of 0.3 ppm, but a STEL of 0.6 ppm. NIOSH's Immediately Dangerous to Life or Health is 20 ppm for formaldehyde [29]. The European Commission classified formaldehyde as 1B carcinogen and mutagen 2 in 2014 in the ordinance EU605/2014, which means that the carcinogenic effect has been demonstrated in animal trials and is probable for humans [21]. The WHO proposed a short-term HCHO exposure limit (30 min) of $0.1 - 0.2\,\mathrm{mg^{-1}\,m^3}$ for the protection of sensory irritation and other long-term health effects [41]. A comprehensive list of standards and guidelines for formaldehyde exposure for various countries is documented [42].

3.2.4 Sources of formaldehyde in the environment

Formaldehyde ubiquitously exists in the environment from both natural sources and anthropogenic activities (Figure 3.2). Natural formaldehyde is a product of photochemical reactions in the earth's troposphere. Photochemical degradation of methane and nonmethane hydrocarbons releases a large amount of formaldehyde. An example is the ozonolysis of alkenes and the reaction of alkanes and alkenes with hydroxyl radicals and nitric oxide. Additionally, natural disasters such as forest and bush fires lead to biomass combustion, and biomolecular decomposition produces an enormous amount of formaldehyde into the atmosphere.

Transportation and industries are the major sources of man-made indoor formaldehyde. The processing of hydrocarbons and combustion of petroleum products, and releases from vehicles, power plants, and manufacturing factories which are made of combustible fuels, are major sources of formaldehyde and other greenhouse gases in the atmosphere. Exhaust gases from automobiles are responsible for over 70% of transport greenhouse gas emissions and much of air pollution in the European Union (EU) [43]. Composite wood products containing UF resin have also been identified as a major source of indoor formaldehyde. In particular, building materials and many household products, such as particleboard, plywood, and fiberboard, glues and adhesives, permanent-press fabrics, paper product coatings, and certain insulation materials,

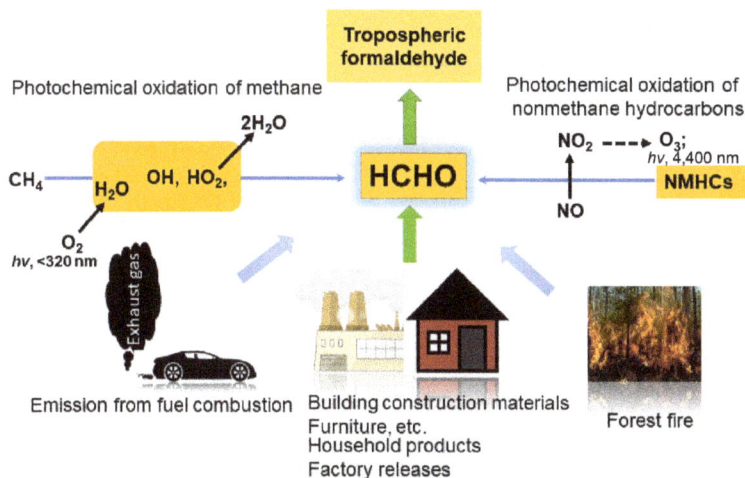

Figure 3.2: Sources of formaldehyde in the atmosphere.

contribute greatly to indoor formaldehyde existence [30]. Formaldehyde is also a by-product of human metabolism and certain anthropogenic activities (e.g., smoking tobacco) [44, 45].

Indoor formaldehyde exposures in residential and renovated apartments, workplaces, and shopping malls can be extremely dangerous. Humans get exposed to formaldehyde from these sources through inhalation, gastric intestinal digestion, or direct skin contact. The surrounding air temperature and relative humidity can influence the emission rate of formaldehyde into the environment, but this can be controlled by effective ventilation [22]. Without doubt, formaldehyde is an integral constituent of the indoor and outdoor air due to its ubiquitous presence in the environment from both natural and anthropogenic sources and its discharge adversely deteriorate the air quality, and therefore, the need for removal. The following section will focus on the removal of formaldehyde from air streams.

3.3 Methods for removal of formaldehyde

There are several methods available for removing VOCs from the indoor air, which a good number of them apply to formaldehyde removal as well. These include source removal, adsorption, condensation, absorption, membrane separation, biological methods, thermal oxidation (incineration), and catalytic oxidations and their advantages and disadvantages are enumerated in Table 3.2. The most suitable method to adopt for removing formaldehyde is dependent on the environmental conditions. However, in this chapter, removal by catalytic oxidation will be elaborated.

Table 3.2: Advantages and disadvantages of different methods for formaldehyde removal.

Removal method	Advantages	Disadvantages
Adsorption	(i) High removal efficiency (80–95%) (ii) Low cost involvement (iii) Cheap cost of adsorbents	(i) Requires frequent regeneration of adsorbent (ii) Problems of adsorbent poisoning and clogging of pores
Biological method (Biofiltration)	(i) Low cost involvement (ii) Release of nonhazardous products (iii) No external energy requirement (ambient temperature) (iv) It is a destructive method (v) Process is highly selective	(i) Process is time consuming (ii) Highly selective to type of pollutant molecules (iii) Generation of secondary wastes (biomass) (iv) Low removal efficiency (60–95%) (v) Controlling of biological parameters
Membrane separation	(i) High removal efficiency (>90%) (ii) No further treatment is required	(i) High cost of membrane (ii) Implementation is complicated
Absorption	(i) High removal efficiency (85–95%)	(i) May require pretreatment step (ii) Difficulty of design due to lack of equilibrium data (iii) It needs rigorous maintenance (iv) Generation of waste water
Catalytic oxidation	(i) High removal efficiency up to 90–98% (ii) Release of non-hazardous products, usually carbon dioxide and water (iii) Requires less or no energy (ambient temperature oxidation) (iv) It is a destructive method	(i) Removal efficiency is sensitive to operating conditions and properties of catalyst (ii) Problems of catalyst poisoning and deactivation (iii) High cost of catalyst (iv) Possibly requires additional control equipment

Data extracted form Ref. [4] with permission from Elsevier.

3.3.1 Adsorption

Adsorption on the surface of solid materials offers a promising method for removal of low-concentration formaldehyde. This method transfers pollutants from gas phase to the adsorbent, and the adsorbent often needs frequent regeneration to degrade the pollutant molecules [46]. Compared with other methods, adsorption is a low-cost process and is relatively easy to carry out. Materials utilized as adsorbents must possess certain inherent properties that promote adsorption of pollutants and offer excellent removal efficiency, and can be regenerated. Some desirable physical properties include high surface area and large pore volume. Efficient adsorbents must possess high adsorption capacity and high diffusion rate, affordable price, and high abundance. Porous materials with these properties include mesoporous silicas, zeolites, and carbonaceous materials such as activated carbons [14].

Activated carbon is the most widely used adsorbent due to its high adsorption capacity and great affinity for both polar and nonpolar compounds. However, this material suffers from high flammability, pore clogging, hydroscopicity, and regeneration problems. Molecular sieves such as zeolite could be applied as an adsorbent for formaldehyde removal despite their lower adsorption capacity. Zeolites are highly stable (for high-temperature process), incombustible, and are easily regenerable at low temperatures, hence suitable as an adsorbent for a broad range of VOCs. Regeneration is typically done by heating or vacuum treatment to desorb the adsorbed gases; the bed is then cooled and dried, and the concentrated vapors are channeled to a recovery system. A major limitation encountered in adsorption processes is the effect of water partial pressure.

Adsorption can be divided into (1) physical and (2) chemical adsorption based on the type of interaction between adsorbate and adsorbent. Generally, formaldehyde molecules are removed from air by physical adsorption. In physical adsorption, organic molecules are held on the surface and in the pores of the adsorbent by the weak Van der Waals forces and are generally characterized by low heat of adsorption and reversible adsorption equilibrium. Since a very low pollutant concentration in indoor air is expensive to treat by other processes such as catalytic oxidation, adsorption becomes an auxiliary method to increase the concentration to a level that is more feasible to clean up. Generally, the design of an adsorption system mostly depends on the chemical characteristics of the pollutant to be removed, the physical properties of the adsorbent, and the inlet stream (temperature, pressure, and volumetric flow-rate).

3.3.2 Biological method

Biological control technique involves the use of biological species, that is, living organisms for the removal or prevention of gaseous indoor pollutants from the environment by converting them under aerobic conditions to carbon dioxide, water, and biomass [20, 47, 48]. This method is inexpensive and environmentally benign, suitable for removal of very low concentrations of highly water-soluble VOCs, such as formaldehyde. Air can be purified by passing through a planted soil or directly on the plants or a packed bed of solid support colonized by microorganisms and the contaminants are then degraded by the attached microorganisms. It was reported that several plants could remove formaldehyde at $19,000 - 46,000 \, \mu g^{-1} \, m^3$ to levels lower than $2,500 \, \mu g^{-1} \, m^3$ (detection limit) in 24 h [49]. Interior plants, such as Hedera helix (English ivy), Chrysanthemum morifolium (pot mum), Dieffenbachia compacta (dump cane), and Epipremnum aureum (golden pathos), grown in growstone were shown to remove formaldehyde [47, 50]. The different parts of plants removed formaldehyde at different rates, and the root zone was found to remove formaldehyde more rapidly than aerial parts of the plant or the entire plant within 24 h, and fast removal was observed under darkness than light conditions [50]. Biological processes can be operated under ambient conditions without the requirement of further treatment process; however, the need for controlling biological parameters poses a challenge to its application, and the treatment period is often longer than other processes. The biological control methods are highly selective and do not generally transfer the pollution problem to another environmental compartment (gas to solid and/or gas to liquid), which is often the case with many other purification methods. An example of a biological VOCs remediation method is *biofiltration*. Biofiltration is a process in which contaminated air is passed through a porous packed medium that supports the living of a microorganism population [51]. The contaminants are first adsorbed from the air to the water/biofilm phase of the medium. The moisture content of the medium should be maintained at optimum to support microbial growth without clogging the pores of the biofilter. Biofiltration success is dependent upon the degradability of the contaminants. An efficient biofilter must also provide a benign environment for microorganisms.

3.3.3 Membrane separation

Pollutants are passed through a membrane into another fluid by affinity separation. This method is normally recommended for highly loaded streams and has yet to be proven efficient at low VOCs levels. If the separated VOCs are not reused, membrane filtration must be completed with a destruction step. Gas permeation and reverse osmosis are the available membrane separation techniques. Membrane separation recovery technology is applied only to low-flow waste gas streams. It has high removal efficiency (90–99%) and generates waste that requires no further treatment. Its wide applicability is limited by cost and complexity.

3.3.4 Absorption

In absorption, the contaminated air is contacted with a liquid solvent and the pollutant is selectively removed in the process by transferring to the liquid phase. It is often referred to as *air scrubbing*. This occurs in an absorber tower designed to provide the appropriate liquid vapor contact area necessary to facilitate mass transfer. The absorber has the same design as that for process application using the vapor/liquid equilibrium data, liquid and vapor flux rates, material balances, and so on [4]. The process efficiency and cost rely on proper choice of solvent, availability of vapor/liquid equilibrium data for the design of the absorber, and required regeneration of the VOC containing solvent by stripping. This method is capable of achieving VOCs removal efficiencies of 95 to 98% [4].

3.3.5 Source removal

Source removal may be described as the elimination of indoor air pollutants from their origins. The feasibility of source removal will sometimes depend on the circumstances. Formaldehyde removal may be an easy venture in the case of wooden products, such as furniture but a much more difficult scenario in the case of a built structure. The surface treatment by the use of reactive or diffusion resistant coatings and fumigation with ammonia were mainly applied in the case of very high formaldehyde levels in the 1980s and 1990s [46]. The reaction with ammonia that yields hexamethylene tetramine was in particular used in prefabricated houses. It is a drastic but very effective and long-lasting solution. The application of natural compounds such as urea, catechin, and vanillin suppresses formaldehyde emission from plywood [52]. Intelligent housing construction and increased ventilation systems have also been proposed to reduce indoor formaldehyde [53].

3.3.6 Catalytic oxidation

Catalytic oxidation is a promising air treatment technology that offers high destructive efficiency of pollutants under mild operating conditions, for example, lower operating temperatures (RT-200 °C) and cheaply available oxygen as oxidant. It is a versatile and efficient method for reducing low concentration organic pollutants to water and carbon dioxide, which are not harmful, with the full control of reaction selectivity. With associated economy, catalytic oxidation has been successfully applied to the removal of a broad range of VOCs, including formaldehyde. Oxidation at low temperatures avoids unwanted side reactions, guarantees environmental friendliness, and saves high energy costs, translating into huge economic benefits. The increase in removal rate by catalysts at low-temperature singles out this method as the most reliable technique for the treatment of indoor air pollution. Among the important parameters to consider for a successful operation are the type of catalyst and its selectivity and stability and the presence of other molecules or contaminants in the air stream to be treated. The presence of contaminants can poison catalysts, leading to fast deactivation, which can significantly hamper the efficiency of the process. Catalytic oxidation shows high operation similarity to thermal oxidation but with much enhanced reaction rate and lower processing temperatures, thus able to reduce energy requirements and guarantee the safety of the surrounding environment.

3.4 Formaldehyde oxidation catalysts

As applicable to other gaseous pollutants (e.g., CO), formaldehyde can be catalytically removed at low temperatures, and the development of highly efficient and stable catalysts that can completely oxidized formaldehyde to harmless compounds at ambient temperatures is critical. Generally, low-temperature oxidation catalysts are divided into two categories: (i) supported metal and (ii) transition metal oxide catalysts (Figure 3.3). Supported metal catalysts are often prepared from noble or precious metals, of which the key players are platinum (Pt), palladium (Pd), gold (Au), ruthenium (Ru), and silver (Ag), usually supported on nonreducible metal oxides (e.g., SiO_2, Al_2O_3, and molecular sieves) or reducible metal oxides, such as MnO_2, Fe_2O_3, CeO_2, Co_3O_4, $MnO_x–CeO_2$, and $Co_3O_4–CeO_2$ [54–59]. Formaldehyde oxidation over supported noble metal catalyst was first reported by Imamura et al. [60] in a study in which water and carbon dioxide were found as products of catalytic oxidation over Ru, Rh, and Pt nanoparticles supported on CeO_2 at low temperatures. The transition metal oxides are also an important group of low-temperature oxidation catalysts, which could be a single transition metal oxide (e.g., MnO_2) or transition metal oxide composites (e.g., $Co_3O_4–MnO_2$ and $CuO–CeO_2$).

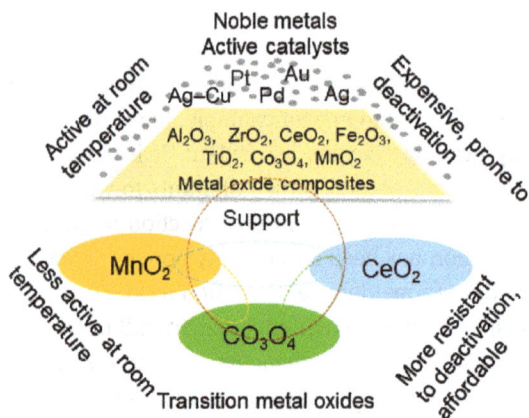

Figure 3.3: Formaldehyde oxidation catalysts.

3.4.1 Oxidation mechanisms and kinetics

The mechanism of catalytic oxidation at low temperature of some gaseous pollutants (CO, VOCs, etc.) exhibits similarity in the general principle [61–64]. A number of different mechanisms have been proposed, but the most popular and

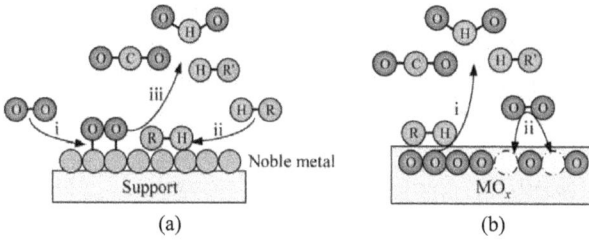

Figure 3.4: Schematic diagrams of (a) Langmuir–Hinshelwood (L-H) mechanism and (b) Mars–van Krevelen (MvK) mechanism. Reproduced from Ref [66] with permission from Elsevier.

generally accepted are Langmuir–Hinshelwood (L-H), Eley–Rideal (E-R), and Mars–van Krevelen (MvK) mechanisms. The validity of each model is dependent on the properties of catalyst and reaction conditions. To this end, the mechanism of formaldehyde oxidation is usually described with respect to the catalyst active sites and the reaction conditions. The MVK model is commonly applied to the kinetic modeling of oxidation of organic molecules including formaldehyde over supported metal or metal oxide catalysts [65].

(i) L-H mechanism: The L-H model assumes that reaction occurs between the adsorbed oxygen species and adsorbed reactant molecules, and the controlling step is the surface reaction between these two adsorbed molecules on different or similar active sites (Figure 3.4a).

(ii) E-R mechanism: According to this model, oxidation reaction occurs between adsorbed oxygen species and gas-phase reactant molecules (i.e., between adsorbed reactant and gas-phase oxygen). The controlling step is the reaction between adsorbed molecules and reactant molecules in the gas phase.

(iii) MvK mechanism: This mechanism involves a redox process in which reaction occurs between a reactant molecule and oxygen-rich sites (oxygen vacancies) on catalyst, assuming a constant oxygen concentration on the surface (Figure 3.4b). A reaction results from an interaction between a reactant molecule and an active site of the catalyst in which the transfer of oxygen atoms from the solid to the gas phase occurs. The first step is the adsorption and dissociation of oxygen molecules from the gas phase into active oxygen species (O), which reacts with the reactant molecule from the air in the second step, and then the reduction of the oxidized catalyst (reoxidation of the intermediate) occurs [67]. The reduction and oxidation rates must be equal in a steady-state, and the oxidation rate can be expressed according to the MvK model equation:

$$O_2 + (\ ) \xrightarrow{k_o} (O_2) \xrightarrow{()} 2(O) \tag{3.1}$$

$$R + (O) \xrightarrow{k_i} CO_2 + H_2O + (\ ) \tag{3.2}$$

where k_o is the kinetic constant for the nonequilibrium dissociative adsorption of oxygen on the catalyst surface, k_i the kinetic constant for reactant oxidation, and R is the reactant molecule. The notation ( ) represents an oxygen adsorption site. The rate of the above reaction is given as:

$$- r_i = \frac{k_o k_i C_o C_i}{k_o C_o + \delta k_i C_i} \tag{3.3}$$

where $-r_i$ is the reaction rate $(\text{mol}^{-1} \, \text{m}^3 \text{s}^1)$, C_o is oxygen concentration, C_i is reactant concentration, and δ is the stoichiometric coefficient for the total oxidation [65, 67].

The kinetic constants follow the Arrhenius dependence on temperature:

$$k_o = k_{oo} \exp\left(- \frac{E_o}{RT} \right) \tag{3.4}$$

$$k_r = k_{or} \exp\left(- \frac{E_r}{RT} \right) \tag{3.5}$$

where k_{oo} and k_{or} are the pre-exponential factors, and E_o and E_r are the activation energies for the two reactions.

The MvK mechanism has been applied to study formaldehyde removal on supported metal catalysts. The oxidation of formaldehyde over supported Pt catalyst based on the MvK is described as follows [68]:

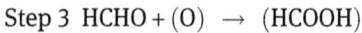

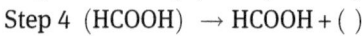

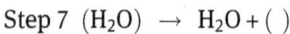

Step 1 $O_2 + (\ ) \rightarrow (O_2)$
Step 2 $(O_2) + (\ ) \rightarrow 2(O)$
Step 3 $HCHO + (O) \rightarrow (HCOOH)$
Step 4 $(HCOOH) \rightarrow HCOOH + (\ )$
Step 5 $(HCOOH) + (O) \rightarrow (COO) + (H_2O)$
Step 6 $HCHO + (O) \rightarrow (HCHO \cdots O) \rightarrow (COO) + (H_2O)$
Step 7 $(H_2O) \rightarrow H_2O + (\ )$
Step 8 $(COO) \rightarrow CO_2 + (\ )$

The aforementioned reactions are represented on a simple network (Figure 3.5).

This mechanism involves both consecutive and parallel surface oxidation reactions. Accordingly, formaldehyde is first oxidized into surface dioxymethylene species and subsequently formate species, which decompose to form surface-bound COO species (rate-determining step) before finally oxidized to CO_2. The surface intermediate (HCOOH) could either be desorbed into the gas phase or be further oxidized into (CO_2), which then desorbs. Similarly, the mechanism of formaldehyde removal on Pt/TiO$_2$ was proposed [55]. On the basis of this mechanism, HCHO oxidizes to formate surface species, which then directly decompose into adsorbed CO species and H_2O, and the CO species finally reacts with O_2 to produce CO_2 (Figure 3.6).

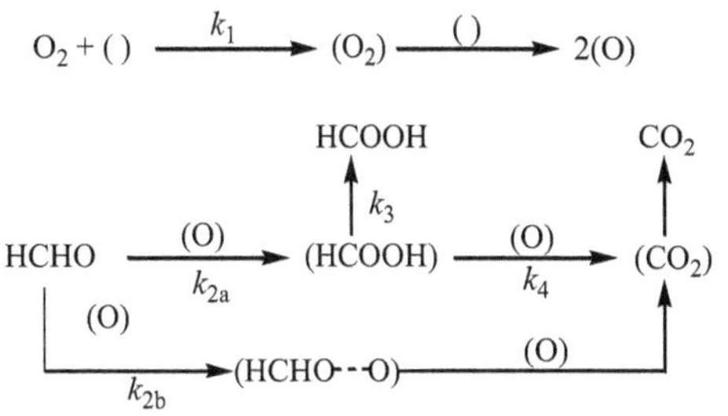

$$O_2 + () \xrightarrow{k_1} (O_2) \xrightarrow{()} 2(O)$$

$$HCHO \xrightarrow[k_{2a}]{(O)} (HCOOH) \xrightarrow{k_3} HCOOH$$

$$(HCOOH) \xrightarrow[k_4]{(O)} (CO_2) \to CO_2$$

$$HCHO \xrightarrow[k_{2b}]{(O)} (HCHO\text{--}O) \xrightarrow{(O)} (CO_2)$$

Figure 3.5: Possible reaction mechanism network for oxidation of formaldehyde on Pt/hydrophobic carbon: steps 1 to 8. Reproduced from Ref. [68] with permission from Elsevier.

M: Pt/TiO$_2$

$$HCHO \xrightarrow{O\text{-}M} HCOO + H \longrightarrow CO + H_2O \xrightarrow{O_2} CO_2 + M$$

(with M beneath HCOO and beneath CO + H$_2$O)

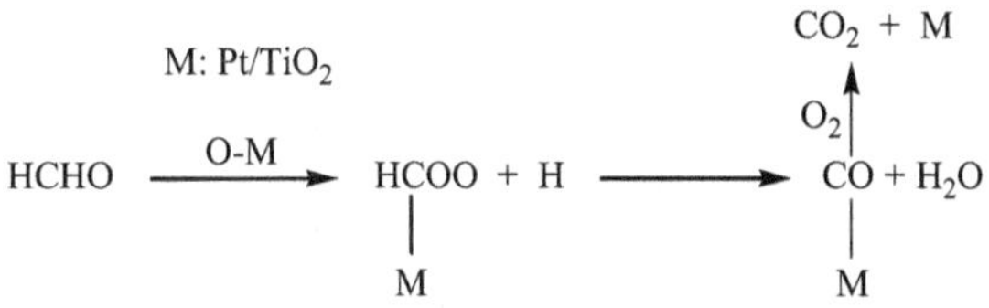

Figure 3.6: Possible reaction pathway of formaldehyde over Pt/TiO$_2$ catalyst. Reproduced Ref. [55] with permission from Elsevier.

Several other mechanisms have been developed by researchers, recognizing formate as an intermediate species in the oxidation of formaldehyde. The formation of formate and OH on the surface on adsorption of oxygen "O" from an oxide catalyst at room temperature is reported [69]:

$$HCHO_{(g)} + 2\text{"O"} \to HCOO_{(ads)} + OH_{(ads)}$$

$$2OH_{(ads)} \to H_2O_{(g)} + \text{"O"}$$

$$HCOO_{(ads)} \to CO_{(g)} + OH_{(ads)}$$

$$HCOO_{(ads)} \to CO_{2(g)} + H_{(ads)}$$

$$OH_{(ads)} + H_{(ads)} \to H_2O_{(g)}$$

In this mechanism, it is believed that the formation of water product is due to surface OH species from the formation of formate species ($HCOO_{ads}$) upon reaction of formaldehyde with active oxygen species from the catalyst surface. However, at a higher temperature, the formation of oxidation products could result in the oxidation of formate species with the participation of oxygen species from the catalyst and desorption of water in the reaction between $OH_{(ad)}$ and $H_{(ad)}$. The formation of formate and OH species as intermediates during oxidation of formaldehyde over layered-MnO$_2$ catalysts was verified via in situ diffuse reflectance infrared Fourier-transform spectroscopy (DRIFTS) experiments in which (COO), (CH), (COO), and n(CH) of formate (HCOO$^-$) species as well as OH species, appeared (Figure 3.7). However, the appearance of negative peaks of OH species shows that the formed species was not consumed in the process; thus, no reaction existed between it and the formate species. Reaction between surface-active oxygen and water regenerates surface hydroxyl groups consumed in the oxidation of HCHO [71–73].

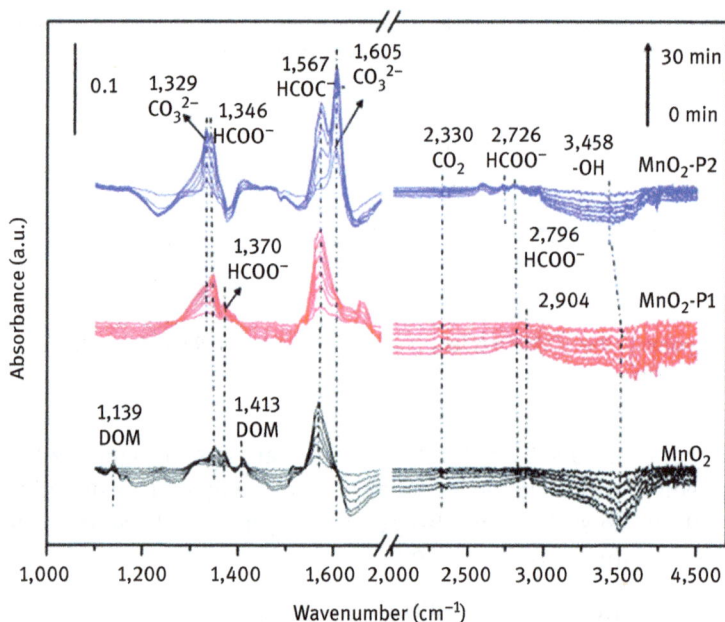

Figure 3.7: In situ DRIFTS spectra of different birnessite samples exposed to the flow of HCHO/O$_2$ at room temperature. Reproduced from Ref. [70] with permission from Elsevier.

The generation of intermediates was confirmed by theoretical studies [74, 75]. Formaldehyde oxidation on the Pt/TiO$_2$ (101) surface was investigated by DFT in the presence of adsorbed oxygen or hydroxyl species [74]. The primary intermediates are CH$_2$O, CH$_2$O$_2$, CHO$_2$, CHO, and CO$_2$. HCHO directly dehydrogenates through the pathway of CH$_2$O → –CHO → CO, while the presence of oxygen promotes oxidation of CH$_2$ → CH$_2$O$_2$ → CHO$_2$ → CO$_2$. The decomposition of formate is the rate-limiting step due to its relatively high-energy barrier. This mechanism reveals that HCHO first oxidizes to dioxymethylene, followed by successive dehydrogenation to carbon dioxide (HCHO → HCHO$_2$ →CHO$_2$ → CO$_2$) as in Figure 3.8. The calculated energies associated with each elementary state are given in Table 3.3.

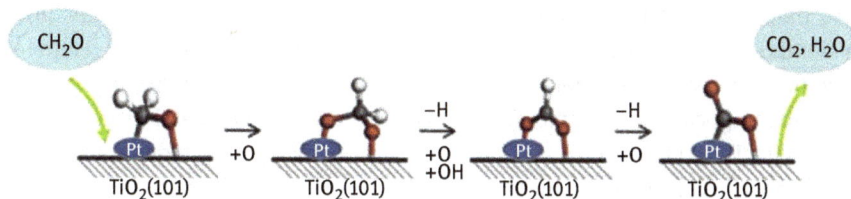

Figure 3.8: Simplified pathway for oxidation of formaldehyde on Pt/TiO_2 (101) surface via dehydrogenation. Reproduced from Ref. with permission from Elsevier.

Table 3.3: Calculated energy barriers E_a (eV) and reaction energies ΔE(eV) for the elementary reactions involved in formaldehyde oxidation on Pt/TiO_2 (101) surface.

Elementary reactions	E_a	ΔE
$CH_2O^* \rightarrow CHO^* + H^*$	1.65(1.47)	−0.58(−0.67)
$CH_2O^* + O^* \rightarrow CHO^* + OH^*$	1.28(1.22)	−1.89(−1.85)
$CH_2O^* + OH^* \rightarrow CHO^* + H_2O^*$	0.91(0.71)	−1.38(−1.34)
$CH_2O^* + O^* \rightarrow CH_2O_2^*$	0.13(0.14)	−1.17(−1.09)
$CH_2O2^* \rightarrow CHO_2^* + H^*$	2.21(1.95)	−0.63(−0.60)
$CH_2O_2^* + O^* \rightarrow CHO_2^* + OH^*$	0.43(0.35)	−4.11(−4.12)
$CH_2O_2^* + OH^* \rightarrow CHO_2^* + H2O^*$	0.24(0.11)	−2.29(−2.20)
$CHO_2^* \rightarrow CO_2^* + H^*$	1.76(1.65)	0.67(0.58)
$CHO_2^* \rightarrow CHO^* + O^*$	1.84(1.78)	0.96(0.92)
$CHO_2^* + O^* \rightarrow CO_2^* + OH^*$	1.35(1.16)	−0.81(−0.86)
$CHO_2^* + OH^* \rightarrow CO_2^* + H_2O^*$	1.95(1.73)	−0.20(−0.20)
$CHO^* \rightarrow CO^* + H^*$	1.06(0.91)	−0.82(−0.84)
$CHO^* + O^* \rightarrow CO^* + OH^*$	1.23(1.08)	−1.80(−1.78)
$CHO^* + OH^* \rightarrow CO^* + H_2O^*$	1.24(1.07)	−1.69(−1.67)
$CO^* + O^* \rightarrow CO_2^*$	1.10(1.07)	−0.39(−0.38)

Values in parentheses are energies with zero point energy corrections. Data adapted from Ref. [74] with permission from Elsevier.

3.4.2 Noble metal-based catalysts

Noble metal-based catalysts possess good catalytic activities, and complete oxidation can be achieved at very low operating temperatures due to their high electron transfer capability. Most studied noble metal catalysts are Pt, Pd, Au, and Ag. These metals exhibit very high oxidation activity for the removal of formaldehyde from air stream and can achieve complete oxidation even at room temperature. However, the associated high cost of procuring these metals hinders large-scale applications and calls for the study of transition metal oxide-based catalysts as an alternative. Moreover, these catalysts are easily deactivated by coke deposition and poisons, such as halogens and water vapor. Despite the high cost and fast deactivation, noble metals are the preferred catalysts for formaldehyde removal at room temperature due to their ease of regeneration. Supported noble metal catalysts are synthesized by loading calculated amounts of the active catalytic metal (e.g., Pt) on supports (e.g., TiO_2) (Table 3.4). The catalytic performance strongly depends on the intrinsic properties of the noble metal, precursor, preparation method, support, size and morphology of particles, and so on [76]. For reducible oxide supports, the metal–support interface plays an important role [77]. The support and particle size, which will be discussed in sections 3.4.5 and 3.5.1 of this chapter, play critical roles in the overall performance of noble metal catalysts.

Table 3.4: Examples of noble metal-based catalysts for oxidation of formaldehyde.

Catalyst	HCHO conc. (ppm)	GHSV (mL/gh)	T100$_{conv}$ (°C)	Ref.
Pt/TiO$_2$	100	50,000/ h	RT	[55]
Pt/Fe$_2$O$_3$	100–500	60,000	RT	[56]
Pt/MnO$_2$	460	20,000	RT	[57]
Pd/TiO$_2$	10	120,000	RT	[58]
Au/CeO$_2$	500	35,400/ h	<40	[78]
2D Au/Co$_3$O$_4$–CeO$_2$	200	55,000/ h	RT	[54]
Pd/CeO$_2$	600	10,000/ h	<20	[72]

Supported Pt and Pd catalysts

Supported Pt and Pd catalysts are efficient for the complete oxidation of formaldehyde at very low temperatures, even at room temperature. Both Pt and Pd show good formaldehyde oxidation activity. Pt/TiO$_2$ is considered as the earliest and efficient formaldehyde oxidation catalyst at room temperature [55, 79]. TiO$_2$ possesses excellent ability as support to activate molecular oxygen by creating oxygen vacancies for activation of formaldehyde molecules [55]. Although Pd/TiO$_2$ exhibits even a better catalytic activity for formaldehyde oxidation at room temperature, and has higher catalytic efficiency at low Pd loading than Pt, it is less studied than Pt/TiO$_2$ catalyst. The dispersion and oxidation states of Pt and Pd play crucial roles, with the metallic outperforming the oxidized form. The promotion of TiO$_2$-supported Pt or Pd catalyst with alkali metal increases the oxidation performance at room temperature by forming Pt-O-alkali metal-active species and providing surface OH species for enhanced formaldehyde adsorption [80–84]. For example, the addition of 2 wt.% Na$^+$ to Pt/TiO$_2$ stabilized the atomically dispersed Pt-O(OH)$_x$–alkali metal species on the catalyst surface. The modified catalysts exhibited enhanced formaldehyde oxidation activity by activating reaction between surface OH and formate species and altering the reaction temperature [80]. A number of other materials are also studied as supports for Pt and Pd catalysts, including Al_2O_3, MnO_2, ZrO_2, Fe_2O_3, Co_3O_4, MnO_x–CeO_2, zeolite, and activated carbon [56, 57, 59]. On a variety of supports such as oxides (TiO_2, Al_2O_3, Mn–CeO_2) and zeolites (Beta, Zeo-13X, USY, ZSM-5, and HM) for Pd catalyst, Park et al. [59] found that Beta zeolite-supported Pd exhibited the highest activity, related to high formaldehyde adsorption capacity of the support and fast formaldehyde oxidation kinetics, in spite of a low (10%) Pd dispersion on the support. Modification of the Pd/Beta catalyst with MnO_x enhanced activity of the catalyst at a very high space velocity (200,000 h^{-1}) and a temperature of 40 °C. The enhanced activity is possibly due to synergistic interaction between Pd and MnO_x. For Pt and Pd catalysts, the large-scale application is limited by high cost, low resistance to poisoning, and fast deactivation.

Supported Au and Ag catalysts

Au, the most inert metal in nature, became popular as oxidation catalyst following the landmark discovery by Haruta et al. [85] in which they demonstrated the ability of Au to oxidize CO below 0 °C. Supported Au catalysts also show excellent catalytic activities for formaldehyde oxidation at low temperatures. Unlike Pt, Au supported on CeO_2 (Au/CeO_2) is most widely studied for formaldehyde oxidation and shows lower activity than Pt/TiO_2 at room temperature. However, numerous other supports including SiO_2, ZrO_2, MnO_2, Co_3O_4–CeO_2, and FeO_x can be utilized [78, 86]. Increasing the surface area of CeO_2 support generates more oxygen vacancies as active sites, leading to higher specific catalytic activity than the low surface area supported Au/CeO_2. This catalyst induced the formation of $Au_xCe_{1-x}O_{2-\delta}$ solid solution favorable for adsorbed O_2 activation [78]. Exposing the CeO_2 reactive facets where oxygen vacancies can easily form enhances the activity of Au/CeO_2. Furthermore, modifying the Au/CeO_2 with some other metal oxides further improves its formaldehyde oxidation performance. Comparatively, formaldehyde oxidation activity of three-dimensional ordered mesoporous Au/Co_3O_4–CeO_2 was higher than that of Au/CeO_2 catalyst [87]. The higher intrinsic performance of the catalyst was attributed to the interaction between Co_2O_3 and CeO_2, which improved surface-active oxygen migration and formation of activated Au species. In comparison with Pt, Pd, and Au, Ag has lower formaldehyde oxidation activity; however, it is relatively cheaper. Supported Ag catalyst exhibits good activities for formaldehyde oxidation at a temperature > 50 °C. For example, 3 wt.% Ag/MnO_x–CeO_2 completely oxidized 580 ppm formaldehyde at 100 °C and 30,000 $mLg^{-1}h^{-1}$ gas hourly space velocity (GHSV), while 100% formaldehyde conversion temperatures of 3 wt.% Ag/CeO_2 and

3 wt.% Ag/MnO$_x$ catalysts were 120 and 140 °C, respectively. At 40 °C, the formaldehyde conversion of the 3 wt.% Ag/MnO$_x$–CeO$_2$ catalyst was only 12% [88].

Bimetallic catalysts

Bimetallic catalysts in this context refer to catalysts that contain two metals such as
Pt–Au/TiO$_2$ [89]. These catalysts may exhibit superior physiochemical properties
and performances than the monometallic catalysts, attributed to synergistic interactions (electronic and geometric effects) between the two component metals, from
the change of the electronic structure to the decrease in the active metal ensemble
size at the interface of bimetallic nanoparticles [90]. For example, Au–Pt bimetals
can often form alloy or intermetallic nanoparticles showing high efficiency for formaldehyde oxidation [89, 91]. It is worthy of note that active sites distribution on bimetallic catalysts depends on its mixing configuration and preparation method,
and metal precursors have noticeable effects on the properties of the bimetallic
metal particle, either enhancing or weakening interactions; weakly interacted bimetallic metal particles form isolated metal sites for reaction [84, 92]. The activity of
supported bimetallic catalysts is inclined to reaction-induced active site modification, which changes the nature of the metal sites and poisoning [90]. In the catalytic oxidation of formaldehyde, it seems research on bimetallic catalysts has not
been sufficiently explored yet.

3.4.3 Transition metal oxide catalysts

Transition metal-based catalysts are good alternatives to noble metals for low-cost
catalytic oxidation and have higher resistance to catalyst poisons and coke deposition. These catalysts can also be supported on both reducible and nonreducible
metal oxides, such as Al$_2$O$_3$, MnO$_2$, ZrO$_2$, Fe$_2$O$_3$, Co$_3$O$_4$, MnO$_x$–CeO$_2$ as well as molecular sieves and activated carbon [66]. Transition metal oxides such as manga-
nese oxides (MnO$_x$), Co$_3$O$_4$, CeO$_2$, or their composites including MnO$_x$–CeO$_2$, Co$_3$O$_4$–
CeO$_2$, and MnO$_2$–Fe$_2$O$_3$ [54, 93] also show very appealing activities for formaldehyde
oxidation, but are not very active under ambient temperatures (low HCHO conversion at room temperature) [94–96]. Their appreciable performance at temperatures
higher than room temperature in most cases endows them with the advantage of
better thermal stability, making them suitable for applications in the removal of
many VOCs.

MnO$_2$ and MnO$_2$-based catalysts

Manganese dioxide (MnO$_2$) exhibits good catalytic activity for oxidation of formaldehyde and has attracted great research interests in the last decade. MnO$_x$-based catalysts possess good thermal stability and are highly affordable. Attention has been shifted to MnO$_2$ as a low-temperature formaldehyde oxidation catalyst following the pioneer study by Sekine [97], who screened a series of metal oxides (Ag$_2$O, MnO$_2$, TiO$_2$, CeO$_2$, CoO, Mn$_3$O$_4$, PdO, WO$_3$, Fe$_2$O$_3$, CuO, V$_2$O$_5$, ZnO, and La$_2$O$_3$) as active catalysts and found that MnO$_2$ showed the highest formaldehyde removal activity at 25 °C. Currently, MnO$_2$ is studied as the most active transition metal oxide catalyst for formaldehyde oxidation. MnO$_2$ with special crystal structures or morphologies tend to perform better than the regular ones, attributed to their specific structures, high specific surface areas, active phase, and a large amount of active surface oxygen species [98]. The different crystal structures of MnO$_2$ (α-, β-, γ-, and δ-MnO$_2$) were prepared by hydrothermal method and tested as catalysts for formaldehyde oxidation. As opposed to other crystal structures, the δ-MnO$_2$ catalyst exhibited the best activity, achieving nearly complete formaldehyde conversion at 80 °C, mainly due to the formed 2D layer tunnel structure that accelerates adsorption and diffusion of formaldehyde [95]. Selectively exposing a larger fraction of the reactive facets where the active sites could be enriched and tuned up endows with the potential to increase the activity of a catalyst [98]. Rong et al. [94] synthesized single-crystalline α-MnO$_2$ nanowires with largely exposed (310) facets by hydrothermal method. The resulting catalyst exhibited a better activity and stability for 100% formaldehyde oxidation (100 ppm HCHO) at 60 °C; this activity is comparable with or even better than some Ag supported catalysts. Wang et al. [99] designed 2D MnO$_x$ on carbon (C@MnO$_2$) for the simultaneous removal of formaldehyde and O$_3$. The carbon matrix enhanced the formation of highly dispersed MnO during calcination. A close interface formed between carbon and MnO facilitated both adsorption and subsequent catalytic reaction. This composite material showed a good removal efficiency (100% for 60 ppm formaldehyde and 180 ppm O$_3$ simultaneously) and 100% CO$_2$ selectivity under a high space velocity (GHSV) of 60,000 mL^{-1} hg at room temperature. This study reveals an interesting mechanism in which active oxygen could be generated in situ by decomposition of the O$_3$, thus enriching surface oxygen for formaldehyde activation. MnO$_2$ modified with some other metals, for example, Co, also showed good catalytic activity, and supported MnO$_2$ could drastically improve the activity for formaldehyde removal at near room temperature, especially for minute formaldehyde concentrations (10 to 60 ppm) typical of the indoor air concentrations in non-industrial areas [99, 100]. MnO$_2$ can be modified by controlling morphology, crystal structure control, doping with other nontransition metals, defect engineering, and so on to improve its activity for formaldehyde removal. In spite of its promising performance for formaldehyde removal, MnO$_2$ shows poor surface adsorption of active oxygen species and is characterized by short-term stability. Moreover, the performance is still far from acceptable at room or near room temperature. The experimental results for some transition metal-based catalysts available in the literature are listed in Table 3.5. With the promising performance, supported or modified MnO_2 requires more exploration as catalysts for technical applications.

Table 3.5: Examples of transition metal-based catalysts for oxidation of formaldehyde.

Catalyst	HCHO conc. (ppm)	GHSV mL/gh	$T_{100conv}$ (°C)	Ref.
α-MnO_2 (310) facet	100	90 (L/gh)	60	[94]
β-MnO_2/SiO_2	120	30,000	130	[101]
α-MnO_2	170	100,000	125	[95]
β-MnO_2			200	[95]
γ-MnO_2			150	[95]
δ-MnO_2			80	[95]
Birnessite	10	180,000/ h	>25	[102]
Birnessite with Mn vacancy	40	120,000	110	[103]
K-Birnessite	0.5(mg/m^3)	1,200,000/ h	>25	[104]
Layered birnessite-type MnO_2	200	120,000	100	[70]
K-MnO_2	100	90 (L/gh)	90	[105]
W-MnO_2	0.3(mg/m^3)	600(L/gh)	30	[106]
H-MnO_x–CeO_2	400	30,000	>100	[69]
MnO_x–$Co_{3-x}O_4$	100	50,000	75	[96]
MnO_x–Co_3O_4–CeO_2	200	36,000	100	[86]
MnO_2–graphene	100	30,000	65	[107]
C@MnO_x	60(HCHO) + 180(O_3)	60,000	RT	[99]
MnO_x/AC–methanol	10	65,000/ h	25	[100]

Co_3O_4-based catalysts

Co_3O_4 exhibits good activity for oxidation of formaldehyde, comparable to that of MnO_x but generally lower than those of noble metal catalysts at room temperature. The catalytic activity is dependent on factors such as physiochemical properties, morphology of the catalyst, and the availability of reactive oxygen species [108]. Various morphologies, including fibers, sheets, cubes, rods, and belts have been reported to influence the catalytic activity of Co_3O_4 through the exposure of catalytically active surface sites [109–111]. For example, the activity comparison of 2D- and 3D–Co_3O_4 nanostructures for complete oxidation of formaldehyde at low temperature showed that 3D–Co_3O_4 possessed a higher reactivity than 2D–Co_3O_4, attributed to abundant surface-adsorbed oxygen species, large specific surface area with more exposed active Co^{3+} species on the (220) crystal facet of 3D–Co_3O_4. Compared with the dense nano–Co_3O_4, the 2D- and 3D-Co_3O_4 had mesoporous channel structures, which could facilitate diffusion of reactants to the active surface for reaction [108]. The textural properties of Co_3O_4 nanofibers greatly boosted its formaldehyde oxidation performance, which was stable up to 160 h at 98 °C. The catalyst was synthesized by electrospinning method and tested under a space velocity of 30,000 mL^{-1} g h [109]. Moreover, Co_3O_4 catalysts with different concentrations of surface oxygen vacancies showed a direct relationship with the catalytic activity in the oxidation of formaldehyde, attributed to improved mobility of oxygen and formation of abundant reactive oxygen species [111].

CeO_2-based catalysts

Cerium is the most abundant rare earth element in the Earth's crust. CeO_2 has a cubic fluorite crystal structure with a cubic array of fourfold coordinated oxygen ions. Cerium cations can occupy half of the eightfold coordinated cationic interstices in CeO_2, making it highly thermally stable and attractive for many catalytic applications. Due to its outstanding oxygen storage capacity, it is usually used as a structural and electronic promoter of catalysts [77]. In CeO_2-promoted catalysts, there is an easy transfer of oxygen species to the surface of CeO_2 by redox reaction between Ce^{4+} and Ce^{3+} species, and also the insertion of metal ions into its lattice. CeO_2 has been studied as an oxidation catalyst for formaldehyde removal, mostly as a support for noble metals or in combination with other metal oxides [112–114]. Like other transition metal oxides, CeO_2-based catalysts display considerable activity for formaldehyde oxidation at temperatures higher than the room temperature.

3.4.4 Transition metal oxide composite catalysts

Transition metal oxide composites, also referred to as *multimetallic oxides*, contain two or more metal oxide components with proportion that either vary or are defined by stoichiometry [77]. Examples are MnO_x–CeO_2, MnO_2–Fe_2O_3, In_2O_3–Co_3O_4 and Co_3O_4–CeO_2 or Cu–CeO_2, Ce–MnO_2 and so on [54, 93, 114, 115]. A common feature of these materials is the presence of multiple oxidation states in the structure of the transition metals, which result in the ability of the cations to undergo reversible oxidation and reduction. The combination of metal oxides or modification with other metals improves the oxidation efficiency at lower temperatures than in the case of single metal oxide due to synergistic interaction, which enhances oxidation capabilities by creating more oxygen vacancies [114], enhancing surface oxygen mobility or charge transport during redox cycles [107]. Generally, insertion of foreign metal cations into a metal oxide lattice significantly changes its electronic properties, chemical bonding, and surface coordination environment, in which case alters the catalytic performance [116]. The metal/metal oxide that is used as a dopant should be of weaker metal–oxygen bonds, have larger radius, and possess lower electronegativity [115]. Using metal oxide composites as supports for noble metal catalysts increases the dispersion of active metal particles and adsorption of reactants and also reduces the loading of noble metals [54]. Transition metal oxide composite catalysts are generally prepared by cosynthesis of two transition metal oxides or the addition of one metal oxide to the other by coprecipitation as well as by other methods [114]. The preparation method plays crucial roles in the formation of active metal species, which in turn determines the activity. According to Wen et al. [117], the higher valence species of Mn in MnO_x–SnO_2 was more active in formaldehyde oxidation. The prevalent oxidation state of Mn in the composite synthesized using redox coprecipitation was Mn^{4+}, while Mn^{3+} was the dominant state in the coprecipitated composite, which accounted for the difference in the activities. Preparation of TiO_2 supported on MnO_x–CeO_2 by consecutive hydrothermal and colloidal reflux condensation prevented the loss of surface oxygen species and unsuitable defect-site depositions of less active ions [118]. This catalyst showed good activity for room temperature catalytic degradation of formaldehyde due to TiO_2-promoted surface electron transfer. The transfer of electrons was also strongly correlated with the specific surface area, porosity, and oxidation states of transition metals in the catalyst. The higher active site exposure derived from the electron transfer benefited chemisorption of formaldehyde. The superoxide radicals activate oxygen species prompted a nucleophilic attack on carbonyl bonds, leading to the formation of carbon dioxide and water. The most active transition metal oxide composite catalyst is MnO_x–CeO_2, which is active for formaldehyde oxide oxidation at a much lower temperature than each of its single oxide counterpart. It owes its improved performance to the synergistic effect of high activity of Mn and oxygen storage capacity of ceria [119, 120] together with the change in Mn oxidation state from Mn^{3+} to Mn^{4+} [107]. The synergistic effect is possible through the ceria oxygen transfer mechanism [114]; this effect is more pronounced in MnO_x–CeO_2 than in Co_3O_4–CeO_2, but the activity of the latter could be improved by doping with noble metals or decreasing the ratio of Co_3O_4 to CeO_2 in the composite [87, 121].

3.4.5 Catalyst support

Catalyst supports are as important as active components in the design of efficient oxidation catalysts. The properties of support materials including specific surface area, pore structure, acidity and surface hydrophobicity or hydrophilicity, and so on, impact metal dispersion and particle size, reactant molecules adsorption and desorption, metal–support interaction, and generation of active oxygen species [122, 123]. Large pores facilitate mass transfer and large specific surface area aids in high metal dispersion, exposes more active sites, regulates particle size, valence state, and adsorption of reactant molecules. The nonreducible supports that have received considerable attention are silica and alumina. The gamma alumina (γ-Al$_2$O$_3$) is often applied to catalytic oxidation of hydrocarbons and VOCs due to its large surface area, high thermal stability, and affordability. Transition metal oxides, either single or mixed, are also excellent supports for many oxidation reactions [118, 121]. Conventional metal oxide supports for the catalytic oxidation of formaldehyde, such as Al$_2$O$_3$, TiO$_2$, ZrO$_2$, CeO$_2$, Fe$_2$O$_3$, Co$_3$O$_4$, and MnO$_2$, or their composites are commonly prepared by precipitation, coprecipitation, or sol–gel methods [121, 124]. Most of them are reducible under a reducing atmosphere, and this property impacts supplementary catalytic activities, improving the overall efficiency of catalysts. The synergy resulting from the interaction between noble metals and transition metal oxides can have profound impacts on the stability and reactivity of surface oxygen species [72]. Catalysts supported on metal oxides with special morphologies exhibit higher catalytic activities in comparison to those supported on bulk metal oxides. These catalysts can generally be prepared via hydrothermal and hard templating methods, while the traditional bulk metal oxides are prepared using precipitation methods. Noble metals supported on metal oxides with special morphology exhibit excellent performance for formaldehyde removal at room temperature. Special morphologies investigated in the catalytic oxidation of formaldehyde include cubes, rods, sheets, flowers, and spheres. Porous materials with low air resistance can also serve as supports to immobilize active catalysts. These kinds of materials including polyethylene terephthalate polyester particulate filter, and porous cellulose fiber, serving as supports for MnO$_2$ nanoparticle catalysts contribute to the excellent activity for formaldehyde oxidation at room temperature [125].

3.5 Factors affecting oxidation performance of catalysts

Various reaction parameters affect the performance of catalysts for formaldehyde oxidation to water and carbon dioxide, such as contaminants or poisons, temperature, gas space velocity, and reactant concentration. These parameters need to be optimized or controlled, for example, space velocity, for the realization of acceptable oxidation performance for practical applications. Catalytic activity is controlled by the catalyst structural and physicochemical properties, which is determined by the deployed method of synthesis. The optimization of the reaction parameters and properties of catalysts is vital for attaining efficient catalytic oxidation of gaseous pollutants at low temperatures.

3.5.1 Catalyst factor

This comprises method of preparation and the active catalyst loading on the support. Commonly adopted preparation methods used to synthesize catalysts for formaldehyde oxidation are sol–gel, precipitation, coprecipitation, and hydrothermal synthesis [126]. The preparation method affects the physicochemical properties, such as surface area, pore volume, number of oxygen species, and oxygen vacancies and oxidation states, noting that the oxidation state of metals and lattice oxygen species are essential factors determining the performance of oxidation catalysts. Moreover, the structure and morphology are also determined during the initial synthesis. Catalysts with improved performance have been synthesized by advanced methods including electrodeposition, chemical vapor deposition, and assisted preparation methods using ultrasound and microwave [127–130]. For instance, ultrasound-assisted nano-casting of 3D–Cr_2O_3 using KIT-6 as template helps in enhancing the penetration of precursor materials into the mesoporous structure of KIT-6, which results in the large specific surface area, pore volume, and mesoporosity of 3D–Cr_2O_3 compared to that synthesized without the aid of ultrasound [131]. Also, MnO_x–CeO_2 formaldehyde oxidation catalysts prepared by modified co-precipitation performed better than that prepared by simple coprecipitation or sol–gel method [114]. A paltry 10% conversion was obtained for catalysts prepared by simple coprecipitation against 90% conversion of catalysts prepared by a modified coprecipitation technique [125].

The composition is also an important factor that determines the optimum performance of a catalyst. The catalytic performance of Pt/TiO_2 catalyst showed an increase in formaldehyde removal efficiency from 0 to 100% when Pt loading increased from 0.3 to 1.0 wt% [55]. However, it should be kept in mind that noble metals are highly expensive. Too much loading of metal catalyst onto a support not only increases the catalyst cost, but also results in agglomeration and poor dispersion, or large particle size. In principle, a catalyst with a relatively small particle size of active metal can outperform that with large particles if consideration is placed only on the exposed active sites. Typically, metal particle sizes ranging 1 to 5 nm form good noble metal oxidation catalysts. One of the persisting challenges is to obtain high catalytic efficiency of these catalysts at a relatively low metal loading.

3.5.2 Catalyst deactivation

For industrial applications, the stability and durability of catalysts are as important as activity and selectivity. Catalysts with low stability are prone to deactivation and must be prevented as possible. Several mechanisms can often lead to deactivation: (i) poisoning, (ii) volatilization of catalyst species, (iii) sintering, and (iv) formation of coke deposits [132, 133]. Chlorinated and sulfur-containing compounds, as well as water vapor, are poisons to noble metals (such as Pt and Pd) and metal oxides frequently employed as formaldehyde oxidation catalysts. Chlorinated compounds poison metals and metal oxides active sites, causing deactivation to the catalysts, as chloride atoms may react with the metal ion to form volatile species that lead to the loss of metal. An example is the catalyst deactivation owing to the loss of Cr from the formation of a volatile compound CrO_2Cl_2 [134]. Generally, poisoning may reduce the activity by reducing the intrinsic rate of formation of reaction products. Furthermore, deactivation of catalysts occurs at low temperatures by water vapor deposition caused by capillary condensation and slow water desorption. For example, the accumulation of water molecules or hydroxylation of the surface of $MnO_x/$ AC is responsible for its deactivation, which was also observed upon the formation of bidentate carbonate $(b-CO_3^{2-})$ [135]. This effect can be addressed by the use of hydrothermally stable catalysts. The formation of coke (carbonaceous deposits) inside the pores or on the surface of catalysts becomes an important consideration when porous materials such as molecular sieves or activated carbons are used as supports for metal catalysts in a number of reactions. Coking obstructs the pores of catalysts and can cause undesired alteration in reaction selectivity as a result of decreased diffusion of reactants and products. The origin of this is the adsorption and accumulation of reactant species on the support, depending on the type and adsorption properties of the support and reaction temperature. For example, a support capable of adsorbing organic molecules such as VOCs at temperatures higher than the reaction temperature exhibited oscillations in conversions during oxidation reactions [136, 137]. A catalyst can also be deactivated by loss of active sites or changing the active site distribution attributed to structural variations of the catalyst caused by thermal sintering. The stability against deactivation thus becomes an important consideration in selecting catalysts for industrial applications, and this is why many catalyst vendors always evaluate their catalysts for stability before putting them up for sale.

3.5.3 Presence of contaminants

The presence of water vapor, often studied as relative humidity in laboratory experiments, and poisons such as chlorine and sulfur affects chemisorption on available sites for catalysis. A trace amount of water vapor is often part of flue gases emitted from various industries; moreover, water is a product of oxidation of VOCs [138]. Water can act either as a poison or a promoter to oxidation catalysts. At high relative humidity, poisonous effect of water is attributed to competitive adsorption of water molecules with the reactant molecules on catalyst active sites on catalysts due to polarity differences [139], leading to blockage of the active sites on catalysts, which prevents free access to reactant molecules. This effect could be prevented by using hydrophobic materials as catalyst supports. Supports with low hydrophobicity require higher reaction temperature to reach high-performance levels [68, 140]. On the other hand, water vapor reacts with the surface-active oxygen (O_2^- or O^-), supplying the surface-active hydroxyl species, which are consumed during formaldehyde oxidation [102, 141]. By competitive adsorption also, it is beneficial in the recovery of catalyst activity. For example, the competitive reaction of water vapor desorbs carbonates from the catalyst surface, enabling the recovery of catalytic activity and preventing the catalyst from fast deactivation [102]. Moreover, over CO_3O_4/TiO_2 catalyst, the presence of water removed Cl radicals from blocking active sites, which accelerated catalyst deactivation [142].

3.5.4 Concentration

Pollutant concentration is an important factor affecting the performance of an oxidation catalyst. Oxidation efficiency tends to decrease with an increase in the concentration of a given pollutant at the same temperature. The relationship between pollutant concentration and reaction rate can be described by the L-H model or the power-law models [143–145]. The pollutant concentrations used in the models are those on the reaction surface and not the concentration of the inlets. Usually, there is an optimal pollutant concentration that maximizes the catalytic reaction rate when the other conditions are kept constant. A very low concentration of formaldehyde exists in the indoor air, and therefore the removal of very low concentrations is extremely important. In literature reports, the concentrations range from 5 to above 1,000 ppm, and the removal efficiencies at low concentrations are higher at ambient temperature probably due to unsaturation of catalyst active sites at low temperatures [138, 146]. The reaction time also influences formaldehyde removal as a function of concentration. For example, MnO_2/AC completely lost its activity after exposed to 5 mg^{-1} m^3 formaldehyde feed concentration for 32 h; the formaldehyde oxidation efficiency was up to 70% for 80 h when a much smaller HCHO feed concentration of 0.5 mg^{-1} m^3 was used [135].

3.5.5 Temperature

Temperature not only affects the catalytic oxidation kinetics, but also the adsorption of the gas-phase compounds onto catalysts. As applicable to most reactions, temperature is also an important parameter, and generally the efficiency of formaldehyde oxidation over catalysts improves with increasing temperature. Increasing temperature positively affects the catalytic oxidation reaction kinetics and the adsorption of the gas-phase molecules on to the surface of catalysts. During adsorption process, the coverage of catalyst's surface by the pollutant molecules decreases progressively with increasing temperature [147]. Correlations between oxidation performance and temperature for formaldehyde removal generally show appreciable levels of removal with an increase in temperature, in most cases, no matter experimental conditions and type of catalyst that are used. Most transition metal-based catalysts achieve complete oxidation of formaldehyde at temperatures above 100 °C, for example, MnO_x-based catalysts, unlike the case of supported noble metal catalysts that can achieve total oxidation (100%) removal at ambient conditions [125, 148]. Some highly active supported noble metal catalysts (e.g., Pt/TiO_2) are able to achieve complete oxidation at room temperature. In this case, removal efficiency becomes a function of time and complete oxidation positively correlates with time, that is, a longer time is required to achieve complete oxidation [125].

3.5.6 Space velocity

GHSV is typically applied to the catalytic gas phase reactions and is calculated as the airflow rate divided by the volume or weight of catalyst or reactor. It affects the efficiency of a fixed bed reaction system by enhancing external mass transfer at high velocity, which leads to high conversion. Simultaneously, residence time reduces which should result in a low conversion. However, available literature report contrasting findings. It thus appears that the controlling factor depends on the oxidation mechanism, affected by the catalyst type, design of the bed, and other operation conditions [149]. Increasing the external mass transfer should lead to high conversion; however, the formaldehyde conversion at room temperature decreases with increasing velocity [30]. For example, conversions of 100%, 85%, and 12% were achieved at 30,000 h^{-1}, 100,000 h^{-1} and 300,000 h^{-1} at 25 °C, respectively, under similar operating conditions [150]. This is because the residence time of formaldehyde molecules decreases with increasing velocity, which results in shorter contact time between the reactant molecules and the catalyst. With this experimental evidence, external mass transfer exhibits an opposing influence on the efficiency of formaldehyde oxidation reaction, and future investigations in this direction will complement the available experimental findings.

3.6 Conclusion

In this chapter, we have mainly presented the origin of formaldehyde pollutant and its hazardous effects on the environment and humans, and the catalytic approach to removing it. Obviously, the catalytic oxidation approach has the capability of remedying most of the challenges associated with the other methods for removing formaldehyde from the indoor environment, such as adsorption and biological methods, and is highly efficient and promising for practical applications.

Supported noble metals are better catalysts at room temperature than the transition metal oxides. MnO_2 and Co_3O_4 can catalyze formaldehyde oxidation, although at temperatures near 100 °C. Due to their affordability and stability in comparison with the noble metals, more research efforts should be made to develop highly efficient transition metal oxide catalysts for formaldehyde oxidation, which may work at low reaction temperatures such as room temperature. In addition to good control of catalyst structure and synthesis in catalyst development, identification of the key or the most active oxygen intermediates for the oxidation reaction and the ability to generate these needed oxygen species are critical. However, it seems activation of molecular oxygen remains a fundamental challenge to the development and application of novel catalysts, and has not been fully explored. In addition, contaminants such as water vapor, halogenated, and sulfur-containing compounds can poison catalyst active sites and reduce long-term catalytic activity, and the reaction condition optimization or reaction and reactor engineering should be paid sufficient attention. It is believed that great breakthrough can be achieved in the future through a comprehensive investigation including catalyst design and development, oxygen activation, and engineering of the reaction and reactor.

References

[1]   World Health Organization releases new global air pollution data Climate & Clean Air Coalition, Climate and Clean Air Coalition. Available at http://ccacoalition.org/en/news/world-health-organization-releases-new-global-air-pollution-data.
[2]   Heinsohn R.J., Kabel R.L. Sources and control of air pollution: Engineering principles, 1998.
[3]   Chung W-C., Mei D-H., Tu X., Chang M-B. Removal of VOCs from gas streams via plasma and catalysis, Catal. Rev., 2018, 1–62.
[4]   Khan F.I., Ghoshal A.K. Removal of volatile organic compounds from polluted air, J. Loss Prev. Process Indust., 2000, 13, 527–545.
[5]   Hester R.E, Harrison R.M. Volatile organic compounds in the atmosphere, Royal Society of Chemistry, 1995.
[6]   Wei W., Cheng S., Li G., Wang G., Wang H. Characteristics of ozone and ozone precursors (VOCs and NOx) around a petroleum refinery in Beijing, China, J. Environ. Sci., 2014, 26, 332–342.
[7]   Shin H-M., McKone T.E., Bennett D.H. Volatilization of low vapor pressure–volatile organic compounds (LVP–VOCs) during three cleaning products-associated activities: Potential contributions to ozone formation, Chemosphere, 2016, 153, 130–137.
[8]   Shin H-M., McKone T.E., Bennett D.H. Contribution of low vapor pressure-volatile organic compounds (LVP-VOCs) from consumer products to ozone formation in urban atmospheres, Atmos. Environ., 2015, 108, 98–106.

[9]   Caillol S. Fighting global warming: The potential of photocatalysis against CO_2, CH_4, N_2O, CFCs, tropospheric O_3, BC and other major contributors to climate change, J. Photochem. Photobiol. C, 2011, 12, 1–19.

[10] Carpenter L.J., Archer S.D., Beale R. Ocean-atmosphere trace gas exchange, Chem. Soc. Rev., 2012, 41, 6473–6506.

[11] Roda C., Kousignian I., Guihenneuc-Jouyaux C., Dassonville C., Nicolis I., Just J., Momas I. Formaldehyde exposure and lower respiratory infections in infants: Findings from the PARIS cohort study, Environ. Health Perspect., 2011, 119, 1653–1658.

[12] Bruce N., Perez-Padilla R., Albalak R. Indoor air pollution in developing countries: A major environmental and public health challenge, Bull. W.H.O., 2000, 78, 1078–1092.

[13] Cao H., Fu T-M., Zhang L., Henze D.K., Miller C.C., Lerot C., Abad G.G., Smedt I.D., Zhang Q., Roozendael Mv. Adjoint inversion of Chinese non-methane volatile organic compound emissions using space-based observations of formaldehyde and glyoxal, Atmos. Chem. Phys., 2018, 18, 15017–15046.

[14] Zhang X., Gao B., Creamer A.E., Cao C., Li Y. Adsorption of VOCs onto engineered carbon materials: A review, J. Hazard. Mater., 2017, 338, 102–123.

[15] Cheng K., Hao W-W., Yi P., Zhang Y., Zhang J-Y. Volatile organic compounds emission from Chinese wood furniture coating industry: Activity-based emission factor, speciation profiles, and provincial emission inventory, Aerosol Air Qual. Res., 2018, 18, 2813–2825.

[16] Zhang J., Xiao J., Chen X., Liang X., Fan L., Ye D. Allowance and allocation of industrial volatile organic compounds emission in China for year 2020 and 2030, J. Environ. Sci., 2018, 69, 155–165.

[17] China's VOC Control Policies, https://www.saiindustrial.com/chinas-voc-control-policies/.

[18] Kamal M.S., Razzak S.A., Hossain M.M. Catalytic oxidation of volatile organic compounds (VOCs)–A review, Atmos. Environ., 2016, 140, 117–134.

[19] Shen L., Xiang P., Liang S., Chen W., Wang M., Lu S., Wang Z. Sources profiles of Volatile Organic Compounds (VOCs) Measured in a typical industrial process in Wuhan, Central China, Atmosphere, 2018, 9, 297.

[20] Salthammer T., Mentese S., Marutzky R. Formaldehyde in the indoor environment, Chem. Rev., 2010, 110, 2536–2572.

[21] Salthammer T. The formaldehyde dilemma, Int. J. Hyg. Environ. Health, 2015, 218, 433–436.

[22] Salthammer T. Formaldehyde in the ambient atmosphere: from an indoor pollutant to an outdoor pollutant?, Angew. Chem. Int. Ed., 2013, 52, 3320–3327.

[23] Cheremisinoff N.P. Photochemical smog, in: Cheremisinoff N.P. (Ed.) Pollution Control Handbook for Oil and Gas Engineering, John Wiley & Sons, Inc, 2016.

[24] Tulpule K., Dringen R. Formaldehyde in brain: An overlooked player in neurodegeneration?, J. Neurochem., 2013, 127, 7–21.

[25] Metz B., Kersten G.F., Hoogerhout P., Brugghe H.F., Timmermans H.A., De Jong A., Meiring H., ten Hove J., Hennink W.E., Crommelin D.J. Identification of formaldehyde-induced modifications in proteins reactions with model peptides, J. Biol. Chem., 2004, 279, 6235–6243.

[26] Metz B., Kersten G.F., Baart G.J., de Jong A., Meiring H., ten Hove J., van Steenbergen M.J., Hennink W.E., Crommelin D.J., Jiskoot W. Identification of formaldehyde-induced modifications in proteins: Reactions with insulin, Bioconjugate Chem., 2006, 17, 815–822.

[27] Tang X., Bai Y., Duong A., Smith M.T., Li L., Zhang L. Formaldehyde in China: Production, consumption, exposure levels, and health effects, Environ. Int., 2009, 35, 1210–1224.

[28] Formaldehyde – Chemical Economics Handbook (CEH), https://ihsmarkit.com/products/form aldehyde-chemical-economics-handbook.html.

[29] Dugheri S., Bonari A., Pompilio I., Colpo M., Mucci N., Arcangeli G. An integrated air monitoring approach for assessment of formaldehyde in the workplace, Saf. Health Work, 2018, 9, 479.

[30] Nie L., Yu J., Jaroniec M., Tao F.F. Room-temperature catalytic oxidation of formaldehyde on catalysts, Catal. Sci. Technol., 2016, 6, 3649–3669.

[31] Tilley S.K., Fry R.C. Chapter 6 – Priority environmental contaminants: Understanding their sources of exposure, biological mechanisms, and impacts on health, in: Fry R.C. (Ed.) Systems Biology in Toxicology and Environmental Health, Academic Press, Boston, 2015, pp. 117–169.

[32] Walker J.F. Formaldehyde. Reinhold Publishing Corporation, New York, 1964.

[33] Wittmann O. Die nachträgliche Formaldehydabspaltung bei Spanplatten, Holz als Roh-und Werkstoff, 1962, 20, 221–224.

[34] Kerns W.D., Pavkov K.L., Donofrio D.J., Gralla E.J., Swenberg J.A. Carcinogenicity of formaldehyde in rats and mice after long-term inhalation exposure, Cancer Res., 1983, 43, 4382–4392.

[35] Swenberg J.A., Kerns W.D., Mitchell R.I., Gralla E.J., Pavkov K.L. Induction of squamous cell carcinomas of the rat nasal cavity by inhalation exposure to formaldehyde vapor, Cancer Res., 1980, 40, 3398–3402.

[36] World Health Organisation. Indoor Air Pollutants – Exposure and Health Effects. EURO Reports and Studies, Copenhagen, 1983.

[37] World Health Organisation. Air Quality Guidelines for Europe. WHO Regional Office for Europe, Copenhagen, 2010.

[38] World Health Organisation. WHO Guidelines for Indoor Air Quality: Selected Pollutants. WHO Regional Office for Europe, Copenhagen, 2010.

[39] Nielsen G.D., Wolkoff P. Cancer effects of formaldehyde: A proposal for an indoor air guideline value, Arch. Toxicol., 2010, 84, 423–446.

[40] Wolkoff P., Nielsen G.D. Non-cancer effects of formaldehyde and relevance for setting an indoor air guideline, Environ. Int., 2010, 36, 788–799.

[41] Organization WH. WHO guidelines for indoor air quality: Selected pollutants, 2010.

[42] Abdul-Wahab S.A., En S.C.F., Elkamel A., Ahmadi L., Yetilmezsoy K. A review of standards and guidelines set by international bodies for the parameters of indoor air quality, Atmos. Pollut. Res., 2015, 6, 751–767.

[43] Suarez-Bertoa R., Mendoza-Villafuerte P., Riccobono F., Vojtisek M., Pechout M., Perujo A., Astorga C. On-road measurement of NH_3 emissions from gasoline and diesel passenger cars during real world driving conditions, Atmos. Environ., 2017, 166, 488–497.

[44] Sekine Y., Katori R., Tsuda Y., Kitahara T. Colorimetric monitoring of formaldehyde in indoor environment using built-in camera on mobile phone, Environ. Technol., 2016, 37, 1647–1655.

[45] Hoffman E.A., Frey B.L., Smith L.M., Auble D.T. Formaldehyde crosslinking: A tool for the study of chromatin complexes, J. Biol. Chem., 2015, 290, 26404–26411.

[46] Quiroz Torres J., Royer S., Bellat J.P., Giraudon J.M., Lamonier J.F. Formaldehyde: Catalytic oxidation as a promising soft way of elimination, ChemSusChem, 2013, 6, 578–592.

[47] Dela Cruz M., Christensen J.H., Thomsen J.D., Müller R. Can ornamental potted plants remove volatile organic compounds from indoor air? – A review, Environ. Sci. Pollut Res., 2014, 21, 13909–13928.

[48] Teiri H., Pourzamani H., Hajizadeh Y. Phytoremediation of VOCs from indoor air by ornamental potted plants: A pilot study using a palm species under the controlled environment, Chemosphere, 2018, 197, 375–381.

[49] Guieysse B., Hort C., Platel V., Munoz R., Ondarts M., Revah S. Biological treatment of indoor air for VOC removal: Potential and challenges, Biotechnol. Adv., 2008, 26, 398–410.

[50] Aydogan A., Montoya L.D. Formaldehyde removal by common indoor plant species and various growing media, Atmos. Environ., 2011, 45, 2675–2682.

[51] Luengas A., Barona A., Hort C., Gallastegui G., Platel V., Elias A. A review of indoor air treatment technologies, Rev. Environ. Sci. Bio., 2015, 14, 499–522.

[52] Uchiyama S., Matsushima E., Kitao N., Tokunaga H., Ando M., Otsubo Y. Effect of natural compounds on reducing formaldehyde emission from plywood, Atmos. Environ., 2007, 41, 8825–8830.

[53] Hayashi M., Osawa H. The influence of the concealed pollution sources upon the indoor air quality in houses, Build. Environ., 2008, 43, 329–336.

[54] Ma C., Wang D., Xue W., Dou B., Wang H., Hao Z. Investigation of formaldehyde oxidation over Co_3O_4–CeO_2 and Au/Co_3O_4–CeO_2 catalysts at room temperature: Effective removal and determination of reaction mechanism, Environ. Sci. Technol., 2011, 45, 3628–3634.

[55] Zhang C., He H., Tanaka K-i. Catalytic performance and mechanism of a Pt/TiO_2 catalyst for the oxidation of formaldehyde at room temperature, Appl. Catal. B: Environ., 2006, 65, 37–43.

[56] An N., Yu Q., Liu G., Li S., Jia M., Zhang W. Complete oxidation of formaldehyde at ambient temperature over supported Pt/Fe_2O_3 catalysts prepared by colloid-deposition method, J. Hazard. Mater., 2011, 186, 1392–1397.

[57] Yu X., He J., Wang D., Hu Y., Tian H., He Z. Facile controlled synthesis of Pt/MnO_2 nanostructured catalysts and their catalytic performance for oxidative decomposition of formaldehyde, J. Phy. Chem C, 2011, 116, 851–860.

[58] Huang H., Leung D.Y. Complete oxidation of formaldehyde at room temperature using TiO_2 supported metallic Pd nanoparticles, ACS Catal., 2011, 1, 348–354.

[59] Park S.J., Bae I., Nam I-S., Cho B.K., Jung S.M., Lee J-H. Oxidation of formaldehyde over Pd/ Beta catalyst, Chem. Eng. J., 2012, 195, 392–402.

[60] Imamura S., Uematsu Y., Utani K., Ito T. Combustion of formaldehyde on ruthenium/cerium (IV) oxide catalyst, Ind. Eng. Chem. Res., 1991, 30, 18–21.

[61] Zhang C., Hu P., Alavi A. A general mechanism for CO oxidation on close-packed transition metal surfaces, J. Am. Chem. Soc., 1999, 121, 7931–7932.

[62] Kim H.Y., Lee H.M., Henkelman G. CO oxidation mechanism on CeO_2-supported Au nanoparticles, J. Am. Chem. Soc., 2012, 134, 1560–1570.

[63] Madey T.E., Engelhardt H.A., Menzel D. Adsorption of oxygen and oxidation of CO on the ruthenium (001) surface, Surf. Sci., 1975, 48, 304–328.

[64] Valden M., Lai X., Goodman D.W. Onset of catalytic activity of gold clusters on titania with the appearance of nonmetallic properties, Science, 1998, 281, 1647–1650.

[65] He C., Cheng J., Zhang X., Douthwaite M., Pattisson S., Hao Z. Recent advances in the catalytic oxidation of volatile organic compounds: A review based on pollutant sorts and sources, Chem. Rev., 2019, 119, 4471–4568.

[66] Zhang Z., Jiang Z., Shangguan W. Low-temperature catalysis for VOCs removal in technology and application: A state-of-the-art review, Catal. Today, 2016, 264, 270–278.

[67] Ordóñez S., Bello L., Sastre H., Rosal R., Díez F.V. Kinetics of the deep oxidation of benzene, toluene, n-hexane and their binary mixtures over a platinum on γ-alumina catalyst, Appl. Catal. B: Environ., 2002, 38, 139–149.

[68] Chuang K.T., Zhou B., Tong S. Kinetics and mechanism of catalytic oxidation of formaldehyde over hydrophobic catalysts, Ind. Eng. Chem. Res., 1994, 33, 1680–1686.

[69] Quiroz J., Giraudon J-M., Gervasini A., Dujardin C., Lancelot C., Trentesaux M., Lamonier J-Fo. Total oxidation of formaldehyde over MnO_x-CeO_2 catalysts: The effect of acid treatment, ACS Catal., 2015, 5, 2260–2269.

[70] Wang J., Zhang G., Zhang P. Layered birnessite-type MnO_2 with surface pits for enhanced catalytic formaldehyde oxidation activity, J. Mate. Chem. A, 2017, 5, 5719–5725.

[71] Zhang C., Liu F., Zhai Y., Ariga H., Yi N., Liu Y., Asakura K., Flytzani-Stephanopoulos M., He H. Alkali-metal-promoted Pt/TiO_2 opens a more efficient pathway to formaldehyde oxidation at ambient temperatures, Angew. Chem. Int. Ed., 2012, 51, 9628–9632.

[72] Tan H., Wang J., Yu S., Zhou K. Support morphology-dependent catalytic activity of Pd/CeO_2 for Formaldehyde oxidation, Environ. Sci. Technol., 2015, 49, 8675–8682.

[73] Rong S., Zhang P., Yang Y., Zhu L., Wang J., Liu F. MnO_2 framework for instantaneous mineralization of carcinogenic airborne formaldehyde at room temperature, ACS Catal., 2017, 7, 1057–1067.

[74] Li S., Lu X., Guo W., Zhu H., Li M., Zhao L., Li Y., Shan H. Formaldehyde oxidation on the Pt/TiO_2 (101) surface: A DFT investigation, J. Organomet. Chem., 2012, 704, 38–48.

[75] Teng B-T., Jiang S-Y., Yang Z-X., Luo M-F., Lan Y-Z. A density functional theory study of formaldehyde adsorption and oxidation on CeO_2 (1 1 1) surface, Surf. Sci., 2010, 604, 68–78.

[76] Huang H., Xu Y., Feng Q., Leung DYC. Low temperature catalytic oxidation of volatile organic compounds: A review, Catal. Sci. Technol., 2015, 5, 2649–2669.

[77] Li Y., Shen W. Morphology-dependent nanocatalysts: Rod-shaped oxides, Chem. Soc. Rev., 2014, 43, 1543–1574.

[78] Li H-F., Zhang N., Chen P., Luo M-F., Lu J-Q. High surface area Au/CeO_2 catalysts for low temperature formaldehyde oxidation, Appl. Catal. B: Environ., 2011, 110, 279–285.

[79] Zhang C., He H., Tanaka K-i. Perfect catalytic oxidation of formaldehyde over a Pt/TiO_2 catalyst at room temperature, Catal. Commun., 2005, 6, 211–214.

[80] Zhang C., Liu F., Zhai Y., Ariga H., Yi N., Liu Y., Asakura K., Flytzani-Stephanopoulos M., He H. Alkali-metal-promoted Pt/TiO_2 opens a more efficient pathway to formaldehyde oxidation at ambient temperatures, Angew. Chem. Int. Ed., 2012, 51, 9628–9632.

[81] Wang L., Yue H., Hua Z., Wang H., Li X., Li L. Highly active Pt/Na_xTiO_2 catalyst for low temperature formaldehyde decomposition, Appl. Catal. B: Environ., 2017, 219, 301–313.

[82] Zhang C., Li Y., Wang Y., He H. Sodium-promoted Pd/TiO_2 for catalytic oxidation of formaldehyde at ambient temperature, Environ. Sci. Technol., 2014, 48, 5816–5822.

[83] Li Y., Zhang C., He H., Zhang J., Chen M. Influence of alkali metals on Pd/TiO_2 catalysts for catalytic oxidation of formaldehyde at room temperature, Catal. Sci. Technol., 2016, 6, 2289–2295.

[84] Guo J., Lin C., Jiang C., Zhang P. Review on noble metal-based catalysts for formaldehyde oxidation at room temperature, Appl. Surf. Sci., 2018, 475, 237–255.

[85] Haruta M., Kobayashi T., Sano H., Yamada N. Novel gold catalysts for the oxidation of carbon monoxide at a temperature far below 0 °C, Chem Lett, 1987, 16, 405–408.

[86] Lu S., Li K., Huang F., Chen C., Sun B. Efficient MnO_x-Co_3O_4-CeO_2 catalysts for formaldehyde elimination, Appl. Surf. Sci., 2017, 400, 277–282.

[87] Liu B., Liu Y., Li C., Hu W., Jing P., Wang Q., Zhang J. Three-dimensionally ordered macroporous Au/CeO_2-Co_3O_4 catalysts with nanoporous walls for enhanced catalytic oxidation of formaldehyde, Appl. Catal. B: Environ., 2012, 127, 47–58.

[88] Tang X., Chen J., Li Y., Li Y., Xu Y., Shen W. Complete oxidation of formaldehyde over Ag/MnO_x–CeO_2 catalysts, Chem. Eng. J., 2006, 118, 119–125.

[89] Na H., Zhu T., Liu Z. Effect of preparation method on the performance of Pt–Au/TiO_2 catalysts for the catalytic co-oxidation of HCHO and CO, Catal. Sci. Technol., 2014, 4, 2051–2057.

[90]   Filez M., Redekop E.A., Dendooven J., Ramachandran R.K., Solano E., Olsbye U., Weckhuysen
       B.M., Galvita V.V., Poelman H., Detavernier C. Formation and functioning of bimetallic
       nanocatalysts: The power of X-ray probes, Angew. Chem. Int. Ed., 2019. http://dx.doi.org/
       10.1002/anie.201902859
[91]   Yu X., He J., Wang D., Hu Y., Tian H., Dong T., He Z. Au–Pt bimetallic nanoparticles supported
       on nest-like MnO$_2$: Synthesis and application in HCHO decomposition, J. Nanopart. Res.,
       2012, 14, 1260.
[92]   Huang J., Mensi M., Oveisi E., Mantella V., Buonsanti R. Structural sensitivities in bimetallic
       catalysts for electrochemical CO$_2$ reduction revealed by Ag-Cu nanodimers, J. Am. Chem.
       Soc., 2019, 141, 6, 2490–2499.
[93]   Wang Z., Pei J., Zhang J. Catalytic oxidization of indoor formaldehyde at room
       temperature–Effect of operation conditions, Build. Environ., 2013, 65, 49–57.
[94]   Rong S., Zhang P., Liu F., Yang Y. Engineering crystal facet of α-MnO$_2$ nanowire for highly
       efficient catalytic oxidation of carcinogenic airborne formaldehyde, ACS Catal., 2018, 8,
       3435–3446.
[95]   Zhang J., Li Y., Wang L., Zhang C., He H. Catalytic oxidation of formaldehyde over manganese
       oxides with different crystal structures, Catal. Sci. Technol., 2015, 5, 2305–2313.
[96]   Shi C., Wang Y., Zhu A., Chen B., Au C. Mn$_x$Co$_{3-x}$O$_4$ solid solution as high-efficient catalysts
       for low-temperature oxidation of formaldehyde, Catal. Commun., 2012, 28, 18–22.
[97]   Sekine Y. Oxidative decomposition of formaldehyde by metal oxides at room temperature,
       Atmos. Environ., 2002, 36, 5543–5547.
[98]   Miao L., Wang J., Zhang P. Review on manganese dioxide for catalytic oxidation of airborne
       formaldehyde, Appl. Surf. Sci., 2019, 466, 441–453.
[99]   Wang H., Huang Z., Jiang Z., Jiang Z., Zhang Y., Zhang Z., Shangguan W. Trifunctional
       C@MnO catalyst for enhanced stable simultaneously catalytic removal of formaldehyde and
       ozone, ACS Catal., 2018, 8, 3164–3180.
[100] Fang R., Huang H., Ji J., He M., Feng Q., Zhan Y., Leung D.Y. Efficient MnO$_x$ supported on
       coconut shell activated carbon for catalytic oxidation of indoor formaldehyde at room
       temperature, Chem. Eng. J., 2018, 334, 2050–2057.
[101] Averlant R., Royer S., Giraudon J.M., Bellat J.P., Bezverkhyy I., Weber G., Lamonier J.F.
       Mesoporous silica-confined manganese oxide nanoparticles as highly efficient catalysts for
       the low-temperature elimination of formaldehyde, ChemCatChem, 2014, 6, 152–161.
[102] Wang J., Zhang P., Li J., Jiang C., Yunus R., Kim J. Room-temperature oxidation of
       formaldehyde by layered manganese oxide: Effect of water, Environ. Sci. Technol., 2015,
       49, 12372–12379.
[103] Wang J., Li J., Jiang C., Zhou P., Zhang P., Yu J. The effect of manganese vacancy in birnessite-
       type MnO$_2$ on room-temperature oxidation of formaldehyde in air, Appl. Catal. B: Environ.,
       2017, 204, 147–155.
[104] Wang J., Li D., Li P., Zhang P., Xu Q., Yu J. Layered manganese oxides for formaldehyde-
       oxidation at room temperature: The effect of interlayer cations, RSC Adv, 2015, 5,
       100434–100442.
[105] Gao J., Huang Z., Chen Y., Wan J., Gu X., Ma Z., Chen J., Tang X. Activating inert alkali-metal
       ions by electron transfer from manganese oxide for formaldehyde abatement, Chem. Eur. J,
       2018, 24, 681–689.
[106] Liu F., Cao R., Rong S., Zhang P. Tungsten doped manganese dioxide for efficient removal of
       gaseous formaldehyde at ambient temperatures, Mater. Design, 2018, 149, 165–172.
[107] Lu L., Tian H., He J., Yang Q. Graphene–MnO$_2$ hybrid nanostructure as a new catalyst for
       formaldehyde oxidation, J. Phy. Chem C, 2016, 120, 23660–23668.

[108] Bai B., Arandiyan H., Li J. Comparison of the performance for oxidation of formaldehyde on nano-Co_3O_4, 2D-Co_3O_4, and 3D-Co_3O_4 catalysts, Appl. Catal. B: Environ., 2013, 142, 677–683.

[109] Wu Y., Ma M., Zhang B., Gao Y., Lu W., Guo Y. Controlled synthesis of porous Co_3O_4 nanofibers by spiral electrospinning and their application for formaldehyde oxidation, RSC Adv, 2016, 6, 102127–102133.

[110] Zeng L., Li K., Huang F., Zhu X., Li H. Effects of Co_3O_4 nanocatalyst morphology on CO oxidation: Synthesis process map and catalytic activity, Chin. J. Catal., 2016, 37, 908–922.

[111] Wang Z., Wang W., Zhang L., Jiang D. Surface oxygen vacancies on Co_3O_4 mediated catalytic formaldehyde oxidation at room temperature, Catal. Sci. Technol., 2016, 6, 3845–3853.

[112] Shen Y., Yang X., Wang Y., Zhang Y., Zhu H., Gao L., Jia M. The states of gold species in CeO_2 supported gold catalyst for formaldehyde oxidation, Appl. Catal. B: Environ., 2008, 79, 142–148.

[113] Xu Q., Lei W., Li X., Qi X., Yu J., Liu G., Wang J., Zhang P. Efficient removal of formaldehyde by nanosized gold on well-defined CeO_2 nanorods at room temperature, Environ. Sci. Technol., 2014, 48, 9702–9708.

[114] Tang X., Li Y., Huang X., Xu Y., Zhu H., Wang J., Shen W. MnO_x–CeO_2 mixed oxide catalysts for complete oxidation of formaldehyde: Effect of preparation method and calcination temperature, Appl. Catal. B: Environ., 2006, 62, 265–273.

[115] Lou Y., Cao X.-M., Lan J., Wang L., Dai Q., Guo Y., Ma J., Zhao Z., Guo Y., Hu P. Ultralow-temperature CO oxidation on an In_2O_3–Co_3O_4 catalyst: a strategy to tune CO adsorption strength and oxygen activation simultaneously, Chem Commun, 2014, 50, 6835–6838.

[116] McFarland E.W., Metiu H. Catalysis by doped oxides, Chem. Rev., 2013, 113, 4391–4427.

[117] Wen Y., Tang X., Li J., Hao J., Wei L., Tang X. Impact of synthesis method on catalytic performance of MnO_x–SnO_2 for controlling formaldehyde emission, Catal. Commun., 2009, 10, 1157–1160.

[118] Li H., Huang T., Lu Y., Cui L., Wang Z., Zhang C., Lee S., Huang Y., Cao J., Ho W. Unraveling the mechanisms of room-temperature catalytic degradation of indoor formaldehyde and its biocompatibility on colloidal TiO_2-supported MnO_x–CeO_2, Environ. Sci. Nano, 2018, 5, 1130–1139.

[119] Xuesong L., Jiqing L., Kun Q., HUANG W., Mengfei L. A comparative study of formaldehyde and carbon monoxide complete oxidation on MnO_x-CeO_2 catalysts, J. Rare Earth, 2009, 27, 418–424.

[120] Imamura S., Shono M., Okamoto N., Hamada A., Ishida S. Effect of cerium on the mobility of oxygen on manganese oxides, Appl. Catal. A: Gen., 1996, 142, 279–288.

[121] Tang X., Chen J., Huang X., Xu Y., Shen W. Pt/MnO_x–CeO_2 catalysts for the complete oxidation of formaldehyde at ambient temperature, Appl. Catal. B: Environ., 2008, 81, 115–121.

[122] Sakata Y., Tamaura Y., Imamura H., Watanabe M. Preparation of a new type of $CaSiO_3$ with high surface area and property as a catalyst support, in: Gaigneaux EM, Devillers M, De Vos DE, Hermans S., Jacobs P.A., Martens J.A., Ruiz P. (Eds.) Stud. Surf. Sci. Catal., Elsevier, 2006, pp. 331–338.

[123] Zhang J., Li Y., Zhang Y., Chen M., Wang L., Zhang C., He H. Effect of support on the activity of Ag-based catalysts for formaldehyde oxidation, Sci. Rep., 2015, 5, 12950.

[124] Bai B., Qiao Q., Li J., Hao J. Progress in research on catalysts for catalytic oxidation of formaldehyde, Chin. J. Catal., 2016, 37, 102–122.

[125] Yusuf A., Snape C., He J., Xu H., Liu C., Zhao M., Chen G.Z., Tang B., Wang C., Wang J. Advances on transition metal oxides catalysts for formaldehyde oxidation: A review, Catal. Rev., 2017, 59, 189–233.

[126] Gunawan P., Xu R., Zhong Z. Heterogeneous gold catalysts for selective oxidation reactions, in: Heterogeneous Gold Catalysts and Catalysis, Royal Society of Chemistry, 2014, pp. 288–400.

[127] Padilla R.H., Priecel P., Lin M., Lopez-Sanchez J.A., Zhong Z. A versatile sonication-assisted deposition–reduction method for preparing supported metal catalysts for catalytic applications, Ultrason. Sonochem., 2017, 35, 631–639.

[128] Gedanken A. Doping nanoparticles into polymers and ceramics using ultrasound radiation, Ultrason. Sonochem., 2007, 14, 418–430.

[129] Wang H., Xu J-Z., Zhu J-J., Chen H-Y. Preparation of CuO nanoparticles by microwave irradiation, J. Cryst. Growth, 2002, 244, 88–94.

[130] Lee Y.H., Zhang X.Q., Zhang W., Chang M.T., Lin C.T., Chang K.D., Yu Y.C., Wang J.T.W., Chang C.S., Li L.J. Synthesis of large-area MoS$_2$ atomic layers with chemical vapor deposition, Adv. Mater., 2012, 24, 2320–2325.

[131] Xia Y., Dai H., Zhang L., Deng J., He H., Au CT. Ultrasound-assisted nanocasting fabrication and excellent catalytic performance of three-dimensionally ordered mesoporous chromia for the combustion of formaldehyde, acetone, and methanol, Appl. Catal. B: Environ., 2010, 100, 229–237.

[132] Argyle M., Bartholomew C. Heterogeneous catalyst deactivation and regeneration: A Review, Catalysts, 2015, 5, 145–269.

[133] Bartholomew C.H. Mechanisms of catalyst deactivation, Appl. Catal. A: Gen., 2001, 212, 17–60.

[134] Oliveira L., Lago R.M., Fabris J., Sapag K. Catalytic oxidation of aromatic VOCs with Cr or Pd-impregnated Al-pillared bentonite: Byproduct formation and deactivation studies, Appl. Clay Sci., 2008, 39, 218–222.

[135] Li J., Zhang P., Wang J., Wang M. Birnessite-Type manganese oxide on granular activated carbon for formaldehyde removal at room temperature, J. Phy. Chem C, 2016, 120, 24121–24129.

[136] Tsou J., Magnoux P., Guisnet M., Órfão J., Figueiredo J. Oscillations in the catalytic oxidation of volatile organic compounds, J. Catal., 2004, 225, 147–154.

[137] Tsou J., Magnoux P., Guisnet M., Órfão J., Figueiredo J. Oscillations in the oxidation of MIBK over a Pt/HFAU catalyst: Role of coke combustion, Catal. Commun., 2003, 4, 651–656.

[138] Wang Z., Pei J., Zhang J. Catalytic oxidization of indoor formaldehyde at room temperature – Effect of operation conditions, Build. Environ., 2013, 65, 49–57.

[139] Beauchet R., Mijoin J., Batonneau-Gener I., Magnoux P. Catalytic oxidation of VOCs on NaX zeolite: Mixture effect with isopropanol and o-xylene, Appl. Catal. B: Environ., 2010, 100, 91–96.

[140] Wu JC-S., Chang T-Y. VOC deep oxidation over Pt catalysts using hydrophobic supports, Catal. Today, 1998, 44, 111–118.

[141] Chen B-b., Zhu X-b., Crocker M., Wang Y., Shi C. FeO$_x$-supported gold catalysts for catalytic removal of formaldehyde at room temperature, Appl. Catal. B: Environ., 2014, 154, 73–81.

[142] Krishnamoorthy S., Rivas J.A., Amiridis M.D. Catalytic oxidation of 1, 2-dichlorobenzene over supported transition metal oxides, J. Catal., 2000, 193, 264–272.

[143] Peral J., Ollis D.F. Heterogeneous photocatalytic oxidation of gas-phase organics for air purification: Acetone, 1-butanol, butyraldehyde, formaldehyde, and m-xylene oxidation, J. Catal., 1992, 136, 554–565.

[144] Obee T.N., Brown R.T. TiO$_2$ photocatalysis for indoor air applications: Effects of humidity and trace contaminant levels on the oxidation rates of formaldehyde, toluene, and 1, 3-butadiene, Environ. Sci. Technol., 1995, 29, 1223–1231.

[145] Yang R., Zhang Y-P., Zhao R-Y. An improved model for analyzing the performance of photocatalytic oxidation reactors in removing volatile organic compounds and its application, J. Air Waste Manage. Assoc., 2004, 54, 1516–1524.

[146] Peng J., Wang S. Performance and characterization of supported metal catalysts for complete oxidation of formaldehyde at low temperatures, Appl. Catal. B: Environ., 2007, 73, 282–291.

[147] Serpone N., Pelizzetti E. Photocatalysis: Fundamentals and applications, Wiley-Interscience, 1989.

[148] Chen T., Dou H., Li X., Tang X., Li J., Hao J. Tunnel structure effect of manganese oxides in complete oxidation of formaldehyde, Micropor. Mesopor. Mater., 2009, 122, 270–274.

[149] Pei J., Zhang J.S. Critical review of catalytic oxidization and chemisorption methods for indoor formaldehyde removal, Hvac&R Research, 2011, 17, 476–503.

[150] An N., Zhang W., Yuan X., Pan B., Liu G., Jia M., Yan W., Zhang W. Catalytic oxidation of formaldehyde over different silica supported platinum catalysts, Chem. Eng. J., 2013, 215, 1–6.

Cheng Zhang, Yanshan Gao, Liang Huang and Qiang Wang

4 Recent advances in catalytic oxidation of NO to NO$_2$

4.1 Introduction

Comparing the stoichiometric engines, lean burn gasoline and diesel engines are normally more fuel efficient, resulting in longer travel distances per unit of fuel and thereby reducing CO$_2$ emissions [1]. However, the diesel engine emission standards are becoming more and more stringent. For example, in 2005, according to Euro IV regulations, the production of soot and NO$_x$ was limited to 25 mg·km^{-1} and 250 mg·km^{-1}, respectively, which is twice as low as Euro III standard [2]. In order to meet the NO$_x$ emission levels, three technologies have been developed recently, which include NO$_x$ storage and reduction (NSR), selective catalytic reduction (SCR), and continuously regenerating trap (CRT) [3–11].

Using these technologies, more and more people believe that the oxidation of NO to NO$_2$ is an important prerequisite for waste-gas treatment, because the produced NO$_2$ contributes to the oxidation of soot and promotes the SCR reaction of NO$_x$ [12–16]. For the NSR technique, NO is first oxidized to NO$_2$ on platinum and then stored as nitrate on BaO [17]. For the SCR technique, it has been demonstrated that when a portion of NO in the exhaust gas is converted to NO$_2$, the SCR reaction rate can be significantly increased. This effect is more obviously at lower temperatures (200–300 °C) when the reaction mixture contains equimolar amounts of NO and NO$_2$ (fast SCR process) [12, 18, 19]. For the CRT technique, NO$_2$ acts as a strong oxidant to oxidize the soot collected on the particulate filter at relatively low temperatures. These temperatures are much lower than the ignition temperature (around 550 °C) of soot in air [20].

However, the nitrogen oxides (NO + NO$_2$) produced by combustion engines are mainly composed of NO (> 90%). Therefore, the oxidation of NO to NO$_2$ is an important step in the posttreatment reaction. With the development of De-NO$_x$ and soot combustion technologies, lots of works have been carried out on NO oxidation. Until now, various types of NO oxidation catalysts have been developed. However, there is lack of a review paper that provides very good insights into the progress in this area. Therefore, it is necessary to review the recent advances in the catalysts for NO oxidation to NO$_2$. By addressing their pros and cons and the potential applications, this chapter can be used as reference to further guide and assist the future development of NO oxidation technology. Furthermore, in this chapter, we not only summarize the latest progress in the synthesis and applications of various catalysts, but also

Cheng Zhang, Yanshan Gao, Liang Huang and Qiang Wang, College of Environmental Science and Engineering, Beijing Forestry University, Beijing, China

https://doi.org/10.1515/9783110544183-004

organize these catalysts into two groups: (1) Pt-based catalysts and (2) metal oxides-based catalysts. Based on the fact that NO oxidation is an important prerequisite step for many De-NO_x technologies, the integration of NO oxidation catalysts into different De-NO_x processes will also be discussed.

4.2 NO oxidation mechanism and reaction equilibrium

As an important part in the NO oxidation model, the reaction mechanism has been extensively studied. Although a large number of papers on the oxidation of NO to NO_2 have been published, the exact reaction mechanism remains controversial. The Eley–Rideal (E-R)-type mechanism, which was first proposed for NO oxidation over Pt/Al_2O_3 by Burch et al. [21], has been generally accepted. They assume the oxygen is the main substance at the active Pt sites, whereas NO reacts from the gas phase. Reaction (4.1) is the dissociative adsorption of oxygen on the platinum to generate atomic oxygen adsorbed on the surface of the Pt particles. Reaction (4.2) is the oxidation of NO, which involves the reaction between NO and adsorbed oxygen according to the E-R mechanism [22]:

$$2Pt + O_2(g) \leftrightarrow 2Pt-O \tag{4.1}$$

$$NO(g) + Pt-O \leftrightarrow Pt + NO_2 \tag{4.2}$$

This mechanism was confirmed by Bai et al. [23] and Crocoll et al. [24] with Pt/Al_2O_3 catalyst and Després et al. [22] with Pt/SiO_2 catalyst. Crocoll et al. [24] studied the mechanism and kinetics of Pt/Al_2O_3-catalyzed NO oxidation in the exhaust gas of the actual engine model. According to the experimental results, a kinetic mean field model was established, including six surface species and 16 basic reactions to simulate catalytic results. The E-R mechanism was again confirmed by their kinetic and spectroscopic studies.

In another report, the E-R mechanism was also confirmed by Li et al. [25] with Pt/TiO_2 catalyst. According to their in situ FTIR analysis, the NO oxidation pathways were proposed as illustrated in Figure 4.1. In the first step, oxygen is adsorbed on the surface of Pt and then disassociated to form O:

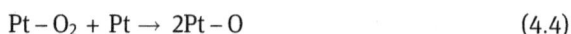

$$Pt + O_2(g) \rightarrow Pt-O_2 \tag{4.3}$$

$$Pt-O_2 + Pt \rightarrow 2Pt-O \tag{4.4}$$

Gaseous NO can react with surface oxygen to form gaseous NO_2:

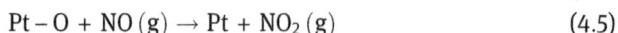

$$Pt-O + NO(g) \rightarrow Pt + NO_2(g) \tag{4.5}$$

Figure 4.1: Proposed NO oxidation pathways on Pt/TiO$_2$ catalysts. Reprinted from Ref. [26], Copyright 2010, Elsevier B.V.

Gaseous NO can also react with surface oxygen to form nitrates, which then migrate to adsorption sites of TiO$_2$ support, similar to a NO storage process:

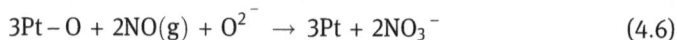

$$3Pt-O + 2NO(g) + O^{2-} \rightarrow 3Pt + 2NO_3^{-} \tag{4.6}$$

At relative high temperatures, the formed nitrates may decompose to produce gaseous NO or NO$_2$:

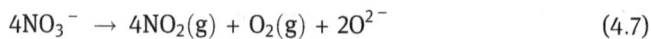

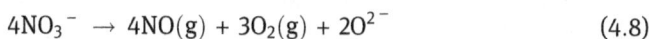

$$4NO_3^{-} \rightarrow 4NO_2(g) + O_2(g) + 2O^{2-} \tag{4.7}$$

$$4NO_3^{-} \rightarrow 4NO(g) + 3O_2(g) + 2O^{2-} \tag{4.8}$$

Based on catalytic results, the gaseous product NO$_2$ was detected up to 150 °C, which indicates that the direct oxidation of gaseous NO to NO$_2$ (reaction (4.5)) does not occur below 150 °C. The surface nitrate species are stable up to 250 °C and does not decompose below 250 °C. Hence, the reactions under the experimental conditions below 150 °C are reaction (4.3), (4.4), and (4.6). In the temperature range of 150–250 °C, NO$_2$ is detected due to the direct oxidation of gaseous NO (reaction (4.5)) and the decomposition of surface nitrates still does not occur. Therefore, the reactions at 150–250 °C are reaction (4.3–4.6). It should be pointed out that once the adsorption of nitrates reaches saturation at below 250 °C, reaction (4.6) will be suspended. At above 250 °C, surface nitrates decompose to NO (a very small proportion to NO$_2$) and gaseous NO is oxidized to NO$_2$ or adsorbed nitrates. So, the reactions at above 250 °C include reaction (4.3–4.8) [26].

On the contrary, Mulla et al. [27, 28] proposed Langmuir–Hinshelwood (L-H)-type mechanism for NO oxidation to NO_2 over Pt/Al_2O_3, which involves the following reaction steps:

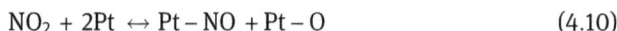

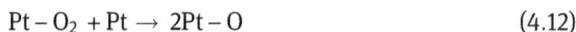

$$NO + Pt \leftrightarrow Pt - NO \tag{4.9}$$

$$NO_2 + 2Pt \leftrightarrow Pt - NO + Pt - O \tag{4.10}$$

$$O_2 + Pt \rightarrow Pt - O_2 \tag{4.11}$$

$$Pt - O_2 + Pt \rightarrow 2Pt - O \tag{4.12}$$

Reaction (4.11) was proposed as the rate-determining step and Pt-O as the most abundant surface intermediate, with reactions (4.9) and (4.10) in quasiequilibrium. Olsson et al. [29] who used E-R mechanism, in a later report, mentioned that they could not confirm which mechanism (L-H, E-R, or both) is the most likely for NO oxidation because all of them seem equally suitable for experimental data [30]. Calculation studies by density functional theory (DFT) also indicated that the existence of surface-adsorbed NO species, Pt-NO, is energetically and geometrically feasible [31, 32]. Getman et al. [31] studied the energy, charge distribution, and vibrational spectra of the stable and metastable of adsorbed NO, NO_2, and NO_3 on Pt(111). At low coverage, the binding energy sequence is NO > NO_3 > NO_2, and the oxidative endotherm of adsorbed NO to NO_2 is 0.78 eV. It was found that NO and NO_2 have many metastable adsorption configurations at low coverage where they occupy a lower co-ordination adsorption sites and reorientation is associated with the Pt surface. At higher NO_x and oxygen surface coverage associated with catalytic oxidation conditions, competition for metal surface charge density tends to favor these metastable configurations. Besides, the lateral interaction between the adsorbates weakens the Pt(111)-O bond to promote the formation of NO_2.

According to another report, Marques et al. [2] investigated the NO oxidation over two different Pt catalysts. The first is the reduced Pt^0 supported on SiO_2, and the second is the oxidized cation Pt_x^+ supported on $CeZrO_2$. With Pt/SiO_2 catalyst, they considered both the E-R and L-H elementary steps, and finally concluded that the kinetic studies cannot distinguish between mechanisms that only consider the basic steps of the E-R sequence or the basic steps of the L-H sequence. However, they noticed that the charge state of supported Pt influences the reaction mechanism. With $Pt_x^+/CeZrO_2$, platinum cation is stable on the surface of the cerium oxide–zirconia support and is surrounded by two oxygen vacancies to form a unique complex active site $\square Pt_x^+\square$. The detailed mechanism is proposed in Figure 4.2 and reactions (4.13–4.17).

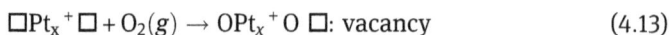

$$\square Pt_x^+\square + O_2(g) \rightarrow OPt_x^+O \quad \square: \text{vacancy} \tag{4.13}$$

$$OPt_x^+O + NO(g) \rightarrow OPt_x^+O(NO) \tag{4.14}$$

$$OPt_x^+O(NO) \rightarrow NO_2(g) + \square Pt_x^+O \tag{4.15}$$

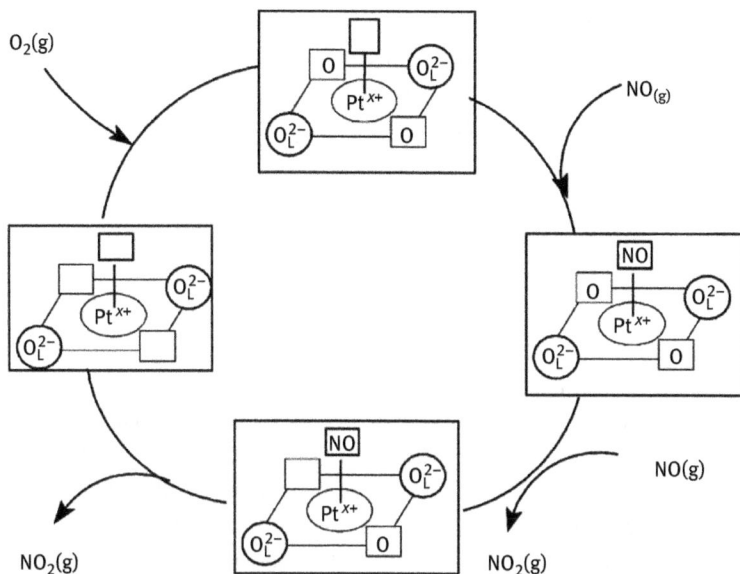

Figure 4.2: Catalytic cycles for NO oxidation on Pt$_x$$^+$/CeZrO$_2$, □: vacancy, OPt$_x$$^+$O(NO): two oxygen anions and one NO as ligands in the coordinative sphere of the platinum cation; □Pt$_x$$^+$O: one oxygen vacancy and one oxygen anion in the complex active site. Reprinted from Ref. [2], Copyright 2004, Elsevier B.V.

$$\Box Pt_x{}^+O + NO(g) \rightarrow NO_2(g) + \Box Pt_x{}^+\Box \qquad (4.16)$$

$$2NO(g) + O_2(g) = 2NO_2(g) \qquad (4.17)$$

In fact, it is now well accepted that the local surface structure as well as the chemical environment can significantly influence the composition of the catalyst surface, and the energetics and kinetics of the individual steps [33–36]. At different reaction conditions or with different supported Pt species, the mechanism might vary.

In addition, the thermodynamic equilibrium of the reaction NO$_2$ $\leftrightarrow$ NO + ½ O$_2$ is also important. Després et al. [22] studied the thermodynamic stability of NO$_2$ as a function of temperature at various partial pressures of O$_2$. NO$_2$ is stable at low temperatures, and starts to dissociate into NO and O$_2$ at temperatures above 200 °C. Consequently, in the presence of catalyst, the NO conversion will generally show a parabolic curve. The maximum NO conversion and the corresponding temperature are important criteria for evaluating the activity of catalysts.

In contrast to Pt-based catalysts, there is clearly no mechanism and kinetics studies on metal oxides-based catalysts. One reason for this might be the fact that metal oxides-based NO oxidation catalysts have just been developed for years. Moreover, most of the time, the mechanism and kinetics of metal oxides that catalyze the oxidation of NO are more dependent on the metal oxides themselves, and such kinds of researches are still going on. However, due to the superior perfor-mance and low price of metal oxides, their practical applications should be very promising. Therefore, the authors believe that the mechanism and kinetics studies on metal oxides-based catalysts are of great importance for searching for the best catalyst for NO oxidation to NO_2.

4.3 Pt-based catalysts

4.3.1 Effect of Pt particle size

The fact that the Pt-containing catalyst exhibits a higher NO oxidation activity than the corresponding support material, which means that Pt is the active phase of the reac-tion. Due to its high cost, Pt has traditionally been deployed as highly dispersed nano-particles to increase the active surface area. Naturally, the dependence of the catalytic activity on the Pt loading and on the Pt particle size would be of great interest to researchers.

In early 1996, Xue et al. [37] studied the effect of Pt loading and the exposed Pt surface area on the catalytic oxidation of NO over Pt/SiO_2, $Pt/\gamma-Al_2O_3$, and Pt/ZrO_2 catalysts. It was observed that the activity of all three catalysts for oxidizing NO and SO_2 increased with the Pt loading within the range from 0.1 to 5 wt% and increased as the amount of Pt exposed on the surface increased. However, the dependence of the specific activity on the particle size of the surface Pt depends on the catalyst. On Pt/SiO_2, it was found that the specific activity has a strong size dependence, and the larger Pt particles exhibited a higher specific activity than the smaller Pt particles. As the Pt loading increases to 2 wt.%, the NO oxidation reaction rate increases; how-ever, as the Pt loading continues to increase to 5 wt.%, the rate remains almost con-stant. Over $Pt/\gamma-Al_2O_3$, the size dependence seems to be far less important than that of the Pt/SiO_2. Although similar trends were also observed, the 2 wt.% $Pt/\gamma-Al_2O_3$ cat-alyst showed a higher rates than 5 wt.% $Pt/\gamma-Al_2O_3$. However, the results of the Pt/ZrO_2 catalyst indicate that the catalytic reaction is likely to be independent of size. A significant increase in the reaction rates was still observed even in the Pt loading range of 2 to 5 wt.%. Later, several other groups have also observed similar phenom-ena [38–42]. For example, Lee et al. [43] synthesized two $Pt/\gamma-Al_2O_3$ catalysts with 4.4% and 82% dispersion of Pt. For the oxidation of NO at 265 °C, the turnover fre-quency (TOF) of the low dispersion catalyst was >50 min^{-1}, which is over 100 times higher than that of the high dispersion sample (0.3 min^{-1}). Olsson et al. [41] investi-gated the particle size sensitivity by pretreating the catalyst at different temperatures (400 °C, 550 °C, 650 °C, 750 °C). As expected, the Pt dispersion decreased with the increase of pretreatment temperature. The effect of different pretreatment tempera-tures on NO oxidation is shown in Figure 4.3. Both the low-temperature activity and the maximum conversion of NO to NO$_2$ increased as the catalyst became more sintered.

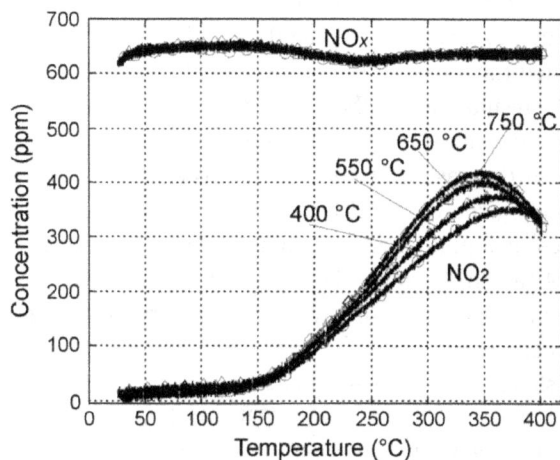

Figure 4.3: The outlet concentration of NO$_2$ and NO$_x$ as a function of temperature for four Pa/Al$_2$O$_3$ catalysts sintered at different temperatures (400 °C, 550 °C, 650 °C, 750 °C). Test condition: 620 ppm NO, 8% O$_2$ in Ar. Reprinted from Ref. [41], Copyright 2002, Elsevier Science (USA).

As mentioned above, large particles are more active than small particles for NO oxidation. By investigating the influence of Pt oxide formation and Pt dispersion on reactivity of Pt/Al$_2$O$_3$ and Pt/BaO/Al$_2$O$_3$, Olsson et al. [41] concluded that the reason for the high activity of large platinum particles for NO oxidation may be that they are more stable and less susceptible to be oxidized. Small Pt particles contain higher concentrations of lower coordination sites than larger particles exposed to more closely packed (111) sites. Consequently, these more-open surfaces of small particles are easier to be oxidized, resulting in a decreased activity [44]. The results reported by Smeltz et al. [34] are consistent with the view that large, closed-packed crystalline Pt particles predominating in the closed-packed (111) plane are kinetically resistant to complete Pt oxidation under NO oxidation reaction conditions, while the small amorphous Pt particles will be oxidized to bulk oxides and thus become inactive.

However, in some very recent reports, some different opinions have also been proposed. Liang et al. [45] investigated the size effect of PtPd nanoparticles on NO oxidation. A series of bimetallic PtPd/SiO$_2$–Al$_2$O$_3$ catalysts of different sizes were controlled by hydrothermal treatment at different times. The experimental results showed that the average precious metal particle size gradually became larger as the aging time increased. An increase in the average particle size resulted in a corresponding reduction in the NO TOFs. Therefore, a new perspective on the relationship between particle size and NO oxidation activity is proposed. During the aging process, the overall trend of precious metal particles increases with the decrease of NO oxidation performance, but in a period of time, a few small particles showed more bridged/chelating nitrate and ionic nitrate species, further delaying its rate of loss activity. As we expected, for noble metal catalysts, the most desirable state is to achieve monoatomic dispersion to achieve 100% atomic utilization, thereby greatly increasing the economics of the catalyst [46]. Narula et al. [47] evaluated the effect of θ-Al$_2$O$_3$-supported single platinum atoms on NO oxidation activity. The results showed that the catalyst exhibited remarkable NO oxidation activity, which is in contrast to the known decrease in NO oxidation activity of supported platinum with decreasing Pt particle size. The first principles density functional theoretical modeling shows that NO oxidation is available on fully oxidized single θ-Al$_2$O$_3$ supported platinum atoms by a modified L-H pathway. Therefore, the overall oxidation of NO on the supported Pt is that as the Pt particle size decreases, the NO oxidation activity decreases, but when Pt exists only as a single atom, the NO oxidation activity is accelerated.

In order to have a clear understanding of the effect of cluster size on the reactivity of small Pt$_x$ clusters toward the oxidation of NO, Xu et al. [33] performed a detailed investigation using periodic DFT calculations. It was proved that the binding of all the relevant reactant species is significantly enhanced compared to the corresponding Pt bulk levels (represented by Pt(111)), especially on Pt$_{1-5}$. In the range of $x < 5$, a large number of species- and size-dependent changes in binding energy are apparent. At Pt$_{4-5}$, the combination of oxygen and NO$_2$ is the smallest (strongest adsorption). And the binding energies of all species on Pt$_{10}$ have already approached the Pt(111) levels. Due to the strong interaction of the reactants and products species with the Pt$_x$ clusters, a deep energy sink is formed on the potential energy surface of the oxidation process, indicating that the reaction energy is worse than on Pt(111). Therefore, this may be the reason why the smallest Pt cluster is less effective for catalyzing NO oxidation than bulk Pt.

4.3.2 Effect of support

Besides the effect of the properties of supported Pt, it is believed that the support influences the NO oxidation activity as well. In the available studies in the literature, various metal oxides including Al$_2$O$_3$ [21, 24, 29], SiO$_2$ [22], TiO$_2$ [25, 26], and CeO$_2$–ZrO$_2$ [2] have been studied as catalyst support. It is commonly accepted that the metal oxides support influences the dispersion and thermal stability of Pt particles, and hence determines the performance of catalysts. Bourges et al. [48] investigated the oxidation of NO over various noble metal catalysts supported on γ-Al$_2$O$_3$. Pt catalysts supported on γ-Al$_2$O$_3$ showed the highest activity. Xue et al. [37] investigated the influence of platinum loading, metal oxide support, and the presence of SO$_2$ on the oxidation of NO. Pt/SiO$_2$ exhibited a higher activity than Pt/Al$_2$O$_3$ and Pt/ZrO$_2$ under the same Pt loading.

Similar support effects were also observed by Oi-Uchisawa et al. [38]. To date, it could be concluded that Pt loaded on different supports show an overall trend with respect to catalyst activity in NO oxidation: SiO$_2$ > Al$_2$O$_3$ > ZrO$_2$ [40]. This ordering can be reasonably performed based on the fact that NO is mainly adsorbed on the carrier and then migrates to Pt where it is oxidized. Considering the weak adsorption of NO on silica and the fact that NO$_2$ is not stored as nitrate, the use of silica as a support facilitates the rapid migration of adsorbed NO to Pt, followed by oxidation and easy desorption of NO$_2$ [49]. Li et al. [25, 26] reported that Pt/TiO$_2$ catalyst prepared by photodeposition showed significant NO oxidation activity even in the presence of SO$_2$. It exhibited a pretty high activity for NO oxidation to NO$_2$; over 90% NO oxidation efficiency was achieved at below 250 °C and at a high GHSV of 180,000 h^{-1}. In the presence of 320 ppm SO$_2$, the conversion still kept as high as ca. 80% at 350 °C.

4.3.3 Effect of promoter

To further increase the activity for the oxidation of NO to NO$_2$, some promoters have been added and studied over Pt-based catalysts. Potassium is one of the most common promoters used for oxidative reactions. For instance, Kuriyama et al. [50] reported that the activity of preferential CO oxidation on Pt/Al$_2$O$_3$ could be enhanced by the addition of potassium. They explained that the additive effect of potassium weakened the interaction between CO and Pt, and also changed the CO adsorption site. In the same way, Mulla et al. [51] investigated the promotion effect of potassium on the catalytic oxidation of NO to NO$_2$ with Pt/K/Al$_2$O$_3$ catalyst. The variation of the NO oxidation reaction rates with temperature over Pt/Al$_2$O$_3$ and Pt/K/Al$_2$O$_3$ catalysts are shown in Figure 4.4. The apparent activation energies (E_a) for both Pt/Al$_2$O$_3$ and Pt/K/Al$_2$O$_3$ catalysts are 60.4 ± 4.2 and 83 ± 9 kJ·mol^{-1}, respectively. Under the same reaction condition (300 °C, 300 ppm NO, 10% O$_2$, 170 ppm NO$_2$), the turnover rates (TOR) on Pt/K/Al$_2$O$_3$ was about 2.5 times higher than that on Pt/Al$_2$O$_3$. Considering that the Pt/K/Al$_2$O$_3$ catalyst exhibited a higher Pt dispersion (about 19% higher than Pt/Al$_2$O$_3$ catalyst), and may possess a smaller Pt particle size than the Pt/Al$_2$O$_3$ catalyst and the fact that larger Pt particles exhibit higher TOR for NO oxidation than smaller particles, the increase in the TOR observed on Pt/K/Al$_2$O$_3$ catalyst indicates the effect of K on Pt/Al$_2$O$_3$ catalyst for the NO oxidation reaction.

Another type of promoters that have been utilized for NO oxidation is metal oxides, including WO$_3$, MoO$_3$, V$_2$O$_5$, and Ga$_2$O$_3$ [52, 53]. Dawody et al. [53] studied the effect of metal oxide additives (WO$_3$, MoO$_3$, V$_2$O$_5$, Ga$_2$O$_3$) on the oxidation of NO and SO$_2$ over Pt/Al$_2$O$_3$. In the absence of SO$_2$, the catalysts containing WO$_3$ and MoO$_3$ showed a significantly higher NO oxidation activity than the other catalysts. When SO$_2$ was introduced into the reaction gas mixture, the NO oxidation activity of all the catalysts was decreased; however, the MoO$_3$-modified catalyst was less

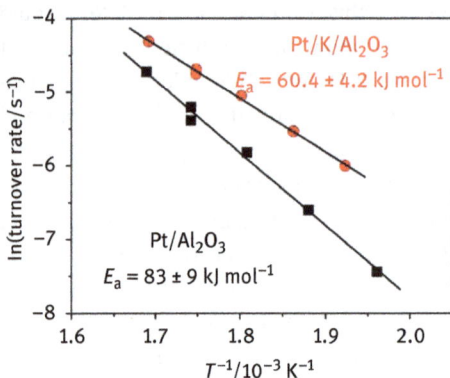

Figure 4.4: Arrhenius plot for NO oxidation on Pt/Al_2O_3 and $Pt/K/Al_2O_3$ assuming a differential reactor. Feed: 300 ppm NO, 10% O_2, 170 ppm NO_2, balance N_2. Reprinted from Ref. [51], Copyright 2002, Elsevier B.V.

affected by the presence of SO_2 and also exhibited the lowest SO_2 oxidation activity. Irfan et al. [54] found that WO_3 could constrain the Pt oxidation, thereby improving the catalytic activity of NO oxidation.

4.3.4 Effect of alloying

Due to the high cost of platinum, the dependence of platinum as the main active element is not ideal. Ideally, people prefer to use relatively less Pt or Pt-free catalysts. One strategy is to develop a bimetallic catalyst that can partially replace the function of Pt and even improve the performance of Pt. Alloying metals have already been utilized as an effective method to enhance the surface reactivity by electronic or holistic effect or both. The electronic effect is a change in the rate constant of the basic steps due to changes in the electronic structure, and the holistic effect is due to changes in the distribution and availability of surface reaction sites [55].

Corro et al. [56] reported that the activity for NO oxidation to NO_2 can be improved by forming Pt–Sn alloy catalyst. The increase in activity was attributed to the enhancement of adsorbed oxygen on the metallic surface resulting from the increase in Sn surface atoms over the catalysts. Jelic et al. [57] initiated a study of Pt (and Pd)-based pseudomorphic monolayer catalysts for NO oxidation using DFT. A series of pseudomorphic monolayer catalysts were prepared based on Pd and Pt monolayers coated on the late transition metal (TM) hosts (Cu, Ru, Rh, Ag, Ir, and Au) and tested for their NO oxidation properties. The activation energies and reaction enthalpies for NO_2 formation on different surfaces are presented in Table 4.1.

It is found that the lattice parameters have a great influence on the activation energy and reaction enthalpy of the reaction. Besides, the thermodynamic stability

Table 4.1: The activation energies (E_{act}) and reaction enthalpies (ΔH) for the NO + O reaction for Pt and Pd pseudomorphic monolayer systems. Reprinted from Ref. [57], Copyright 2008, Elsevier B.V.

	Host metal	Cu	Ru	Rh	Pd	Ag	Ir	Pt	Au
Pt	E_{act}	0.29	0.36	0.54	1.13	1.66	0.41	1.07	1.56
	ΔH	−0.58	−0.81	−0.57	0.25	1.12	−0.55	0.06	0.94
Pd	E_{act}	1.06	0.84	0.94	1.26	1.61	0.98	1.36	1.63
	ΔH	0.04	−0.07	0.13	0.55	1.07	0.10	0.56	1.06

when adsorbing NO and O$_2$ was also investigated. Although pseudomorphic mono-layer separation systems with Pt and Pd on the TM were found to be stable to systems involving Ru, Ir, Rh, and Au, even in the presence of adsorbates, it was found that oxygen can reverse the position of the support monolayer and the host metal for systems involving Ag or Cu [57]. Finally, considering the prediction that Pt will stay on the surface even in contact with reactants, the Pt/Ir system is of great interest due to its small NO oxidation barrier.

By designing proper alloy, the sulfur resistance of Pt-based catalysts can be enhanced. For instance, Tang et al. [55] applied a d-band center weighting model and found that the alloy surface with a surface d-band center of about −2.22 eV possesses the highest NO oxidation selectivity even in the presence of SO$_2$. Figure 4.5 shows the ratio of the prefactor for NO oxidation to that for SO$_2$ oxidation as a function of surface d-band center. The Ir alloyed Pt(111) surfaces showed the highest selectivity (about 9.6) compared to a clean Pt(111) surface (about 2.0), and all others

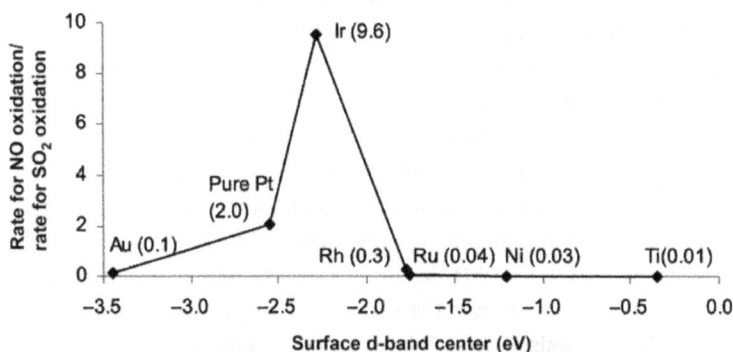

Figure 4.5: Plot of the rate ratio between the oxidation of NO and the oxidation of SO$_2$ at 427 °C vs the surface d-band center of the surfaces (1M^x/Pt(111), with x = a for SO$_2$ and x = d for NO). M of each point is marked in the plot. Reprinted from Ref. [55], Copyright 2006, American Chemical Society.

(e.g., Au, Rh, Ru, Ni, or Ti alloyed surfaces) show much lower selectivities (lower than 0.3).

Pt-based catalysts will agglomerate under high-temperature lean conditions, resulting in rapid loss of Pt surface area, and therefore it has been proposed to alloy Pt with Pd or Rh as a possible solution to this problem [58]. Graham et al. [59] synthesized a series of highly dispersed Pt–Pd catalysts, which were aged, characterized and tested to evaluate the effect of alloy composition on dispersion stability and NO oxidation performance. According to the aging experiments, it could be concluded that alloying Pt with Pd in alumina-based catalysts can stabilize the metal particle size after aging, and it does not adversely affect the TOF compared to pure Pt. In practical applications, these results indicate that Pt can be partially replaced with cheaper Pd while improving catalyst durability and without significantly sacrificing NO oxidation activity. Kaneeda et al. [60] also reported that the addition of Pd significantly increased the NO oxidation activity of the catalyst. The Pd-modified catalysts were active even after a strict heat treatment at 830 °C for 60 h. The optimized Pd-modified Pt/Al$_2$O$_3$ catalyst exhibits a maximum activity limited by chemical equilibrium under the reaction conditions. In addition, Auvray et al. [61] also reported that the addition of Pd into Pt/Al$_2$O$_3$ catalysts prevented the agglomeration of Pt and improved the thermal stability of metallic particles by the formation of Pt–Pd alloy, and the formed alloy structure can improve the NO oxidation activity by restricting the Pt oxidation.

Other novel metals such as Ru, Rh, Au, and Pd were also studied as NO oxidation catalysts [62–67]. With TiO$_2$ as catalyst support, the catalyst NO oxidation activity showed the following order: Ru/TiO$_2$ > Rh/TiO$_2$ > Pd/TiO$_2$ > Au/TiO$_2$. Even with a very high GHSV of 180,000 h^{-1}, Ru/TiO$_2$ achieved a 50% NO conversion to NO$_2$ at 250 °C and a 94% conversion at 275 °C. The effect of different supports on the Ru supported catalyst was also investigated, and the activity followed the order of Ru/TiO$_2$ > Ru/TiO$_2$-R > Ru/ZrO$_2$ > Ru/SiO$_2$ > Ru/Al$_2$O$_3$ > Ru/TiO$_2$-A. It was concluded that TiO$_2$ P25 with both anatase and rutile phase seems to be the most suitable support for Ru and the optimal Ru loading was determined to be ca. 2 wt.% [64]. In another report, Qu et al. [63] synthesized Ru-FAU catalysts for NO oxidation. The results show that the addition of rare earth ions to the FAU structure can improve the NO oxidation performance of the catalyst, while the introduction of alkali metal ions at the cationic sites has a negative impact on the catalyst. Ru-REY exhibited the best activity and achieved a maximum NO conversion of around 94% at 250 °C. However, thermal treatment of Ru-based catalysts at high temperature resulted in a severe deactivation. Moreover, the presence of H$_2$O and/or SO$_2$ in the reaction system also showed obvious negative effects on NO oxidation over the Ru-based catalyst [64]. In a recent report, Adjimi et al. [67] synthesized a series of noble metals modified K-OMS–2 catalysts for NO oxidation. The results showed that the NO conversion followed the order of Ru/K-OMS–2 > Pt/K-OMS–2 ≈ KOMS–2 > Ag/K-OMS–2 ≈ Pd/K-OMS–2. It was concluded that Ru/K-OMS–2 exhibited the highest NO conversion rate, and the reaction reach 5.3 μmol·g^{-1}·s^{-1} at 584 K, which could be attributed to the enhanced catalyst reducibility upon incorporation of Ru.

4.3.5 The main issues toward commercial application

For the commercial application of Pt-based catalysts, two of the most crucial problems should be first solved [24]. One is Pt oxidation by the reaction product, NO$_2$. When Pt was exposed to NO + O$_2$ at 250 °C, its activity was decreased. Based on XPS data, this inactivation was attributed to the oxidation of Pt to PtO or PtO$_2$, which is less active than reduced platinum. The other is sulfur poisoning, which forms sulfate with support alumina (such as Al$_2$(SO$_4$)$_3$), resulting in a decrease in catalytic activity [2]. What's more, nonnoble metal NO oxidation catalysts are highly demanded for economical consideration.

4.4 Transition metal oxides-based catalysts

4.4.1 Cobalt oxide-based catalysts

Considering the high price of Pt and the autoinhibition problem caused by the reaction product NO$_2$, people begin to search for alternatives from TM oxides. The first reported nonnoble metal NO oxidation catalyst was Co$_3$O$_4$-based catalyst [16, 68–70]. In addition to its low price, the existence of NO$_2$ does not inhibit the catalytic activity of Co$_3$O$_4$-based catalysts. Irfan et al. [71] prepared various TM oxide-supported catalysts such as MnO$_x$/TiO$_2$, CuO$_x$/TiO$_2$, Co$_3$O$_4$/TiO$_2$, and Co$_3$O$_4$/SiO$_2$. The MnO$_x$/TiO$_2$, CuO$_x$/TiO$_2$, and Co$_3$O$_4$/TiO$_2$ did not present proper conversion (about 50%), while for Co$_3$O$_4$/SiO$_2$, a remarkable conversion (69%) was obtained at a lower temperature (300 °C). Without catalyst supports, the activity order was Co$_3$O$_4$ > MnO$_x$ > CoMnO$_x$ > CuO$_x$. Since Co$_3$O$_4$ has been considered as the most promising metal oxide catalyst for NO oxidation and lots of works have been done on Co$_3$O$_4$-based catalysts, in the following parts, we will first review all the works on Co$_3$O$_4$-based catalysts (effect of support, effect of prepared method, effect of Co loading, and potassium promotion effect), then introduce the works on other TM oxides-based catalysts.

Effect of support

Kim et al. [72] have tested NO oxidation activity over cobalt oxide supported on various supports such as SiO$_2$, ZrO$_2$, TiO$_2$, CeO$_2$ (H, surface area = 230 m^2/g), and CeO$_2$ (M, surface area = 66 m^2/g). Unsupported Co$_3$O$_4$ and 1 wt.% Pt/γ-Al$_2$O$_3$ were also tested for comparison, as shown in Figure 4.6. The NO oxidation performance decreased in the following order: Co_3O_4/CeO_2 (H) $\approx$ 1 wt.% $Pt/\gamma\text{-}Al_2O_3 \gg Co_3O_4 > CoO_x/SiO_2 > Co_3O_4/CeO_2$ (M) $> Co_3O_4/ZrO_2 > Co_3O_4/TiO_2$. The most active catalyst Co_3O_4/CeO_2 (H) exhibits comparable catalytic activity to well-known noble metal catalyst, $Pt/\gamma\text{-}Al_2O_3$, and a NO conversion of 72% was achieved at 265 °C. This Co_3O_4/CeO_2 (H) also exhibited higher catalytic activity than Co_3O_4 at the same contact time, indicating that the catalytic activity of cobalt oxides catalyst can be further improved by selecting a suitable support. Furthermore, the catalytic activity of Co_3O_4 supported on ceria with different surface areas was also investigated, and the results showed that the catalytic activity for NO oxidation is strongly dependent on the surface area of CeO_2. The larger the specific surface area, the better the NO oxidation activity [72].

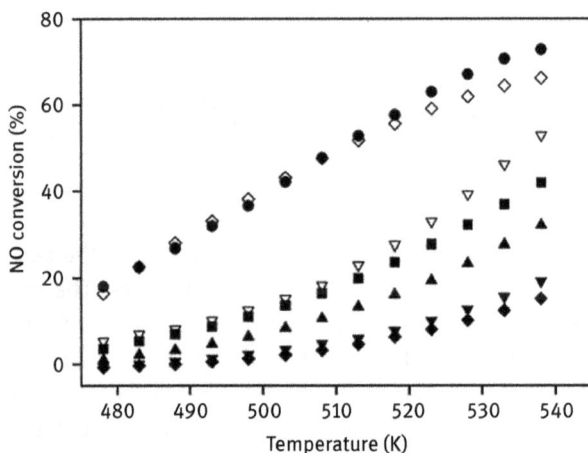

Figure 4.6: NO conversions over cobalt oxide catalysts supported on various supports such as (●) CeO_2 (H), (■) SiO_2, (▲) CeO_2 (M), (▼) ZrO_2, and (◆) TiO_2, (▽) Co_3O_4, (◇) 1 wt.% Pt/Al_2O_3. Gas composition: 500 ppm NO, 5% O_2 in N_2, 200 mL·min^{-1}. Reprinted from Ref. [72], Copyright 2009, Korean Institute of Chemical Engineers, Seoul, Korea.

Irfan et al. [71] compared the role of TiO_2 and SiO_2 as catalyst support for NO oxidation. The activity order of these catalysts is: $Co_3O_4/SiO_2 > Pt\text{-}WO_3/TiO_2 > Pt/TiO_2 > Pt/SiO_2 > CuO_x/TiO_2 > Co_3O_4/TiO_2 > MnO_x/TiO_2 > Pt\text{-}V_2O_5/TiO_2$. At 300 °C and with an SV of 100,000 h^{-1}, the NO conversions over Co_3O_4/SiO_2 and Co_3O_4/TiO_2 are 69% and 36%, respectively. Yung et al. [70] compared the performance of Co_3O_4/TiO_2 and Co_3O_4/ZrO_2 for the oxidation of NO to NO_2 in excess oxygen (10%). When compared to other reports, much higher NO conversions were achieved over these two catalysts, 86% at 280 °C for Co_3O_4/TiO_2 and 92% at 260 °C for Co_3O_4/ZrO_2. In all the tested conditions, the performance of Co$_3$O$_4$/ZrO$_2$ was slightly higher than that of Co$_3$O$_4$/TiO$_2$.

Besides, Wang et al. [73] synthesized CoO$_x$/TiO$_2$ catalysts by sol–gel (SG) method, and investigated the NO oxidation in excess O$_2$. Under the reaction conditions, 9% CoO$_x$/TiO$_2$ showed the highest NO oxidation efficiency of 79.8% at 300 °C. Huang et al. [74] supported Co$_3$O$_4$ on mesoporous SiO$_2$, and the catalyst showed a 82% NO conversion at 300 °C, at a space velocity of 12,000 h^{-1}. Mumar et al. [75] compared the effect of commercial ceria, commercial cobalt oxide, and a laboratory-synthesized nanoceria-supported cobalt oxide on catalytic NO oxidation. It was found that ceria-supported cobalt oxide showed highest catalytic activity for NO oxidation at moderate temperature, exhibiting 80% NO conversion at 300 °C at a space velocity of 150,000 h^{-1}. In another report, Yu et al. [76] prepared a series of cobalt supported on metal-doped ceria catalysts (M = Zr, Sn, and Ti) for NO oxidation. The results showed that the type of doping metal affected the dispersion state of cobalt. The Ce–Co metal-oxide catalyst possessed a relatively large amount of finely dispersed cobalt species, more oxygen vacancies, and excellent redox ability. These characteristics may be responsible for the improved NO oxidation activity.

In summary, it is apparent that there is still no common agreement on the effect of supports. With the available reports, it is probably difficult to conclude which support is the best for Co$_3$O$_4$-based catalysts for NO oxidation. Even with the same catalyst, for example, Co$_3$O$_4$/TiO$_2$, the conversion of NO varies with reports, ranging from 36% to 86% [70, 71]. It suggests that more works are needed to clarify the roles of catalyst supports.

Effect of preparation method

Since Co$_3$O$_4$ catalyst is found to be the most active catalyst among all the metal oxides, many works have been carried out on the preparation parameters, for example, the synthesis method, the cobalt precursor, and the calcination temperature. Yung et al. [70] compared the SG and incipient-wetness impregnation (IWI) methods with Co/TiO$_2$ catalyst, and found that the overall activity of Co/TiO$_2$ (IWI) was higher than that of Co/TiO$_2$ (SG). The reason for their difference in activity was investigated by a temperature-programmed reduction analysis. With Co/TiO$_2$ (SG), the reduction temperature was very high (> 550 °C) and the reduction peak was broad, beginning between 400 and 450 °C and ending between 700 and 750 °C. This indicates that the SG preparation method provides linkages between Ti and Co species and the interaction between them is strong. In contrast, Co/TiO$_2$ (IWI) had a much lower reduction temperature and a much sharper reduction peak, indicating a weaker interaction with the support and a greater homogeneity among the types of supported cobalt species. In another report, Shang et al. [77] investigated the influence of citrate complexation method (CC-series) and impregnation method

(IM-series) on the NO oxidation activity of a series of cobalt supported on $Zr_{1-x}Ce_xO_2$ catalysts. Compared to the IM-series, the CC-series catalysts were found to have easier reducibility, lower crystallite size, and higher relative amount of Ce^{3+}, which in turn led to their higher catalytic activity.

The effect of cobalt precursors, such as cobalt acetate (A) and cobalt nitrate (N), on the NO oxidation performance was investigated over cobalt oxides supported on SiO_2, ZrO_2, and CeO_2 as shown in Figure 4.7. The NO oxidation activity of the catalysts was decreased in the following order: $Co_3O_4(A)/CeO_2 \gg CoO_x(A)/SiO_2 \sim Co_3O_4$ $(N)/CeO_2 > Co_3O_4(N)/SiO_2 > Co_3O_4(N)/ZrO_2 > Co_3O_4(A)/ZrO_2$. The activation energy was measured to be 64.8, 68.2, 74.0, 93.1, 75.7, and 89.8 kJ·mol^{-1} for $Co_3O_4(A)/CeO_2$, $CoO_x(A)/SiO_2$, $Co_3O_4(N)/CeO_2$, $Co_3O_4(N)/SiO_2$, $Co_3O_4(N)/ZrO_2$, and $Co_3O_4(A)/ZrO_2$, respectively. $CoO_x(A)/SiO_2$ exhibited a little higher catalytic activity than $Co_3O_4(N)/SiO_2$ especially at low temperatures. In contrast, Co_3O_4/ZrO_2 prepared from cobalt nitrate presented higher catalytic activity than that prepared from cobalt acetate. A significant improvement of NO oxidation activity was found over Co_3O_4/CeO_2 when cobalt acetate was applied as the cobalt precursor instead of cobalt nitrate [78].

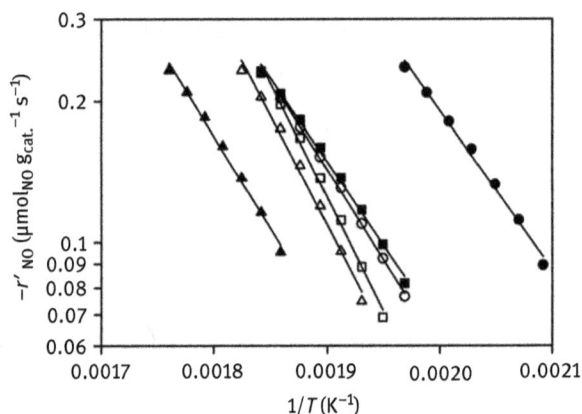

Figure 4.7: The specific NO oxidation rates of supported cobalt oxides prepared from different cobalt precursors as a function of reaction temperature. (●) $Co_3O_4(A)/CeO_2$, (■) $CoO_x(A)/SiO_2$, (▲) $Co_3O_4(A)/ZrO_2$, (O) $Co_3O_4(N)/CeO_2$, (□) $Co_3O_4(N)/SiO_2$, and (△) $Co_3O_4(N)/ZrO_2$. Reactants: 500 ppm NO and 5% O_2 in N_2. Reprinted from Ref. [78], Copyright 2010, Korean Institute of Chemical Engineers, Seoul, Korea.

The effect of calcination temperature on the NO conversion performance of the unsupported Co_3O_4 catalyst is shown in Figure 4.8. The results showed that the activity decreased with the increase of calcination temperature, which might be caused by a decrease in surface area due to high-temperature sintering. The catalyst calcined at 300 °C gave the highest NO to NO_2 conversion of 76% as well as

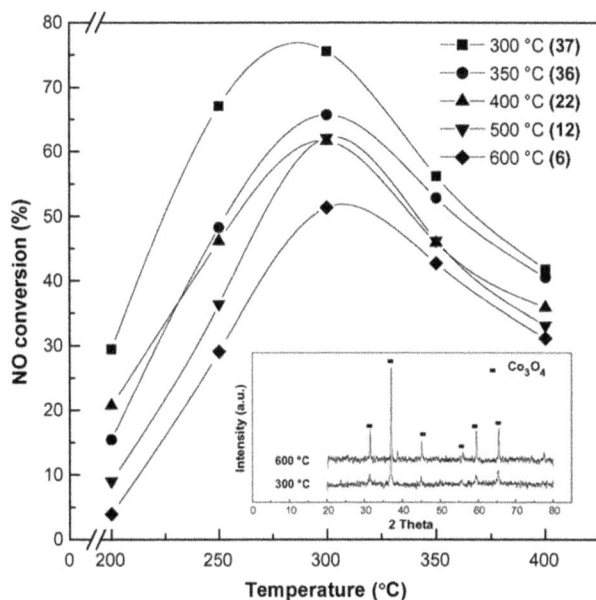

Figure 4.8: Effect of calcination temperature on NO oxidation as a function of reaction temperatures over Co$_3$O$_4$ catalyst; XRD analysis of Co$_3$O$_4$ catalyst at different calcination temperatures (150 ppm NO, SV = 150,000 h^{-1}, 8% H$_2$O, 10% O$_2$ and N$_2$ balance). Values in parenthesis are BET surface areas. Reprinted from Ref. [71], Copyright 2007, Elsevier B.V.

the highest BET surface area (37 m$^2\cdot$g^{-1}) compared to the others. Further increasing the calcination temperature to 500 °C and 600 °C, the conversion rates of catalysts were 61 and 51%, respectively. The inset of Figure 4.8 indicates that the XRD peaks of Co$_3$O$_4$ calcined at low temperature of 300 °C are broader than those calcined at higher temperature of 600 °C. It suggests that high calcination temperature increases the crystalline degree of Co$_3$O$_4$. Therefore, the catalyst calcined at a lower temperature has a smaller particle size and a larger surface area, thus increasing the conversion of NO to NO$_2$ [71]. Yung et al. [70] observed the same trend over Co/TiO$_2$ and Co/ZrO$_2$ samples prepared by IWI. However, with the samples prepared by SG method, a slightly higher calcination temperature (400 °C) is needed. Although the sample calcined at 300 °C had a higher surface area, it was amorphous, lacking TiO$_2$ crystallinity, which may partially explain the different behavior in activity. Besides, Qiu et al. [79] also studied the NO oxidation activity over a series of 10 wt% Mn-5 wt% Co/TiO$_2$ catalysts calcined at different temperatures. The experimental results showed that the catalytic activity increased with the calcination temperature increasing from 573 K to 773 K, and then decreased. The reduced NO oxidation activity can be assigned to the decrease of chemical-adsorbed oxygen amounts and the reduction of Mn^{4+} to Mn^{3+} and Mn^{2+} when the calcination temperature higher than 773 K.

Effect of cobalt loading

Yung et al. [70] investigated the influence of cobalt loading on the activity of Co/ZrO$_2$ catalysts prepared by IWI and calcined at 300 °C, as shown in Figure 4.9(a). The

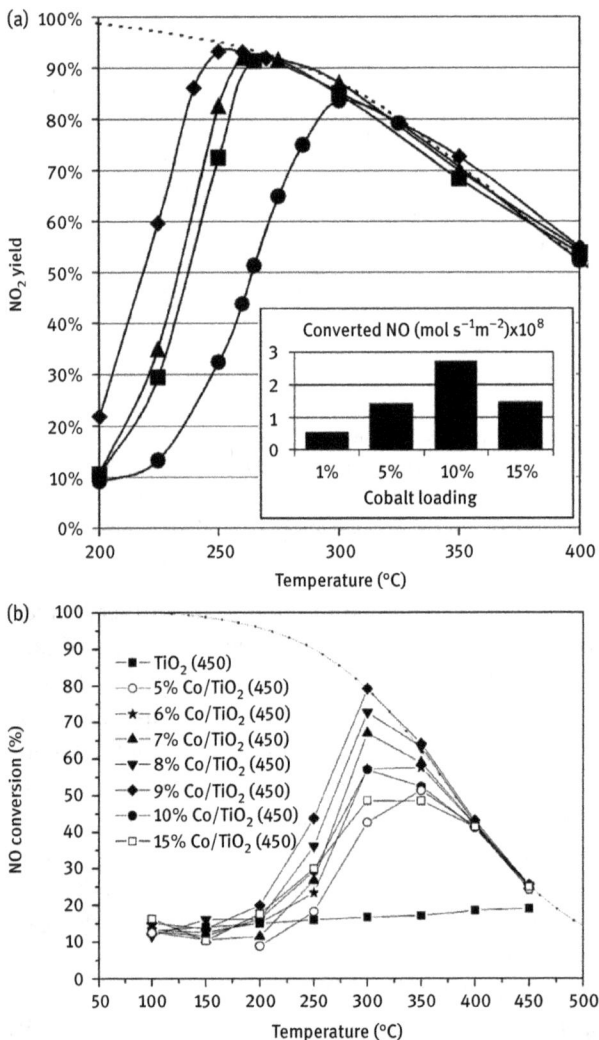

Figure 4.9: (a) Comparison of the activities of Co/ZrO$_2$ (IWI, calcined at 300 °C) catalysts with various cobalt loadings: (■) 15 wt.%, (♦) 10 wt.%, (▲) 5 wt.%, and (●) 1 wt.%. The inset shows the NO conversion rates at 225 °C. (b) NO oxidation efficiency of CoO$_x$/TiO$_2$ catalysts at various cobalt loadings. Reaction conditions: [NO] = 600 ppm, [O$_2$] = 4%, balance N$_2$, total flow rate 1,200 mL/min, GHSV 30,000 h^{-1}. (a) Reprinted from Ref. [70], Copyright 2007, Elsevier Inc. (b) Reprinted from Ref. [73], Copyright 2009, Springer Science Business Media, LLC.

1 wt.% Co/ZrO$_2$ sample presented pretty lower activity than the sample with higher co-balt loadings. The inset compared the conversion rates normalized with respect to sur-face area at 225 °C. It was concluded that 10 wt.% loading is the optimum value among the tested loading levels. In another report, Wang et al. [73] studied the cobalt loading effect with a series of CoO$_x$/TiO$_2$ prepared by SG and calcined at 450 °C, as shown in Figure 4.9(b). For pure TiO$_2$, it had almost no catalytic oxidation ability over the entire temperature range. By addition Co to TiO$_2$, the oxidation activity was improved rapidly, especially when the temperature was higher than 200 °C. For 9 wt.% CoO$_x$/TiO$_2$, it ex-hibited the highest activity in this samples that reached equilibrium at 300 °C, corre-sponding to a NO conversion of 79.8%. Some other studies also used 10 wt.% as the optimal Co loading for NO oxidation reaction [72, 78, 79].

Potassium promotion effect

It is widely accepted potassium promotes the oxidation activity of Pt/Al$_2$O$_3$ [50, 51]. Mulla et al. [51] observed that the Pt/K/Al$_2$O$_3$ catalyst had a lower apparent activation energy and a higher TOR than the Pt/Al$_2$O$_3$ catalyst. The promotion of K on the cata-lyst activity was attributable to the enhancement of O$_2$ absorption on the surface of Pt/K/Al$_2$O$_3$. However, Wang et al. [16] found that the activity of Co/TiO$_2$ could not be promoted by directly impregnating potassium, as shown in Figure 4.10. After impreg-nating 1 wt.% K into 5 wt.%/TiO$_2$ catalyst, the NO conversion was even worse than that of 5 wt.% Co/TiO$_2$ without K.

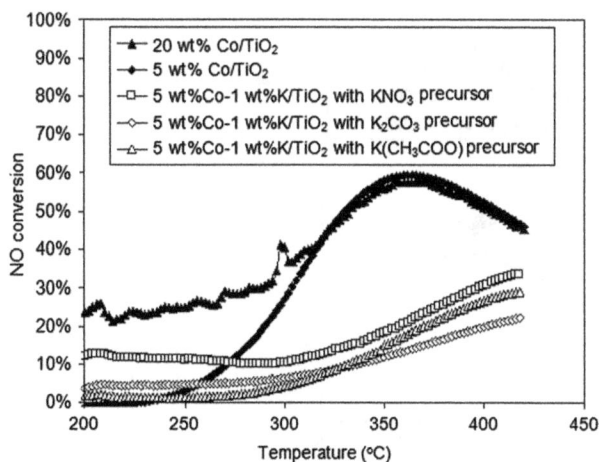

Figure 4.10: NO conversion over 20 wt.% Co/TiO$_2$, 5 wt.% Co/TiO$_2$, and 5 wt.%Co-1 wt.% K/TiO$_2$ with KNO$_3$, K$_2$CO$_3$, and K(CH$_3$COO) as potassium precursors. Reprinted from Ref. [16], Copyright 2007, Elsevier B.V.

In order to further increase the NO oxidation activity of Co_3O_4-based catalyst, Wang et al. [15, 16] proposed a new method to prepare the potassium-containing Co_3O_4-based catalyst, $Co_3O_4/K_xTi_2O_5$. After obtaining $K_2Ti_2O_5$ via solid-state method, it was introduced into $Co(NO_3)_2$ aqueous solution for ion exchange. And the loading of K and Co_3O_4 can be controlled based on the exchange time and the concentration of the Co precursor solution.

It was believed that reactions (4.18) and (4.19) occurred during the immersing of $K_2Ti_2O_5$ in Co^{2+} precursor solution. After thermal pretreatment, the $K_x(H_3O)_{2-x}Ti_2O_5$ was converted to $K_xTi_2O_5$, and $Co(OH)_2$ deposited on surface sites was decomposed to form Co_3O_4.

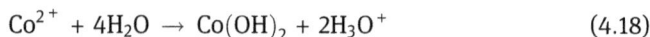

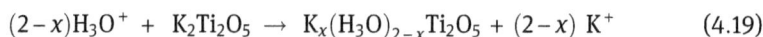

$$Co^{2+} + 4H_2O \rightarrow Co(OH)_2 + 2H_3O^+ \tag{4.18}$$

$$(2-x)H_3O^+ + K_2Ti_2O_5 \rightarrow K_x(H_3O)_{2-x}Ti_2O_5 + (2-x)\,K^+ \tag{4.19}$$

As shown in Figure 4.11, the catalytic activity of $Co_3O_4/K_xTi_2O_5$ for NO oxidation was much higher than that of Pt-based catalysts. When GHSV was maintained at 20,000 h^{-1}, the maximum NO conversion of $Co/K_xTi_2O_5$ was 86%, which was much higher than 41% of Pt/Al_2O_3 or 59% of Pt/TiO_2. The maximum conversion of 86% was achieved at 280 °C, and the temperature was much lower than the temperatures of 420 °C obtained for Pt/Al_2O_3 and 340 °C for Pt/TiO_2. This catalyst also showed very stable performance in a long-term test. In another report, Shang et al. [80] investigated the existing states of potassium species in K-doped Co_3O_4 catalysts and found that there were two kinds of K species existing on Co_3O_4, namely, the "free" K species

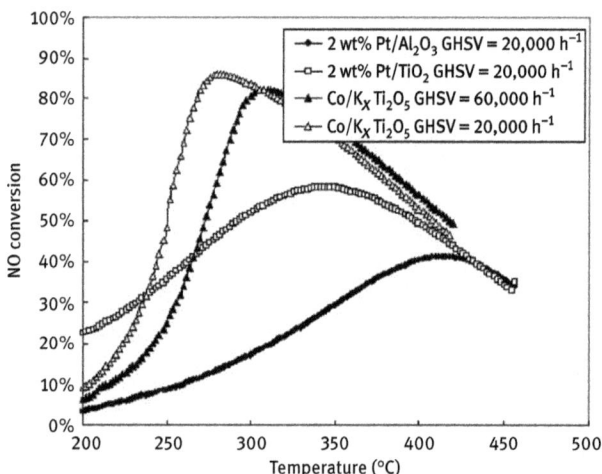

Figure 4.11: NO conversions to NO_2 over $Co/K_xTi_2O_5$, 2 wt.% Pt/Al_2O_3 and 2 wt.% Pt/TiO_2. Feed gas: 700 ppm NO, 10% O_2 in He, 300 mL·min^{-1}; temperature 200–420 °C. Reprinted from Ref. [15], Copyright 2008, Elsevier B.V.

presenting as carbonates and nitrates with high mobility and the "stable" K species entering the lattice or strongly interacting with Co$_3$O$_4$ on the surface or subsurface. The present of "free" K species could improve the contact state of catalyst soot and thus accelerate soot oxidation significantly, but it also covered the active surface of Co$_3$O$_4$ severely and was adverse for NO oxidation. However, the "stable" K species increased the surface Co^{3+} content and accelerated the formation of oxygen vacancies and O$^-$/O^{2-} species, thereby enhancing the intrinsic activities of Co$_3$O$_4$ for soot and NO oxidation.

4.4.2 Manganese oxide-based catalysts

In addition to cobalt-based catalysts, manganese-based catalysts have received increasing attention due to their high activity and low cost. They have been extensively studied due to their good redox properties, low environmental toxicity, and their use in many catalytic oxidation reactions, such as the oxidation of CH$_4$, CO, NO$_x$, and VOCs [9, 81–84]. Many literatures have also reported that Mn-based catalysts, such as Mn-based composite oxide [13, 85, 86], Mn-containing perovskites [87, 88], and supported MnO$_x$ catalysts [89, 90], exhibit good catalytic properties for NO oxidation. In the following parts, we will summarize several major factors affecting the activity of Mn-based catalysts (effect of preparation method, effect of valences states, and effect of crystal structure).

Effect of preparation method

Mn-based catalysts have attracted great attention due to their excellent activities. Many works have been performed on preparation parameters, such as the Mn precursor and the calcination temperature. Park et al. [91] compared the effect of Mn precursor on the surface Mn species, as shown in Figure 4.12. It was found that the low-temperature NO oxidation properties could be improved when manganese acetate precursor was used. According to the analysis results, it can be concluded that the increased in activity was attributed to the fact that the catalyst prepared form the manganese acetate precursor has a well-dispersed Mn species and higher surface area. Abundant Mn^{3+} species, hydroxyl groups, increased oxygen linking to Mn, and high concentrations of amorphous Mn on the surface resulted in an increased NO oxidation activity at lower temperatures.

Chen et al. [92] investigated the influence of calcination temperature on the NO oxidation ability over Mn-based catalysts, the results are presented in Figure 4.13. The NO oxidation activity of the catalysts did not show a significant difference when the calcination temperature was lower than 400 °C; the maximum NO conversion reached 93% at 220 °C; and the catalysts exhibited excellent low-temperature

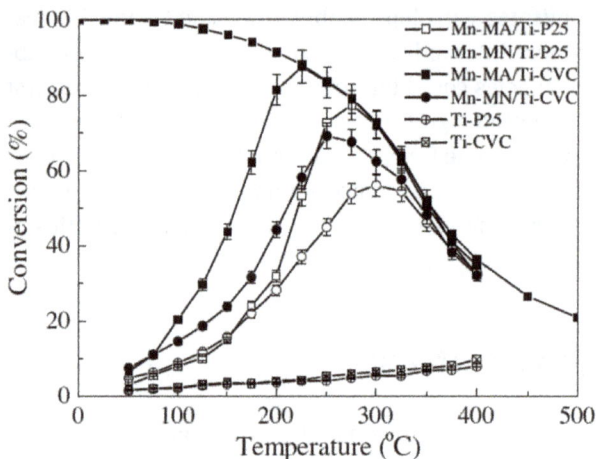

Figure 4.12: NO conversion over TiO$_2$ and Mn/TiO$_2$ nanocomposites. Reprinted from Ref. [91], Copyright 2012, Elsevier Inc.

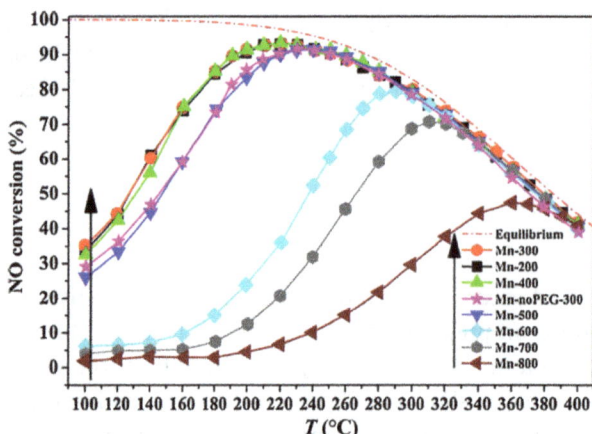

Figure 4.13: NO conversion over different Mn-based catalysts. Reaction conditions: 500 ppm NO, 5% O$_2$, N$_2$ balance, GHSV = 48,000 mL·g^{-1}·h^{-1}. Reprinted from Ref. [92], Copyright 2018, Elsevier B.V.

activity with the conversion of above 50% at 130 °C. When the calcination temperature was 500 °C, the reaction temperature at which the maximum NO conversion was reached was increased to 240 °C, and when the test temperature was lower than 200 °C, the activity was lowered by 10%. After continuing to increase the calcination temperature to 600 °C or higher, the NO conversion decreased rapidly. After calcining the catalyst at 800 °C, the maximum NO conversion at 360 °C was just 47%. It was apparent that catalysts calcined at high temperature (600, 700, and 800 °C) exhibited worse catalytic activity than catalysts calcined at low temperature (200, 300, and 400 °C). From XPS analysis results, it was found that when the calcination temperature was lower than 500 °C, the Mn^{4+} species that promote oxidation of NO to NO_2 could be detect on the surface of the catalyst. Meanwhile, an increase in the calcination temperature lead to a decrease in the O_α/O_β ratio. Therefore, it was concluded that the catalytic activity is closely related to the surface concentration of O_α species. In another report, Meng et al. [86] also reported the same trend.

Effect of valence states

As we know, manganese has different valence states (MnO_2, Mn_2O_3, and Mn_3O_4), and due to the variability and uncontrollability of the valence state of Mn species in the Mn-based catalysts, it is difficult to clarify their roles in NO catalytic oxidation. In a very recent report, to find the difference in NO oxidation activities, Wang et al. [12] prepared a series of Mn-based catalysts with different valences (MnO_2, Mn_2O_3, and Mn_3O_4) by a hydrothermal method, and systematically investigated their NO oxidation performance. As presented in Figure 4.14(a), MnO_2 showed better NO oxidation performance (about 91.4% at 250 °C) than Mn_2O_3 (about 89.9% at 250 °C) and Mn_3O_4 (about 69.0% at 300 °C). In the low temperature range of 100–250 °C, the NO conversion sequence of the three catalysts at the same temperature is as follow: MnO_2 > Mn_2O_3 > Mn_3O_4, which indicates that the Mn^{4+} ion had the highest NO oxidation activity. And then the effects of NO concentrations, O_2 contents, and GHSV were also performed over the three catalysts (Figure 4.14(b), (c) and (d)). The results showed that the effects of NO concentration and O_2 content on MnO_2 catalytic activity are relatively small, and MnO_2 exhibited the most excellent catalytic performance in a wide range of GHSV among the three catalysts.

Effect of crystal structure

According to the above description, Mn^{4+} is more active species for NO oxidation. However, considering the different crystal structure of MnO_2, such as α-MnO_2, β-MnO_2, and γ-MnO_2 with tunnel structure and δ-MnO_2 with layer structure, it is difficult to choose proper MnO_2 as a NO oxidation catalyst. Many researchers have reported the applications of different MnO_2 crystal structure in different reaction processes, such as the deep oxidation of propane, VOCs degradation, and CO oxidation [93–95]. In order to clarify the relationship between catalytic performance and crystal phase or channel structure of MnO_2 for NO oxidation, Gao et al. [96] successfully prepared α-MnO_2 and β-MnO_2 nanorods, and urchin-like γ-MnO_2 catalysts with different tunnel structures by a hydrothermal method and investigated their NO conversion ability. The results are

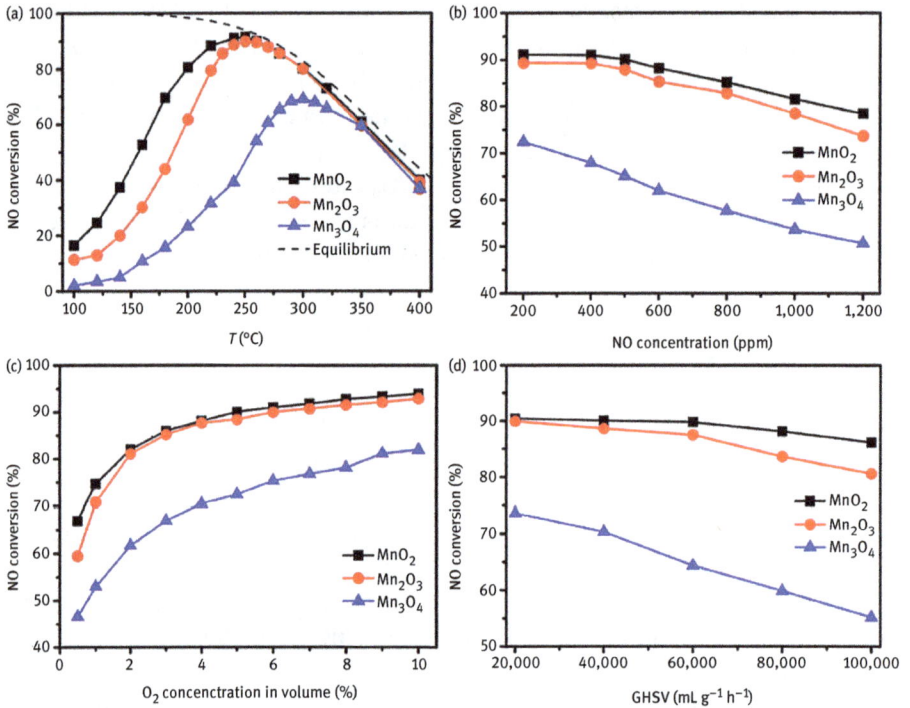

Figure 4.14: (a) NO conversions of the three catalysts. Reaction conditions: 500 ppm NO, 5 vol.% O_2, N_2 balance and GHSV = 48,000 mL·g^{-1}·h^{-1}. (b) NO conversions of the three catalysts at different NO concentrations. (c) NO conversions of the three catalysts at different O_2 concentrations. (d) NO conversions of the three catalysts at different GHSV. Reprinted from Ref. [12], Copyright 2019, Elsevier B.V.

summarized in Figure 4.15. Obviously, the γ-MnO_2 catalyst showed the best catalytic activity among them, with more than 80% NO conversion at 250 °C, and the NO oxidation activities decreased in the order: γ-MnO_2 > β-MnO_2 ≈ α-MnO_2. In another report, Zhao et al. [97] studied the effect of phase structure, morphology on NO catalytic oxidation performance of single-phase MnO_2 catalysts and pointed out the NO oxidation activities decreased in the order γ-MnO_2 > α-MnO_2 > β-MnO_2 > δ-MnO_2.

4.4.3 Chromium-based catalysts

Chromium metal oxides with the Cr^{3+}/Cr^{6+} redox couple are very effective in oxidation reaction process for its strong oxidizing property to absorb the acid gas NO and facilitate complete oxidation [98–102]. The main factors affecting the NO oxidation activity of Cr-based catalysts are the support, chromium loading, and the solvent types. The

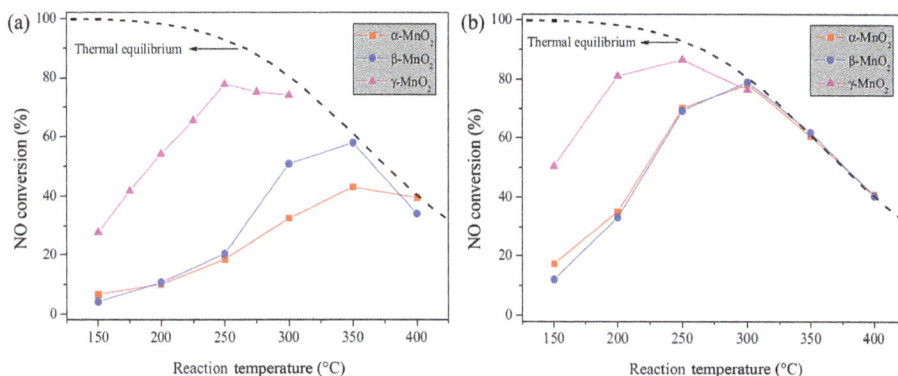

Figure 4.15: NO conversions of α-MnO$_2$, β-MnO$_2$, and γ-MnO$_2$ catalysts for NO oxidation at a GHSV of 48,000 h^{-1} (a) and 24,000 h^{-1} (b) by temperature series. (Reaction conditions: [NO] = 500 ppm, [O$_2$] = 5%, gas feed flow rate = 250 mL·min^{-1}). Reprinted from Ref. [96], Copyright 2017, Elsevier B.V.

most common supports for Cr-based catalysts are CeO$_2$ [102], TiO$_2$ [103], ZrO$_2$ [104], Ce–Ti [98, 99] mixed metal oxide, and Ce–Zr mixed metal oxide [100, 101, 105].

Wang et al. [106] compared the NO oxidation ability of three series of Cr-based mixed oxides (Cr–Co, Cr–Fe, and Cr–Ni) with their pure metal oxide (CrO$_x$, Co$_3$O$_4$, FeO$_x$, and NiO) counterparts, and found that all mixed oxides exhibited improved low concentration NO oxidation activity at room temperature. The best performance catalysts can achieve complete NO conversion in about 30 h at a high space velocity of 45,000 mL·g^{-1}·h^{-1}. The amorphous structure was found to be the key factor to maintaining the high activity and durability of these catalysts. Besides, the Cr loading, nitrate precursor decomposition temperature and catalyst calcination temperature were also key parameters for the synthesis of these highly active catalysts.

Zhang et al. [103] reported chromium oxide as active site over TiO$_2$-pillared clays (TiO$_2$–PILC) for selective catalytic oxidation of NO. It was found that the conversion of NO decreased in the following order: Cr(10)TiP > Cr(12)TiP > Cr(8)TiP > Cr (15)TiP > Cr(5)TiP, and the 10 wt.% chromium doped catalyst exhibited the best catalytic activity. Wang et al. [104] investigated the effect of Cr/Zr atomic ratios on the NO conversion at ambient temperature, as shown in Figure 4.16. Compared to the Zr-300 catalyst, all Cr-containing catalysts showed complete initial NO conversions. However, as the reaction time increased, significant differences were observed: NO oxidation activity increased first with increasing zirconium content and Cr8Zr1-300 was found to have the best catalytic performance. Further increases in Zr loading resulted in a decrease in catalytic performance.

Besides, the authors also studied the influence of calcination temperature on the NO oxidation performance, and the results are presented in Figure 4.17. And the

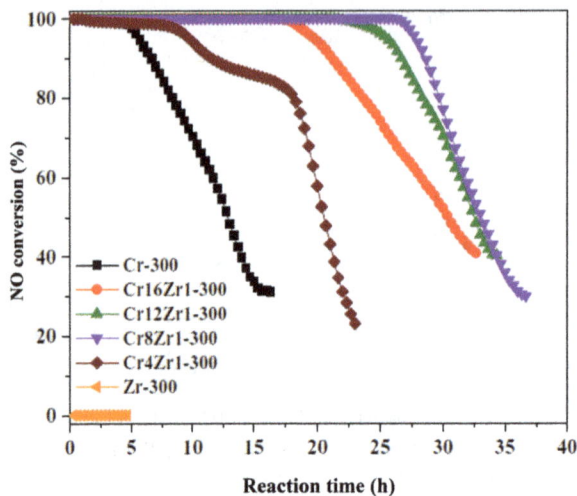

Figure 4.16: NO conversion as a function of reaction time on Cr$_x$Zr$_y$-300 and Cr (or Zr)-300. The reactant feed contains 12 ppm of NO, 21% of O$_2$ balanced with Ar at a GHSV of 45,000 mL·g^{-1}·h^{-1} and a reaction temperature of 25 °C. Reprinted from Ref. [104], Copyright 2016, Elsevier B.V.

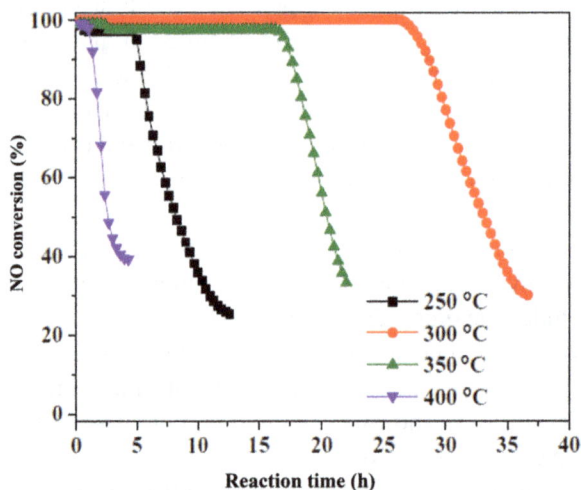

Figure 4.17: NO conversion as a function of reaction time on Cr8Zr1 catalysts calcined at different temperatures. The reactant feed contains 12 ppm of NO, 21% of O$_2$ balanced with Ar at a GHSV of 45,000 mL·g^{-1}·h^{-1}, and a reaction temperature of 25 °C. Reprinted from Ref. [104], Copyright 2016, Elsevier B.V.

calcination temperature has a significant effect on the catalytic performance. Among them, when the calcination temperature was 300 °C, Cr8Zr1-300 exhibited the highest activity, and its NO conversion can be maintained at 100% for 26 h. Higher or lower calcination temperature will result in worse catalytic activity. According to the XRD and BET analysis results, the authors found that with the calcination temperature higher than 400 °C, the sharp Cr$_2$O$_3$ XRD peaks could be clearly observed. At the same time, with increasing calcination temperature, the BET surface areas of the mixed-oxides first increase, reaching the highest value at 300 °C, and then decrease. It can be summarized that the best performance for Cr8Zr1-300 is related to its amorphous structures and highest surface area.

For NO$_x$ removal reaction, specific morphology could enhance the catalytic activity at low temperature [100–102]. Therefore, the selection of a suitable solvent can be an effective means of controlling the morphology or surface properties of the catalyst during the synthesis processes. Cai et al. [100–102] have done a lot of work investigating the effect of solvents on the NO conversion of catalysts. First, the Cr-modified Ce$_{0.2}$Zr$_{0.8}$O$_2$ catalyst was prepared using cinnamic acid as ligand, and four different solvents (water, ethanol, ethylene glycol, and isopropanol) were added into the precursor solution for comparison [100]. As shown in Figure 4.18(a), all catalysts exhibited a broad active temperature window, and the catalyst synthesized with ethylene glycol as solvent presented the highest NO conversion, reaching 66.3% at 350 °C. Based on the analysis results, it was found that one-step preparation can simplify the experimental procedure, resulting in a reduction in catalyst particle size and an increase in catalysts specific surface area. Furthermore, the solvent with a suitable polarity can reduce the band gap of the

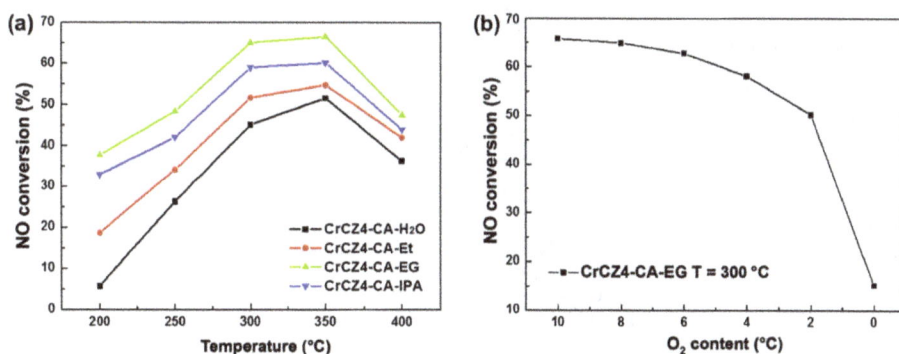

Figure 4.18: (a) Catalytic performance of CrCZ4–CA catalysts with different solvents. Reaction conditions: 390 ppm NO, 8% O$_2$, N$_2$ as balance gas, GHSV = 35,400 h^{-1}; (b) the effect of O$_2$ concentration on NO conversion over CrCZ4-CA-EG. Reaction conditions: 390 ppm NO, N$_2$ as balance gas, GHSV = 35,400 h^{-1}, T = 300 °C. Reprinted from Ref. [100], Copyright 2014, Elsevier B.V.

catalyst, which is favorable for forming more Ce^{3+} and Cr^{6+} on the catalyst surface, thereby improving the adsorption ability of NO and O_2.

Moreover, the influence of O_2 concentration on NO oxidation of CrCZ4–CA-EG catalyst was also measured, and the results were presented in Figure 4.18(b). O_2 concentration showed a significant effect on catalyst performance. Specifically, when the O_2 concentration is in the range of 4–10%, the NO conversion rate was relatively stable, and when the O_2 concentration is 2%, the conversion rate was slightly reduced. Further reducing the O_2 concentration from 2% to 0, the NO conversion rate dropped sharply from 50.2% to 15.2%. The possible reason is that O_2 can increase the ability of the catalyst to adsorb NO, thereby reducing the NO concentration of the gas outlet after the reaction.

In another report, Cai et al. [101] concluded that the structural properties of Cr/$Ce_{0.2}Zr0.8O_2$ were determined by the $Ce_{0.2}Zr_{0.8}O_2$ support, and that the solvents were found to have an effect on the surface properties. The $Ce_{0.2}Zr_{0.8}O_2$ support prepared by using ethylene glycol as the solvent has a smaller particles sizes, a higher surface area, and more surface Ce^{3+} species. Moreover, the present of H_2O during the impregnation process could increase the dispersion of Cr species. Activity tests confirmed that the surface Ce^{3+} and Cr^{6+} ratios are the key factors for NO oxidation.

In a recent report, Cai et al. [102] successfully synthesized 3D spherical Cr–Ce-mixed oxides by the hydrothermal method in different solvents and evaluated their efficiency in catalyzing the oxidation of NO. As shown in Figure 4.19, among all catalysts, CrCeEt (ethanol was used as the solvent) exhibited the best low-temperature NO oxidation performance, which was over 40% at 200 °C and reached 50% at 230 °C. It was found that the particle size and roughness of the catalyst

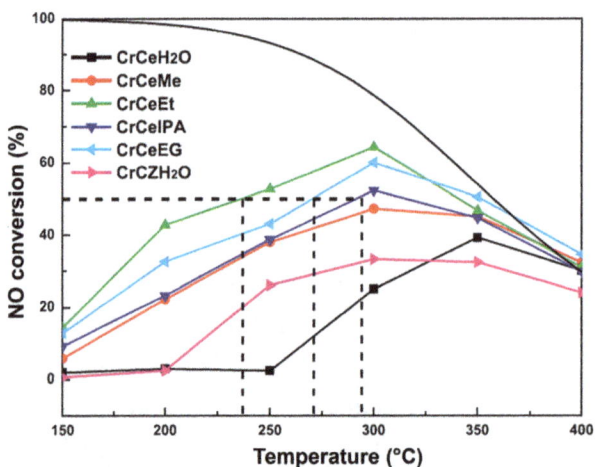

Figure 4.19: Catalytic oxidation of NO over the CrCe(x) and CrCZH$_2$O catalysts. Reprinted from Ref. [102], Copyright 2017, Elsevier B.V.

surface can increase the exposed active sites, thereby improving the catalytic performance at low temperatures. Further, the exposed surface could generate more Cr^{6+} and O_β ratios, which played the key role in NO oxidation. Moreover, CrCeEt catalyst exhibited stable catalytic performance and strong water resistance.

4.4.4 Copper-based catalysts

As reported in many literatures, copper-based catalysts can also be utilized for the NO oxidation to NO_2, using different supports, such as zeolites [107–110], Al_2O_3 [111, 112], TiO_2 [112], MnO_2 [113], perovskites [114–116], and also ceria [117, 118]. Generally speaking, these studies indicate that copper-based catalysts improve the activity toward the NO oxidation to NO_2, even at low temperatures.

López-Suárez et al. [111] reported the effect of copper loading on the NO oxidation performance over Cu/Al_2O_3 catalysts. The results showed that the amount of the most active Cu(II) species that can accelerate NO conversion increases with the copper loading from 1 wt.% to 5 wt.% and remains almost constant for higher copper loading. Similar results were also verified by Akter et al. [107] Moreover, Zhou et al. [116] thoroughly investigated the influence of Cu loading on NO oxidation performance of hexagonal phase of $LaCoO_3$ ($LaCo_{1-x}Cu_xO_3$ (x = 0.1, 0.2, 0.3)) catalysts. As shown in Figure 4.20, the catalytic activities of all prepared catalysts followed the order of $LaCo_{0.9}Cu_{0.1}O_3$ > $LaCoO_3$ > $LaCo_{0.8}Cu_{0.2}O_3$ > $LaCo_{0.7}Cu_{0.3}O_3$, and the $LaCo_{0.9}Cu_{0.1}O_3$ catalyst presented the highest NO conversion at 310 °C, about 82%.

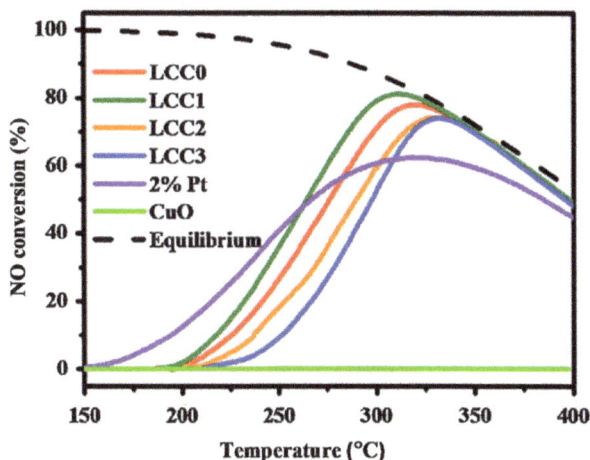

Figure 4.20: The conversion of NO oxidation as a function of temperature and composition: 400 ppm NO, 10% O_2, and a balance of N_2. Republished from Ref. [116], Copyright 2014, Royal Society of Chemistry.

A decreased catalytic performance can be observed by increasing Cu doping content, which can be attributed to the production of isolated CuO on the surface of the oxides. Based on the first principle calculations, the reaction mechanism of NO oxidation on the surface was also proposed. Cu doping could promote the reaction by reducing the energy of oxygen vacancy formation and the NO_2 desorption barrier of Co- or Cu-nitrite.

Albaladejo-Fuentes et al. [114] compared the effects of synthesis method (SG or hydrothermal) on NO oxidation activity over $BaTi_{0.8}Cu_{0.2}O_3$ catalysts. And the hydrothermal catalyst was found to exhibit a high NO to NO_2 oxidation activity, which can be attributed to the highly dispersed CuO species formed on the surface of the catalyst. Lin et al. [117] reported the effect of barium as a promoter on the performance of the catalyst, and found that Ba restrains sintering of $(Cu, Ce)O_x$ after hydrothermal aging of the catalyst at 800 °C for 10 h, resulting in relatively more favorable redox properties and thus higher activity for NO oxidation. In another report, Shi et al. [113] prepared Cu-modified MnO_2 catalysts with improved activity and H_2O resistance for NO oxidation, as presented in Figure 4.21. H_2O exhibited a negative effect on all catalysts activities. While, for CuMnpH6 and CuMnpH8 catalysts, the NO oxidation performance was only slightly lowered, about 60% of the NO oxidation efficiency was obtained at 350 °C. As for the CuO and α-MnO_2 catalysts, the activity decreased significantly compared to other catalysts. At 300 °C, the NO conversion of the α-MnO_2 catalyst was 27% much lower than 82% in the absence of H_2O. This indicated that H_2O can significantly inhibit the NO oxidation activity of CuO and α-MnO_2 catalysts. Based on the analysis results, it can be

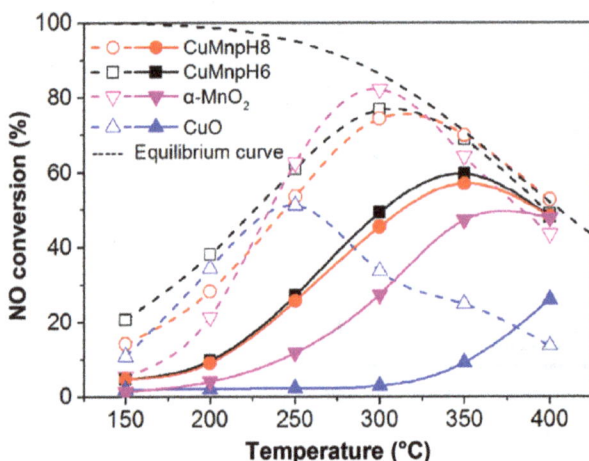

Figure 4.21: NO conversion efficiency over CuO, α-MnO_2, and CuMn catalysts in the absence (dashed lines) and presence (solid lines) of 5% H_2O. Reprinted from Ref. [113], Copyright 2018, American Chemical Society.

concluded that the basicity of α-MnO$_2$ was weakened after the addition of Cu, which might inhibit the adsorption of H$_2$O on the surface oxygen atoms, thereby promoting the adsorption of NO on the CuMnpH6 catalyst.

4.4.5 Other metal oxides

Besides Co$_3$O$_4$, some mixed metal oxides including CeO$_2$–ZrO$_2$ [119], CeO$_2$–MnO$_x$ [120], FeO$_x$–MnO$_x$ [121], MnO$_x$/TiO$_2$ [122], Fe-MFI, and Fe-ferrierite [123] have also been tested for NO oxidation. Among them, CeO$_2$–MnO$_x$ showed a very interesting performance, with NO oxidation activity at a very low-temperature range, 100–200 °C. This catalyst achieved a NO conversion of 60% at 150 °C [120]. In another report, Wu et al. [122] prepared various MnO$_x$/TiO$_2$ composite nano-oxides by deposition–precipitation (DP) method, and compared the activity with samples prepared by conventional wet-impregnation (WI) method. The sample with the Mn/Ti ratio of 0.3 showed excellent NO oxidation activity. For MnO$_x$(0.3)/TiO$_2$(DP) catalyst, the maximum NO conversion could reach 89% at 250 °C, which was much higher than the catalyst prepared by conventional WI method (69% at 330 °C). The higher activity of MnO$_x$(0.3)/TiO$_2$(DP) was attributed to the enrichment of well-dispersed MnO$_x$ on the surface and the abundance of Mn^{3+} species. In addition, DRIFT studies and long-time running experiments indicated that NO$_2$ was derived from the decomposition of adsorbed nitrogen-containing species. There is also one paper reporting La$_{1-x}$Ce$_x$CoO$_3$ perovskite as NO oxidation catalyst [124]. It was reported that the prepared La$_{1-x}$Ce$_x$CoO$_3$ samples exhibited good activities as regular noble metal catalysts, with the highest conversion of 80% at about 300 °C.

In all, there is no doubt now that metal oxides catalysts, especially Co$_3$O$_4$-based materials, are very promising for NO oxidation to NO$_2$. These materials are very cheap and easy to prepare. Through proper synthesis design, metal oxides can provide even higher oxidation activity than Pt-based catalysts, and the produced NO$_2$ has no inhibiting effect on the materials. Therefore, it is easy to predict that more research works on NO oxidation will be focused on metal oxide catalysts in future. With metal oxides catalysts, at least there are several issues remaining unsolved, for example, the mechanism of NO oxidation to NO$_2$ over TM oxides and the long-term stability in the presence of sulfur. Currently, there are a lot of studies on the mechanism of NO oxidation over Pt-based catalysts, but none on TM oxides catalysts. This mechanism study should be very important to further design more efficient and stable metal oxides catalysts.

In addition to the above-mentioned complex metal oxides catalysts, there are some other types of catalysts with special structure that can also be used for NO oxidation, such as perovskite-based catalysts, carbonaceous-based catalysts and zeolite-based catalysts. Therefore, in the following parts, we will briefly review these three types of catalysts.

4.5 Perovskite-based catalysts

Perovskite-based oxides have been proved to be a promising group of NO oxidation catalysts that have received great attention due to their low cost, good activity, and thermal stability [115, 116, 125–127]. These materials have the general formula ABO_3, in which the A site is a rare-earth or alkaline cation with a 12-fold coordination and the B site is a TM with a sixfold coordination with the oxygen anions [128, 129]. Interestingly, the catalytic redox properties of perovskites can be easily adjusted by substituting a small portion of A or B site with other cations. The A or B site-substituted perovskite based on $LaMnO_3$ or $LaCoO_3$ can highly enhance the activity compared with unmodified ones [88, 115, 116, 127, 130–134].

Chen et al. [130] reported a series of structural modified La_xMnO_3 (x = 0.9, 0.95, 1, 1.05, 1.11) perovskite catalysts by an SG method. $La_{0.9}MnO_3$ sample presented excellent activity for NO oxidation. The analysis results indicated that the lower La content in A site induced the conversion of Mn^{3+} to Mn^{4+} to achieve charge balance and structural stability, thereby accelerating the NO oxidation. Zhou et al. [131] synthesized a series of $La_{1-x}Ba_xCoO_3$ (x = 0, 0.1, 0.2, 0.3) perovskites and tested their NO oxidation performance, as presented in Figure 4.22. The results showed that the introduction of Ba at the A site in $LaCoO_3$ greatly improved the activity for NO oxidation. The best performance with NO conversion of 93% at 265 °C was obtained on $La_{0.9}Ba_{0.1}CoO_3$. The enhanced activity could be due to the highly isolated Co species resulting from Ba substitution. It is also important to emphasize that all Ba-doped $LaCoO_3$ had better NO oxidation performance than that of the classical Pt/Al_2O_3 catalyst. This suggests that this kind of

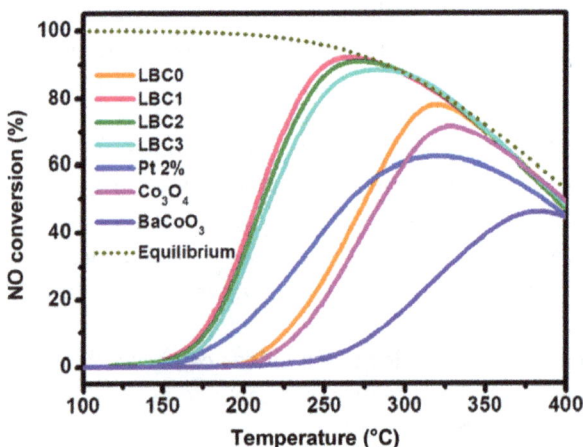

Figure 4.22: NO removal percentage as a function of temperature and composition: 400 ppm NO, 10% O_2, and a balance of N_2. Reprinted from Ref. [131], Copyright 2015, RSC Publishing.

materials could be considered as the potential replacement candidate for noble metal-based catalysts and provides a viable direction for finding materials with low cost and good NO conversion. A similar result was also confirmed by Shen et al. [135]. Besides, Choi et al. [132] investigated the effect of Sr on NO oxidation over La$_{1-x}$Sr$_x$CoO$_3$ perovskite catalysts, and the Sr-substituted catalysts exhibited higher NO oxidation rates than pure LaCoO$_3$. It was pointed out that Sr substitution decreased the amount of NO$_2$ adsorption and enhanced its desorption and improved the total oxygen-exchange capacity.

The B site modification can also be a useful strategy for improving the properties of perovskites-based catalysts. Zhong et al. [133] reported LaNi$_{1-x}$Co$_x$O$_3$ ($x = 0$, 0.1, 0.3, 0.7, 1.0) perovskite catalysts with different Ni/Co ratios and evaluated their catalytic oxidation ability for NO. As presented in Figure 4.23, LaNiO$_3$ showed the lowest activity for the NO oxidation. The addition of cobalt resulted in a substantial increase in the NO oxidation activity and the maximum activity point shifts to lower reaction temperature. The activity of LaNi$_{1-x}$Co$_x$O$_3$ first increased and then decreased with the increase in x value, and the LaNi$_{0.7}$Co$_{0.3}$O$_3$ exhibited the highest activity among all catalysts. According to the analysis results, it can be concluded that the Co doping at B site in LaNiO$_3$ was beneficial for adsorbing more O$_2$ and NO, while Ni was in favor of tuning the adsorbed strength of O$_2$ on the catalysts surface. LaNi$_{0.7}$Co$_{0.3}$O$_3$ displayed the best activity among all samples due to the proper amount of adsorbed O$_2$ and NO and the moderate adsorption strength of O$_2$. In another report, the effect of Co/Mn ratios on NO conversion over LaMnO$_3$ catalysts were also investigated [134]. It was found that partial

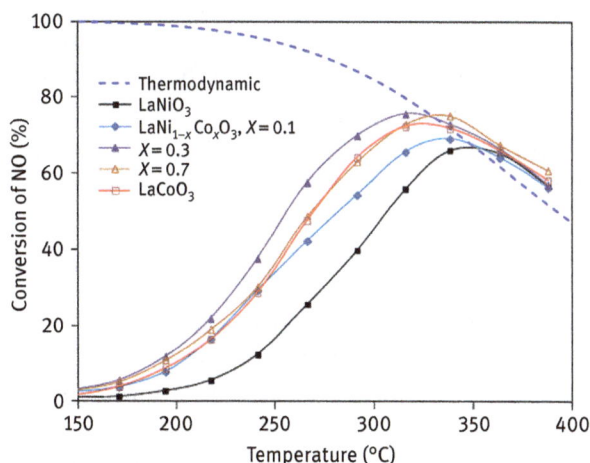

Figure 4.23: NO conversion of LaNi$_{1-x}$Co$_x$O$_3$ for NO oxidation. Reprinted from Ref. [133], Copyright 2015, Elsevier B.V.

substitution of Co in $LaMnO_3$ can improve the stability of Mn^{4+} and activated oxygen species.

Generally, the synthesis methods and synthesis conditions also affect the performance of catalysts. As reported by Wang et al. [136], high surface area mesoporous $LaCoO_3$ oxides (270 $m^2 \cdot g^{-1}$) with well-crystallized perovskite framework were successfully prepared via a simple conanocasting method. The prepared mesoporous $LaCoO_3$ possessed much higher NO oxidation activity due to its high surface area and crystal lattice defects. Zhao et al. [137] reported an acid-etching method for preparing an ultralarge surface area $LaMnO_{3+\delta}$ perovskites. After etching, the NO catalytic oxidation activities and long-term stabilities were significantly improved. The favorable activities of the etched $LaMnO_{3+\delta}$ catalysts were mainly due to their high specific surface area, abundant activated oxygen species and vacancies, as well as the higher valence state of Mn ions resulted from the loss of La^{3+} in the perovskite structure. Similarly, Onrubia et al. [88] used an optimized preparation procedure to enhance textural properties and obtain pure perovskites for optimum NO oxidation performance.

4.6 Carbonaceous-based catalysts

Carbon-based catalysts such as activated carbons (ACs) [138–142], activated carbon fibers (ACFs) [143–147], and carbon xerogels (CXs) [148, 149] have been considered to be ideal catalysts for NO oxidation due to their large surface area, high porosity, catalytic and redox properties, and acid sites. Sousa et al. [139] synthesized a series of ACs with different N loading, in which the ACs treated with aqueous urea solution exhibited the best catalytic activity. It was found that the presence of nitrogen-containing surface groups in the carbon surface could enhance the activity of carbon materials in oxidation reactions. In another paper, Sousa et al. [141] also demonstrated that ACs treated with nitric acid and melamine were the most efficient for NO oxidation, with a conversion of 88%. In a very recent report, the nitric acid hydrothermal treatment of ACs was carried out to adjust pore size, and N-doping resulted in a significant increase in NO oxidation activity at room temperature [140]. After the modification, the catalyst exhibited a more than 400 h stability, which is promising for a continuous NO removal.

In addition, Yu et al. [143] prepared a novel PAN-based carbon nanofibers (PCNFs) by electrospinning using $g\text{-}C_3N_4$ as a sacrificial template for oxidizing NO to NO_2 at room temperature. Compared with PCNFs, the $g\text{-}C_3N_4$-modified catalyst showed much higher activity due to the increased N-containing functional groups, larger pore volume, and higher degree of graphitization. Besides, metal oxides-modified carbon nanofibers were also proposed. As described by Guo et al. [147], $MnO_x\text{-}CeO_2\text{-}Al_2O_3$ mixed oxides (MCAOs) were embedded in PAN precursor to fabricate composite PCNFs (MCAOs–PCNFs) via electrospinning. The ternary complex coupled with PCNFs display remarkable catalytic activities for NO oxidation at room temperature.

CXs have been considered as novel porous carbon materials that can be obtained from carbonization of organic xerogels prepared by SG polycondensation of some specific monomers, such as resorcinol and formaldehyde [148, 149]. Sousa et al. [148, 149] studied the NO oxidation performance of CXs catalyst modified with or without nitrogen. O$_2$ was first adsorbed on the surface of CXs and then reacted with NO to form adsorbed NO$_2$, which played a key role in the NO conversion process. Finally, NO$_2$ can desorb to the gas phase.

4.7 Zeolite-based catalysts

In recent years, zeolite-based catalysts have been shown to be one of the ideal NO oxidation catalysts due to their high and stable activity. Among them, the Fe- or Cu-exchanged zeolites have been extensively investigated [108, 110, 150–153]. In addition, Fe-exchanged zeolite catalysts, in general, are more active than Cu-exchanged zeolite catalysts for NO oxidation to NO$_2$ [154].

Ellmers et al. [150] prepared a series of Fe-ZSM-5 by different methods and investigated the influence of the Fe content on NO oxidation activity. It was found that deliberate introduction of even very low amounts of iron can effectively promote the NO oxidation activity. Ruggeri et al. [152] presented direct evidence for the formation of nitrites/HONO in the oxidative activation of NO over Fe-zeolites. Based on chemical trapping techniques, it can be concluded that the nitrites/HONO are intermediates in the oxidation of NO to NO$_2$. In another report, Ruggeri et al. [108] systematically investigated the mechanism of NO oxidation over a commercial Cu-CHA catalyst by in suit DRIFTS technique. During the experiment, both prereduced and preoxidized catalyst samples were prepared to investigate mechanism of NO oxidation to NO$_2$. The results showed that the NO$^+$ species were the key surface intermediates in the process of NO oxidation to NO$_2$ and nitrates. When the catalyst was exposed to NO$_2$, both NO$^+$ and nitrate are formed. After the catalyst was exposed to NO + O$_2$, the nitrate continuously evolves to NO$^+$, suggesting that nitrite-like species, rather than NO$_2$, were formed as the major products of the NO oxidative activation over Cu-CHA. When the catalyst was only in contact with NO, NO$^+$ and nitrates were formed on a preoxidized sample but not on a prereduced one, which demonstrated the red-ox nature of the NO oxidation mechanism.

Moreover, Metkar et al. [154] compared the NO oxidation properties of Fe- and Cu-zeolite. Based on the steady-state integral measurement results, Fe-zeolite exhibited a more active NO oxidation performance than Cu-zeolite. The NO oxidation reaction was inhibited by NO$_2$, and this inhibition was more pronounced on Cu-chabazite than Fe-ZSM-5 due to the strongly adsorbed NO$_2$ on the Cu-chabazite. Unfortunately, both catalysts exhibited very poor H_2O resistance. Besides the Fe- or Cu-exchanged zeolites, some H- and Na-exchanged zeolites were also investigated. Loiland et al. [155] reported that H- and Na-exchanged zeolite frameworks (BEA, MFI, and CHA) can achieve rapid catalytic NO oxidation rates (up to 1,000 ×) at temperatures above 423 K. According to the experiment results, two possible reaction regimes for the oxidation of NO over H- and Na-zeolites were proposed: Activity increases with decreasing temperature below 423 K, and increases with increasing temperature above 423 K.

In a recent report, Selleri et al. [109] studied the NO oxidation activity of Fe- and Cu-zeolites mixed with BaO/Al_2O_3. It was found that in these combined systems, the initial rate of NO oxidative activation was significantly higher than that of the only metal-promoted zeolite as long as the NO_x storage sites are available, because the inhibition of NO oxidation products (NO_2 or $HONO/N_2O_3$) can be effectively removed from the gas phase and stored in the form of Ba-nitrites.

4.8 Sulfur poisoning

It has been well known that the deactivation by SO_2 poisoning is a serious problem for the application of both Pt-based and metal oxides-based catalysts for NO oxidation. For Pt-based catalysts, quite a few works have been done on Pt/Al_2O_3, Pt/SiO_2, Pt/ZrO_2, Pt/CeO_2, Pt/TiO_2, $Pt-WO_3/TiO_2$, and $Pt-MO_x/Al_2O_3$ (M = W, Mo, V, Ga) [25, 26, 37, 49, 52, 53]. It was reported that the NO oxidation activity of Pt-based catalysts decreased sharply in the presence of SO_2 at ppm level. SO_2 can be oxidized to SO_3 at the Pt sites and then adsorbed on the catalyst to form surface sulfate species. And then the sulfate species were gradually deposited on the catalyst surface, which was the main reason for SO_2 poisoning [25, 37, 49]. For example, to understand the reason why SO_2 deactivated the catalyst during NO oxidation, Ji et al. [49] carried out over DRIFTS measurements over Pt/CeO_2 with a feed gas of NO, O_2, and 20 ppm SO_2. Upon adding SO_2, the intensity of the nitrate bands significantly decreased, and the sulfate species appeared. This is due to the fact that metal sulfates are more stable than metal nitrates and are difficult to decompose. In another work, Li et al. [25] tested the NO conversion to NO_2 over Pt/TiO_2 catalyst in the presence of various amount of SO_2. Results in Figure 4.24 indicate that the SO_2 has a serious adverse effect on NO oxidation. In the presence of 80 ppm SO_2, the maximal NO conversion was reduced from 94% to 79% and the corresponding reaction temperature was raised from 250 °C to 300 °C. Increasing the SO_2 concentration to 320 ppm resulted in severe deactivation of Pt/TiO_2 during NO oxidation. Again, the negative effect of SO_2 was attributed to the oxidation of SO_2 to SO_3. The formed SO_3 may not only occupy the active sites for NO oxidation but also cover the support materials by forming sulfates [25]. Dawody et al. [53] investigated the effect of metal oxide additives (WO_3, MoO_3, V_2O_5, Ga_2O_3) on the oxidation of NO and SO_2 over Pt/Al_2O_3

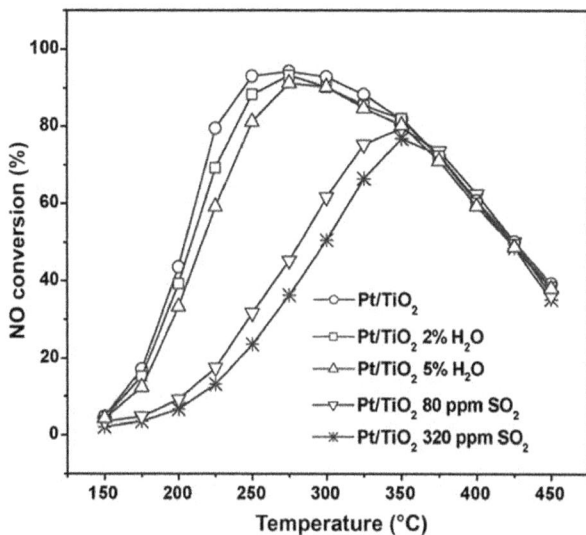

Figure 4.24: Effects of H$_2$O and SO$_2$ on NO oxidation activity over Pt/TiO$_2$. Reaction conditions: 0.15 g catalyst, 400 ppm NO, 10% O$_2$, 0–320 ppm SO$_2$, 0–5% H$_2$O, and the balance He, GHSV = 180,000 h^{-1}. Reprinted from Ref. [25], Copyright 2009, Elsevier B.V.

catalyst. The NO oxidation activity decreased for all catalysts in the presence of SO$_2$, but the MoO$_3$-modified catalyst was less affected by SO$_2$ and also exhibited the lowest SO$_2$ oxidation activity.

However, Olsson et al. [156] recently observed a beneficial effect of SO$_2$ on platinum migration and NO oxidation over Pt containing monolith catalysts (Pt/Al$_2$O$_3$). They evaluated the NO oxidation activity of Pt/Al$_2$O$_3$ by first exposing the catalyst to 630 ppm NO + 8% O$_2$ for 30 min and then to 630 ppm NO + 8% O$_2$ + 30 ppm SO$_2$ for 22 h. At the beginning of the introduction of SO$_2$, the NO oxidation activity was lowered; however, after about 3 h, the NO$_2$ concentration gradually increased, indicating an increase in NO oxidation activity. Measurements before and after the experiment indicated that the Pt dispersion decreased from 12% to 3.5%. And therefore it was suggested that when SO$_2$ was added, the slow increase in activity was due to the slow migration and sintering of Pt. Because the larger Pt particles are more active for NO oxidation.

The similar negative effect of SO$_2$ was also observed on metal oxides-based catalysts [16, 71, 78, 123]. Wang et al. [16] studied the sulfur-poisoning effect on Co/K$_x$Ti$_2$O$_5$, as shown in Figure 4.25. It was found when the SO$_2$ concentration was kept within 10 ppm, there was only a slight decrease in NO conversion, while the catalyst activity decreased appreciably when increasing the SO$_2$ concentration to 16 ppm. Later, Irfan et al. [71] and Kim et al. [78] further confirmed that SO$_2$ can cause an irreversible deactivation of Co$_3$O$_4$-based catalysts even in the presence of

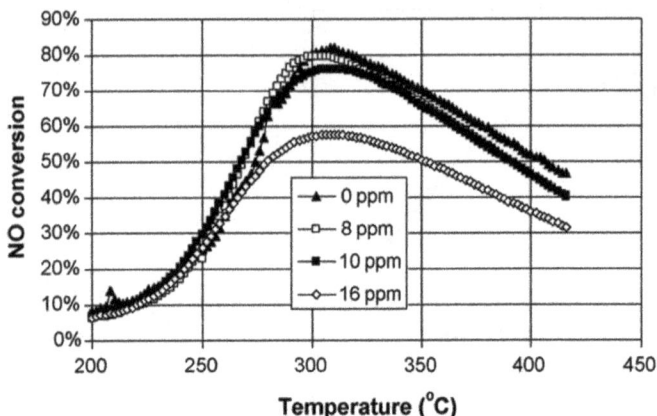

Figure 4.25: Influence of SO_2 on NO conversion over $Co/K_xTi_2O_5$ catalyst as a function of temperature. Feed composition: 700 ppm NO, 10% O_2 in He, GHSV 120,000 h^{-1}. Reprinted from Ref. [16], Copyright 2007, Elsevier B.V.

5 ppm SO_2 in the feed. Giles et al. [123] investigated the effect of SO_2 on Fe-MFI and Fe-ferrierite for NO oxidation. In the presence of 100–500 ppm SO_2, the NO oxidation activity was greatly inhibited. The maximum NO conversion decreased from ca. 68% without SO_2 to 40% with 100 ppm SO_2, with an increase in the optimal temperature from ca. 300 °C to 400 °C. For the metal oxides-based catalysts, apart from the above studies, little information is available on the effects of SO_2 during a long-term operation. Therefore, we believe that more researches are needed to understand not only the SO_2-poisoning mechanism in the NO oxidation process, but also how to design sulfur-tolerant NO oxidation catalysts in the future.

4.9 Practical applications of NO oxidation catalysts

The oxidation of NO to NO_2 plays a key role in many reaction systems, including NO_x abatement using NSR catalysts, the SCR of NO_x using ammonia, urea or hydrocarbon as the reductants, and the oxidation of soot in catalyzed diesel particulate filters (DPFs) [41, 157, 158]. In the following part, we will briefly introduce how the NO oxidation catalysts are integrated into the above technologies.

4.9.1 Practical applications in NSR

NSR is among the most promising De-NO_x technology that is especially designed for vehicle emissions control, in which the diesel or gasoline combustion in the engine

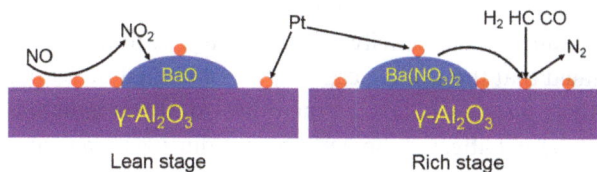

Figure 4.26: The typical mechanism for the NO$_x$ storage and reduction over Pt-Ba/Al$_2$O$_3$ catalyst.

takes place alternately over long oxygen-rich and short fuel-rich periods. As shown in Figure 4.26, a typical NSR catalyst is composed of three parts: (1) a high surface area support, (2) a NO$_x$-storage component, and (3) a noble metal. The NSR technology generally consists of four steps, which include (1) NO oxidation to NO$_2$, (2) NO$_x$ storage, (3) stored NO$_x$ release, and (4) NO$_x$ reduction to N$_2$. During the lean burn stage, NO is first oxidized to NO$_2$ by O$_2$, and then stored in the form of nitrate or nitrite. During the rich burn stage, the excess HC, H$_2$, and CO can react with the stored NO$_x$ in the presence of noble metal catalyst, selectively reducing it to N$_2$ [157, 159]. The whole process can be described by the following reactions (4.20–4.25):
Lean stage:

$$NO(g) + O_2(g) \ \rightarrow \ NO_2\,(g) \tag{4.20}$$

$$NO_2\,(g) + BaO(s) \ \rightarrow \ Ba(NO_3)_2\,(s) \tag{4.21}$$

Rich stage:

$$Ba(NO_3)_2(s) \rightarrow BaO(s) \ + NO/NO_2(g) \tag{4.22}$$

$$NO/NO_2(g) \ + \ H_2(g) \ \rightarrow \ N_2\,(g) \ + \ H_2O(g) \tag{4.23}$$

$$NO/NO_2\,(g) + HC(g) \ \rightarrow \ N_2\,(g) \ + CO_2(g) + H_2O(g) \tag{4.24}$$

$$NO/NO_2(g) + CO(g) \ \rightarrow \ N_2\,(g) + CO_2(g) \tag{4.25}$$

It is apparent that the NO oxidation is a key step for the NSR reactions. It has been proven that the performance of NSR catalyst can be enhanced in the presence of more NO$_2$ [160]. Therefore, the approaches of integrating the NO oxidation catalyst into the NSR catalyst are of great interests to researchers. Generally, there are three methods to further increase the NO oxidation activity, which are (1) controlling the location of Pt [161–163], (2) modifying Pt with other metal oxides, for example, Co$_3$O$_4$, Fe$_2$O$_3$, and so on [164], and (3) replacing Pt directly by other metal oxides, for example, Co$_3$O$_4$, CuO, and so on [165–167]. Büchel et al. [161] investigated the influence of Pt location on BaCO$_3$ or Al$_2$O$_3$ during NO$_x$ storage and reduction. It was found that the Pt on Al$_2$O$_3$ exhibited a better activity for NO oxidation, the limiting step for the overall NO storage process, at low temperatures (300 °C). However, during reduction, Pt on Ba showed much better activity than Pt on Al_2O_3. Storage of NO and NO_2 until saturation revealed that the storage capacity is not limited by Ba but by the loss of Pt oxidation activity during catalyst restructuring. At higher temperature (350 °C), the location of Pt barely affected the performance during storage and reduction. These differences in activity of Pt in contact with Al or Ba can be exploited for the design of NSR catalysts with improved performance at low temperatures [161–163]. For example, in Figure 4.27(a), the Pt-Ba/Al_2O_3 NSR catalyst with Pt locating at the junctions of Al_2O_3 and BaO might show enhanced performance for low temperature NSR process. However, it is apparent that this unique position is very difficult to achieve. Vijay et al. [164] synthesized cobalt-containing NSR catalysts and found that the addition of 5 wt.% Co increased the NO_x storage capacity of a 1 wt.% Rh/15 wt.% Ba catalyst by 50% and that of a 1 wt.% Pt/15 wt.% Ba catalyst by 100%. This promotion was attributed to the high oxidizing ability of Co_3O_4, which provided an additional oxidation sites for the conversion NO to NO_2, and more contact area for NO_2 spillover to Ba storage sites. The introduction of Co_3O_4 can also decrease the loading of noble metal Pt, for instance, it was found that a 0.25 wt.% Pt/5 wt.% Co/15 wt.% Ba-containing catalyst showed better performance than a 1 wt.% Pt/15 wt.% Ba-containing catalyst [164]. In other reports, Vijay et al. [165] reported a noble metal-free NSR catalyst 5 wt.% Co/15 wt.% Ba, which is as efficient as 1 wt.% Pt/15 wt.% Ba. By utilizing Co as an oxidizing metal instead of Pt, the storage capacity of NSR catalysts can be improved without the need for an expensive noble metal oxidizing catalyst. Finally the promotional effect of Co is also attributed to the combination of increased NO to NO_2 oxidation and improved surface area for NO_2 spillover to the Ba storage sites [166]. The Co-containing NSR catalyst was also reported by another group. Milt et al. synthesized Co–Ba–K/CeO_2 as a potential catalyst for simultaneous abatement of soot and NO_x in diesel exhausts. Figure 4.27 shows the three schemes for integrating the NO oxidation step into NSR process.

4.9.2 Practical applications in SCR

The SCR technology is generally recognized as the most effective approach to control the emission of NO_x. In particular, the SCR process with NH_3 is a well-developed and commercially available technology used in the removal of NO_x from stationary sources. Nowadays, many investigations have been done to extend its applications to mobile sources. Besides, NH_3, many other reducing agents including soot, urea, hydrocarbon, and carbon monoxide (CO), and so on have been studied for the SCR technology [159, 168].

Later on, it has been reported by Koebel et al. [18] and Madia et al. [19] that the SCR reaction rate can be significantly increased, when a portion of NO is converted

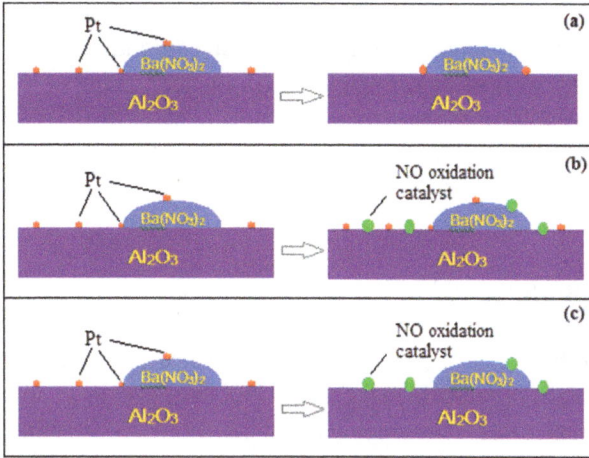

Figure 4.27: Three schemes of integrating the NO oxidation step into NSR process. (a) controlling the position of Pt, (b) modifying Pt with additional NO oxidation catalyst, and (c) directly replacing Pt by other metal oxides catalyst.

to NO$_2$. This effect is most pronounced at lower temperatures (200–300 °C) when the reaction mixture contains equimolar amounts of NO and NO$_2$. This process called *fast SCR process* can be expressed by reaction (4.26) [169]. This implies that the rate of NO$_x$ conversion can be accelerated by use of an oxidation catalyst upstream of the SCR unit, so as to convert ca. 50% of the NO to NO$_2$; this, in turn, enables the SCR catalyst volume to be reduced. The general scheme for integrating the NO oxidation catalyst into SCR process is shown in Figure 4.28.

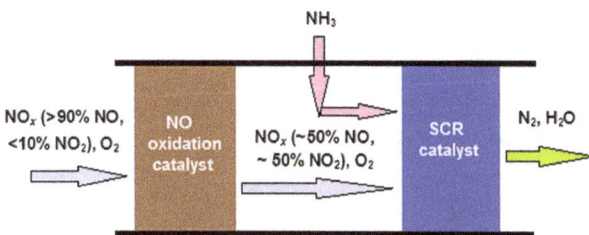

$$2NH_3 + NO + NO_2 \rightarrow 2N_2 + 3H_2O \qquad (4.26)$$

Figure 4.28: The general scheme of integrating NO oxidation catalyst into SCR.

Besides adding one NO oxidation catalyst upstream the SCR, some researchers also tried to modify the SCR catalyst by introducing some metal oxides. For instance, Irfan et al. [170] introduced WO$_3$ into Pt/TiO$_2$ SCR catalyst and noticed that the increase in WO_3 loading on the catalyst increased the NO oxidation more rapidly at lower temperatures. However, it is worthy to mention that when combining the NO oxidation catalyst and the SCR catalyst together, the oxidation/combustion of the reducing agents over the NO oxidation catalyst should be considered. If the NO oxidation catalyst is too oxidative and consumes too much reducing agent, putting the NO oxidation catalyst upstream the SCR process is preferred.

4.9.3 Practical applications in CRT

Particulate matters (PMs) filtration from the diesel engine emissions is gaining increasing attention for both light-duty and heavy-duty vehicles. Passenger cars equipped with DPFs already appeared in the market to meet the low particulate emission standards in Europe. The particulate filter technology is also considered to be the most promising solution for achieving vehicle emissions standards [171].

The diesel filter materials have shown quite high filtration efficiencies, frequently in excess of 90%, as well as acceptable mechanical and thermal durability. Figure 4.29(a) shows one example of DPF. The most important issue with diesel traps is filter regeneration. It is necessary to regenerate the filter, either periodically or continuously, to prevent the accumulation of backpressure of the loaded DPF during regular engine operation.

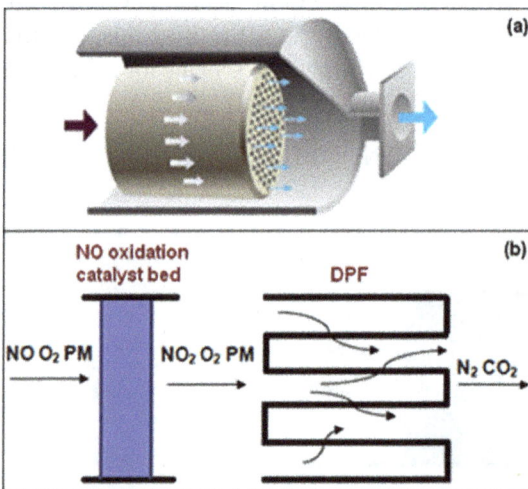

Figure 4.29: (a) One model of DPF. (b) The general working scheme of CRT technology.

CRT is a newly developed technology that combines trapping of soot in a wall-flow monolith, and two chemical processes, that is, catalytic oxidation of NO to NO_2 and the subsequent oxidation of trapped soot with the NO$_2$ produced. The principle of CRT regeneration is due to the fact that diesel PM is easily oxidized by NO$_2$. Carbon in the form of soot can be rapidly oxidized by O$_2$ above 550 °C with noticeable reaction rates. For NO$_2$, the process has begun to occur at 250 °C, which is the temperature that can be encountered in diesel exhaust during normal driving cycles. Therefore, within CRT technology, an efficient NO oxidation catalyst is the key component and it determines the regeneration temperature. The NO oxidation catalyst also functions as catalyst for CO and HC oxidation.

Figure 4.29(b) shows the general scheme of CRT, which consists of an NO oxidation catalyst installed upstream of a typical ceramic wall-flow diesel filter. Apart from converting CO and HC, the main role of the catalyst is to oxidize NO to NO$_2$. The produced NO$_2$ then helps in soot combustion and greatly lowers the ignition temperature from 550 °C to around 250 °C. In the presence of the upstream NO oxidation catalyst, the DPFs could be continuously regenerated. The overall reactions involved in this process include reactions (4.27) and (4.28):

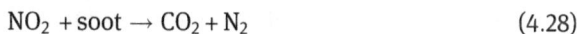

$$NO + O_2 \rightarrow NO_2 \tag{4.27}$$

$$NO_2 + soot \rightarrow CO_2 + N_2 \tag{4.28}$$

4.10 Conclusions and remarks

There is a growing awareness that NO oxidation is a very important prerequisite step for many environmental processes. In this chapter, we provided a detailed introduction to the developed NO oxidation catalysts in literature, the involved reaction mechanisms and reaction kinetics, the sulfur poisoning issues, as well as the practical applications in NSR, SCR, and CRT processes. The NO oxidation catalysts can be roughly classified into two groups according to their based materials: Pt-based catalysts and metal oxides-based catalysts.

With Pt-based catalysts, both E-R and L-H mechanisms were proposed for the NO oxidation reaction. Despite the large number of papers published on the oxidation of NO to NO$_2$, the exact reaction mechanism is still a matter of debate. However, it is now well accepted that the local surface structure as well as the chemical environment can significantly influence the composition of the catalyst surface and the energetics and kinetics of the individual steps. At different reaction conditions or with different supported Pt species, the mechanism might vary. In contrast to Pt-based catalysts, there is almost no mechanism and kinetics studies on metal oxide based catalysts.

With Pt-based catalysts, the NO oxidation activity generally increases with the increase in Pt particle size. The optimal Pt loading is around 2 wt.%. With different catalyst supports, Pt/SiO$_2$ shows the highest activity and the general trend follows the order of Pt/SiO$_2$ > Pt/Al$_2$O$_3$ > Pt/ZrO$_2$. Potassium greatly promotes the NO oxidation activity, for example, the TOR of $Pt/K/Al_2O_3$ is about 2.5 times higher than that of Pt/Al_2O_3. Metal oxides are another type of catalyst promoter. In the absence of SO_2, introducing WO_3 and MoO_3 to Pt/Al_2O_3 significantly increases the NO oxidation activity, while in the presence of SO_2, introducing MoO_3 to Pt/Al_2O_3 enhances the sulfur tolerance of the catalyst. Alloying Pt with other metals (e.g., Ir, Pd, etc.) is believed to be an efficient strategy to improve either the thermal stability or the sulfur resistance of Pt-based catalysts. Although Pt-based catalysts have already been commercially available, its wide application is still limited due to the "self-inhibition" issue caused by the reaction product of NO_2 and the high price of Pt.

(1) With metal oxides-based catalysts, Co_3O_4 is regarded as the most promising candidate catalyst for NO oxidation. With Co_3O_4 as the active species, high NO conversions have been reported with many catalyst supports including CeO_2, SiO_2, TiO_2, and ZrO_2. The optimal loading of Co_3O_4 is around 10 wt.%, and a relatively low calcination temperature (ca. 300 °C) is required for high activity. Directly adding potassium to Co_3O_4-based catalysts showed a negative effect on NO oxidation activity. However, the promoted effect of potassium on Co_3O_4-based catalysts was observed with a novel catalyst $Co/K_xTi_2O_5$ prepared by ion exchange method. Besides Co_3O_4, some mixed metal oxides were also reported as NO oxidation catalysts, for example, CeO_2–ZrO_2, CeO_2–MnO_x, FeO_x–MnO_x, MnO_x/TiO_2, and $La_{1-x}Ce_xCoO_3$. Some of them also showed very promising performance. Based on the facts that the catalytic performances are comparable to Pt-based catalysts, the materials are cheap, and the catalysts are easy to prepare, the authors highly believe that metal oxides-based catalysts will play a great role in NO oxidation catalysis.

(2) With both Pt-based and metal oxides-based catalysts, SO_2 generally shows a negative effect on NO oxidation, and it is attributed to the oxidation of SO_2 to SO_3. The formed SO_3 may not only occupy the active sites for NO oxidation but also cover the support materials by forming sulfates. However, since the presence of SO_2 in the feed helps Pt migration and sintering at low temperatures, SO_2 may show a beneficial effect on NO oxidation under certain conditions. This activity increase is due to the increase of the Pt particles, and it is known that larger Pt particles are more active for NO oxidation. For metal oxides-based catalysts, probably more studies are needed for the understanding of the sulfur-poisoning effect as well as for the future design of sulfur-resistant NO oxidation catalysts.

(3) The NO oxidation catalysts can be integrated into many environmental processes such as NSR, SCR, and CRT to increase the overall conversion of NO_x and/or soot. For all these processes, the NO oxidation catalyst is generally installed upstream of the main converter or added as a component to the main catalysts to fully or partially oxidize NO to NO_2. The current concerns are mainly the activity and stability of the NO oxidation catalysts and the compatibility with the main catalysts.

Acknowledgments: This work is supported by the National Natural Science Foundation of China (U1810209, 51572029, 51622801).

References

[1] Valle S. Palma Del, Marie O., Nguyen H.P., Appl. Catal. B: Environ. 223, 2018, 116–124.
[2] Marques R., Darcy P., Costa P.D., Mellottée H., Trichard J.-M., Djéga-Mariadassou G., J. Mol. Catal. A: Chem. 221, 2004, 127–136.
[3] Yang R., Cui Y., Yan Q., Zhang Q., Qiu L., O'Hare D., Wang Q., Chem. Eng. J. 326, 2017 656–666.
[4] Nguyen H.P., Palma S. Valle S., Marie O., Appl. Catal. B: Environ. 231, 2018,391–399.
[5] Wang Q., Zhu J., Wei S., Chung J., Guo Z., Ind. Eng. Chem. Res., 2010.
[6] Wang Q., Guo Z., Chung J. Shik, Chem Commun (Camb) 2009,5284–5286.
[7] Wang Q., Sohn J.H., Chung J.S., Appl. Catal. B: Environ. 89, 2009.97–103.
[8] Wang Q., Chung J.S., Appl. Catal. A: Gen. 358, 2009, 59–64.
[9] Yan Q., Chen S., Zhang C., Wang Q., Louis B., Appl. Catal. B: Environ. 238, 2018,236–247.
[10] Shah A.N., Ge Y., Tan J., Liu Z., He C., Zeng T., J. Environ. Sci. 24, 2012,1449–1456.
[11] Yamamoto K., Sakai T., Catal. Today 242, 2015,357–362.
[12] Wang H., Chen H., Wang Y., Lyu Y.-K., Chem. Environ. J. 361, 2019,1161–1172.
[13] S.-j. Ma, X.-w. Wang, Chen T., Z.-h. Yuan, Chem. Eng. J. 354, 2018,191–196.
[14] Yuan H., Chen J., Wang H., Hu P., ACS Catal. 8, 2018,10864–10870.
[15] Wang Q., Park S.Y., Duan L., Chung J.S., Appl. Catal. B: Environ. 85, 2008,10–16.
[16] Wang Q., Park S.Y., Choi J.S., Chung J.S., Appl. Catal. B: Environ. 79, 2008,101–107.
[17] Sakamoto Y., Motohiro T., Matsunaga S., Okumura K., Kayama T., Yamazaki K., Tanaka T., Kizaki Y., Takahashi N., Shinjoh H., Catal. Today 121, 2007,217–225.
[18] Koebel M., Madia G., Elsener M., Catal. Today 73, 2002,239–247.
[19] Madia G., Koebel M., Elsener M., Wokaun A., Ind. Eng. Chem. Res. 41, 2002,3512–3517.
[20] Walker A.P., Top. Catal. 28, 2004,165–170.
[21] Burch R., Fornasiero P., Watling T.C., J. Catal. 176, 1998,204–214.
[22] Després J., Elsener M., Koebel M., Kröcher O., Schnyder B., Wokaun A., Appl. Catal. B: Environ. 50, 2004,73–82.
[23] Bai Y., Bian X., Wu W., Appl. Surf. Sci. 463, 2019,435–444.
[24] Crocoll M., Kureti S., Weisweiler W., J. Catal. 229, 2005,480–489.
[25] Li L., Shen Q., Cheng J., Hao Z., Appl. Catal. B: Environ. 93, 2010,259–266.
[26] Li L., shen Q, Cheng J., Hao Z., Catal. Today 158, 2010, 361–369.
[27] Mulla S.S., Chen N., Delgass W.N., Epling W.S., Ribeiro F.H., Catal. Lett. 100, 2005, 267–270.
[28] Mulla S.S., Chen N., Cumaranatunge L., Blau G.E., Zemlyanov D.Y., Delgass W.N., Epling W. S., Ribeiro F.H., J. Catal. 241, 2006, 389–399.
[29] Olsson L., Westerberg B., Persson H., Fridell E., Skoglundh M., Andersson B., Phys J.. Chem. B 103, 1999, 10433–10439.
[30] Olsson L., Persson H., Fridell E., Skoglundh M., Andersson B., J. Phys. Chem. B 105, 2001, 6895–6906.
[31] Getman R.B., Schneider W.F., Phys J. Chem. C 111, 2007, 389–397.
[32] Ge Q., Neurock M., Am J. Chem. Soc. 126, 2004, 1551–1559.
[33] Xu Y., Getman R.B., Shelton W.A., Schneider W.F., Phys. Chem. Chem. Phys. 10, 2008, 6009–6018.
[34] Smeltz A.D., Getman R.B., Schneider W.F., Ribeiro F.H., Catal. Today 136, 2008, 84–92.

[35]   Roldan Cuenya B., Behafarid F., Surf. Sci. Rep. 70, 2015, 135–187.
[36]   Liu X., Tian D., Ren S., Meng C., J. Phys. Chem. C 119, 2015, 12941–12948.
[37]   Xue E., Seshan K., Ross J.R.H., Appl. Catal. B: Environ. 11, 1996, 65–79.
[38]   Oi-Uchisawa J., Obuchi A., Enomoto R., Liu S., Nanba T., Kushiyama S., 38, Applied Catalysis B: Environmental 26, 2000, 17–24.
[39]   Denton P., Giroir-Fendler A., Praliaud H., Primet M., J. Catal. 189, 2000, 410–420.
[40]   Benard S., Retailleau L., Gaillard F., Vernoux P., Giroir-Fendler A., Appl. Catal. B: Environ. 55, 2005, 11–21.
[41]   Olsson L., Fridell E., J. Catal. 210, 2002, 340–353.
[42]   Weiss B.M., Iglesia E., J. Phys. Chem. C 113, 2009, 13331–13340.
[43]   Lee J.-H., Kung H.H., Catal. Lett. 51, 1998, 1–4.
[44]   Altman E.I., Gorte R.J., Surf. Sci. 195, 1988, 392–402.
[45]   Liang Y., Ou C., Zhang H., Ding X., Zhao M., Wang J., Chen Y., Ind. Eng. Chem. Res. 57, 2018, 3887–3897.
[46]   Botao Q., Aiqin W., Xiaofeng Y., Allard L.F., Zheng J., Yitao C., Jingyue L., Jun L., Tao Z., Nat. Chem. 3, 2011, 634–641.
[47]   Narula C.K., Allard L.F., Stocks G.M., Moses-DeBusk M., Sci. Rep. 4, 2014, 7238.
[48]   Bourges P., Lunati S., Mabilon G., Proceedings of the fourth international congress on catalysis and automotive pollution control, Brussels, Belgium, 1997.
[49]   Ji Y., Toops T.J., Graham U.M., Jacobs G., Crocker M., Catal. Lett. 110, 2006, 29–37.
[50]   Kuriyama M., Tanaka H., Ito S.-i, Kubota T., Miyao T., Naito S., Tomishige K., Kunimori K., J. Catal. 252, 2007, 39–48.
[51]   Mulla S.S., Chen N., Cumaranatunge L., Delgass W.N., Epling W.S., Ribeiro F.H., Catal. Today 114, 2006, 57–63.
[52]   Irfan M.F., Goo J.H., Kim S.D., Int. J. Chem. Kinet. 38, 2006, 613–620.
[53]   Dawody J., Skoglundh M., Fridell E., Mol J.. Catal. A: Chem. 209, 2004, 215–225.
[54]   Irfan M.F., Goo J.H., Kim S.D., Hong S.C., Chemosphere 66, 2007, 54–59.
[55]   Tang H., Trout B.L., J. Phys. Chem. B 110, 2006, 6856–6863.
[56]   Corro G., Elizalde M.P., Velasco A., React. Kinet. Catal. Lett. 76, 2002, 117–122.
[57]   Jelic J., Meyer R.J., Catal. Today 136, 2008, 76–83.
[58]   Morlang A., Neuhausen U., Klementiev K.V., Schütze F.W., Miehe G., Fuess H., Lox E.S., Appl. Catal. B: Environ. 60, 2005, 191–199.
[59]   Graham G.W., Jen H.W., Ezekoye O., Kudla R.J., Chun W., Pan X.Q., McCabe R.W., Catal. Lett. 116, 2007, 1–8.
[60]   Kaneeda M., Iizuka H., Hiratsuka T., Shinotsuka N., Arai M., Appl. Catal. B: Environ. 90, 2009, 564–569.
[61]   Auvray X., Olsson L., Appl. Catal. B: Environ. 168–169 2015, 342–352.
[62]   Hernández-Fernández J., Aguilar-Elguezabal A., Castillo S., Ceron-Ceron B., Arizabalo R.D., Moran-Pineda M., Catal. Today 148, 2009, 115–118.
[63]   Qu L., Li J., Hao Z., Li L., Catal. Lett. 131, 2009, 656–662.
[64]   Li L., Qu L., Cheng J., Li J., Hao Z., Appl. Catal. B: Environ. 88, 2009, 224–231.
[65]   Weiss B.M., Artioli N., Iglesia E., ChemCatChem 4, 2012, 1397–1404.
[66]   Deka R.C., Bhattacharjee D., Chakrabartty A.K., Mishra B.K., RSC Adv. 4, 2014, 5399.
[67]   Adjimi S., García-Vargas J.M., Díaz J.A., Retailleau L., Gil S., Pera-Titus M., Guo Y., Giroir-Fendler A., Appl. Catal. B: Environ. 219, 2017, 459–466.
[68]   Wang Q., Chung J.S., Proceedings of the Korean Environmental Sciences Society Conference, Jeju Island, South Korea, 2006.
[69]   Wang Q., Park S.Y., Chung J.S., 11th Korea-Japan symposium on catalysis, Seoul, South Korea, 2007.

[70] Yung M.M., Holmgreen E.M., Ozkan U.S., J. Catal. 247, 2007, 356–367.
[71] Irfan M.F., Goo J.H., Kim S.D., Appl. Catal. B: Environ. 78, 2008, 267–274.
[72] Kim D.S., Kim Y.H., Yie J.E., Park E.D., J Korean. Chem. Eng. 27, 2010, 49–54.
[73] Wang H., Wang J., Wu Z., Liu Y., Catal. Lett. 134, 2010, 295–302.
[74] Huang Y., Gao D., Tong Z., Zhang J., Luo H., J. Nat. Gas Chem. 18, 2009, 421–428.
[75] Kumar Megarajan S., Rayalu S., Teraoka Y., Labhsetwar N., J. Mol. Catal. A: Chem. 385, 2014, 112–118.
[76] Yu Y., Zhong L., Ding J., Cai W., Zhong Q., RSC Adv. 5, 2015, 23193–23201.
[77] Shang D., Zhong Q., Cai W., J. Mol. Catal. A: Chem. 399, 2015, 18–24.
[78] Kim D.S., Kim Y.H., Yie J.E., Park E.D., J Korean . Chem. Eng. 27, 2010, 822–827.
[79] Qiu L., Wang Y., Pang D., Ouyang F., Zhang C., Cao G., Catalysts 6, 2016, 9.
[80] Shang Z., Sun M., Che X., Wang W., Wang L., Cao X., Zhan W., Guo Y., Guo Y., Lu G., Catal. Sci. Technol. 7, 2017, 4710–4719.
[81] Stobbe E.R., Boer D., Geus J.W., Catal. Today 47, 1999, 161–167.
[82] B. Bai, J. Li, J. Hao, Appl. Catal. B: Environ. 164, 2015, 241–250.
[83] Ren Y., Ma Z., Dai S., Materials (Basel) 7, 2014, 3547–3556.
[84] Saputra E., Muhammad S., Sun H., Ang H.-M., Tadé M.O., Wang S., Appl. Catal. B: Environ. 154–155 2014, 246–251.
[85] Wang Z., Lin F., Jiang S., Qiu K., Kuang M., Whiddon R., Cen K., Fuel 166, 2016, 352–360.
[86] Meng L., Wang J., Sun Z., Zhu J., Li H., Wang J., Shen M., J. Rare Earth. 36, 2018, 142–147.
[87] Thampy S., Zheng Y., Dillon S., Liu C., Jangjou Y., Lee Y.-J., Epling W.S., Xiong K., Chabal Y.J., Cho K., Hsu J.W.P., Catal. Today 310, 2018, 195–201.
[88] Onrubia J.A., Pereda-Ayo B., De-La-Torre U., González-Velasco J.R., Appl. Catal. B: Environ. 213, 2017, 198–210.
[89] Zhao B., Ran R., Wu X., Weng D., Wu X., Huang C., Catal. Commun. 56, 2014, 36–40.
[90] You F.-T., Yu G.-W., Wang Y., Xing Z.-J., Liu X.-J., Li J., Appl. Surf. Sci. 413, 2017, 387–397.
[91] Park E., Chin S., Jeong J., Jurng J., Micropor. Mesopor. Mat. 163, 2012, 96–101.
[92] Chen H., Wang Y., Lyu Y.-K., Mol. Catal. 454, 2018, 21–29.
[93] Xie Y., Yu Y., X. Gong, Y. Guo, Y. Guo, Wang Y., Lu G., CrystEngComm 17, 2015, 3005–3014.
[94] Li Y., Fan Z., Shi J., Liu Z., Shangguan W., Chem. Eng. J. 241, 2014, 251–258.
[95] Liang S., Teng F., Bulgan G., Zong R., Zhu Y., J. Phys. Chem. C 112, 2010, 5307–5315.
[96] Gao F., Tang X., Yi H., Chu C., Li N., Li J., Zhao S., Chem. Eng. J. 322, 2017, 525–537.
[97] Zhao B., Ran R., Wu X., Weng,D., Appl. Catal. A: Gen. 514, 2016, 24–34.
[98] Zhong L., Cai W., Zhong Q., RSC Adv. 4, 2014, 43529–43537.
[99] Zhong L., Cai W., Yu Y., Zhong Q., Appl. Surf. Sci. 325, 2015, 52–63.
[100] W. Cai, Q. Zhong, W. Zhao, Chem. Eng. J. 246, 2014, 328–336.
[101] Cai W., Zhong Q., Ding J., Bu Y., Chem. Eng. J. 270, 2015, 1–8.
[102] Cai W., Zhao Y., Chen M., Jiang X., Wang H., Ou M., Wan S., Zhong Q., Chem. Eng. J. 333, 2018, 414–422.
[103] Zhang J., Zhang S., Cai W., Zhong Q., J. Environ. Sci. 25, 2013, 2492–2497.
[104] Wang A., Guo Y., Gao F., Peden C.H.F., Appl. Catal. B: Environ. 202, 2017, 706–714.
[105] Cai W., Bu Y., Zhao Y., Chen M., Yu Y., Zhong Q., Chem. Eng. J. 317, 2017, 376–385.
[106] Wang A., Lin B., Zhang H., Engelhard M.H., Guo Y., Lu G., Peden C.H.F., Gao F., Catal. Sci. Technol. 7, 2017, 2362–2370.
[107] Akter N., Chen X., Parise J., Boscoboinik J.A., Kim T., Korean J. Chem. Eng. 35, 2017, 89–98.
[108] Ruggeri M.P., Nova I., Tronconi E., Pihl J.A., Toops T.J., Partridge W.P., Appl. Catal. B: Environ. 166–167 2015, 181–192.
[109] Selleri T., Gramigni F., Nova I., Tronconi E., Appl. Catal. B: Environ. 225, 2018, 324–331.

174 —— Cheng Zhang et al.

[110] Chen H.-Y., Wei Z., Kollar M., Gao F., Wang Y., Szanyi J., Peden C.H.F., Catal. Today 267, 2016, 17–27.

[111] López-Suárez F.E., Bueno-López A., Illán-Gómez M.J., Appl. Catal. B: Environ. 84, 2008, 651–658.

[112] Suárez S., Jung S.M., Avila P., Grange P., Blanco J., Catal. Today 75, 2002, 331–338.

[113] Shi C., H. Chang, C. Wang, Zhang T., Peng Y., Li M., Wang Y., Li J., Ind. Eng. Chem. Res. 57, 2018, 920–926.

[114] Albaladejo-Fuentes V., López-Suárez F., Sánchez-Adsuar M.S., Illán-Gómez M.J., Top. Catal. 60, 2016, 220–224.

[115] Torregrosa-Rivero V., Albaladejo-Fuentes V., Sánchez-Adsuar M.-S., Illán-Gómez M.-J., RSC Adv. 7, 2017, 35228–35238.

[116] Zhou C., Liu X., Wu C., Wen Y., Xue Y., Chen R., Zhang Z., Shan B., Yin H., Wang W.G., Phys. Chem. Chem. Phys. 16, 2014, 5106–5112.

[117] Lin F., Wu X., Weng D., Catal. Today 175, 2011, 124–132.

[118] Guerrero S., Águila G., Araya P., Catal. Commun. 28, 2012, 183–190.

[119] Atribak I., Guillén-Hurtado N., Bueno-López A., García-García A., Appl. Surf. Sci. 256, 2010, 7706–7712.

[120] Li H., Tang X., Yi H., Yu L., J. Rare Earth. 28, 2010, 64–68.

[121] Zhang J., Huang Y., Chen X., Nat J.. Gas Chem. 17, 2008, 273–277.

[122] Wu Z., Tang N., Xiao L., Liu Y., Wang H., J. Colloid Interf. Sci. 352, 2010, 143–148.

[123] Giles R., Cant N.W., Kögel M., Turek T., Trimm D.L., Appl. Catal. B: Environ. 25, 2000, L75–L81.

[124] Wen Y., Zhang C., He H., Yu Y., Teraoka Y., Catal. Today 126, 2007, 400–405.

[125] X. Liu, Z. Chen, Y. Wen, R. Chen, B. Shan, Catal. Sci. Technol. 4, 2014, 3687–3696.

[126] Kurt M., Say Z., Ercan K.E., Vovk E.I., Kim C.H., Ozensoy E., Top. Catal. 60, 2016, 40–51.

[127] Yoon D.Y., Lim E., Kim Y.J., Kim J.H., Ryu T., Lee S., Cho B.K., Nam I.-S., Choung J.W., Yoo S., J. Catal. 319, 2014, 182–193.

[128] PeñA M.A., Fierro J.L., Chem. Rev. 101 20011981–2017.

[129] Tanaka H., Misono M., Curr. Opin. Solid St. M. 5, 2001, 381–387.

[130] Chen J., Shen M., Wang X., Qi G., Wang J., Li W., Appl. Catal. B: Environ. 134–135 2013, 251–257.

[131] Zhou C., Feng Z., Zhang Y., Hu L., Chen R., Shan B., Yin H., Wang W.G., Huang A., RSC Adv. 5, 2015, 28054–28059.

[132] Choi S.O., Penninger M., Kim C.H., Schneider W.F., Thompson L.T., ACS Catal. 3, 2013, 2719–2728.

[133] Zhong S., Sun Y., Xin H., Yang C., Chen L., Li X., Chem. Eng. J. 275, 2015, 351–356.

[134] Wang J., Su Y., Wang X., Chen J., Zhao Z., Shen M., Catal. Commun. 25, 2012, 106–109.

[135] Shen B., Lin X., Zhao Y., Chem. Eng. J. 222, 2013, 9–15.

[136] Wang Y., Cui X., Y. Li, Shu Z., Chen H., Shi J., Micropor. Mesopor. Mat. 176, 2013, 8–15.

[137] Zhao B., Ran R., Sun L., Guo X., Wu X., Weng D., RSC Adv. 6, 2016, 69855–69860.

[138] Ghafari M., Atkinson J.D., Environ Sci Technol 50, 2016, 5189–5196.

[139] Sousa J.P.S., Pereira M.F.R., Figueiredo J.L., Catal. Today 176, 2011, 383–387.

[140] You F.-T., Yu G.-W., Xing Z.-J., Li J., Xie S.-Y., Li C.-X., Wang G., Ren H.-Y., Wang Y., Appl. Surf. Sci. 471, 2019, 633–644.

[141] Sousa J.P.S., Pereira M.F.R., Figueiredo J.L., Fuel Process. Technol. 106, 2013, 727–733.

[142] Wang X., Yao R., Bai Z., Ma H., Reaction Kinet. Mech. Catal. 120, 2016, 209–217.

[143] Yu Y., Bu Y., Zhong Q., Cai W., Catal. Commun. 87, 2016, 62–65.

[144] Talukdar P., Bhaduri B., Verma N., Ind. Eng. Chem. Res. 53, 2014, 12537–12547.

[145] Atkinson J.D., Zhang Z., Yan Z., Rood M.J., Carbon 54, 2013, 444–453.

[146] Guo Z., Huang Z.-H., Wang M., Kang F., Catal. Sci. Technol. 5, 2015, 827–829.
[147] Guo Z., Liang Q.-H., Yang Z., Liu S., Huang Z.-H., Kang F., Catal. Sci. Technol. 6, 2016, 422–425.
[148] Sousa J.P.S., Pereira M.F.R., Figueiredo J.L., Catalysts 2, 2012, 447–465.
[149] Sousa J.P.S., Pereira M.F.R., Figueiredo J.L., Appl. Catal. B: Environ. 125, 2012, 398–408.
[150] Ellmers I., Pérez Vélez R., Bentrup U., Schwieger W., Brückner A., Grünert W., Catal. Today 258, 2015, 337–346.
[151] Ellmers I., Vélez R.P., Bentrup U., Brückner A., Grünert W., J. Catal. 311, 2014, 199–211.
[152] M.P. Ruggeri, Selleri T., Colombo M., Nova I., Tronconi E., J. Catal. 311, 2014, 266–270.
[153] Verma A.A., Bates S.A., Anggara T., Paolucci C., Parekh A.A., Kamasamudram K., Yezerets A., Miller J.T., Delgass W.N., Schneider W.F., F.H. Ribeiro, J. Catal. 312, 2014, 179–190.
[154] Metkar P.S., Balakotaiah V., Harold M.P., Catal. Today 184, 2012, 115–128.
[155] Loiland J.A., Lobo R.F., J. Catal. 325, 2015, 68–78.
[156] L. Olsson, H. Karlsson, Catal. Today 147, 2009, S290–S294.
[157] EplingW.S., Campbell L.E., Yezerets A., Currier N.W., Parks J.E., Catal. Rev. 46, 2004, 163–245.
[158] van Setten B.A.A.L., Makkee M., Moulijn J.A., Catal. Rev. 43, 2001, 489–564.
[159] Wang Q., Luo J., Zhong Z., Borgna A., Encyclopedia of Semiconductor Nanotechnology, American Scientific Publisher, 2017.
[160] Sedlmair C., Seshan K., Jentys A., Lercher J.A., J. Catal. 214, 2003, 308–316.
[161] Büchel R., Strobel R., Krumeich F., Baiker A., Pratsinis S.E., J. Catal. 261, 2009, 201–207.
[162] Maeda N., Urakawa A., Baiker A., Top. Catal. 52, 2009, 1746.
[163] Büchel R., Strobel R., Baiker A., Pratsinis S.E., Top. Catal. 52, 2009, 1709.
[164] Vijay R., Snively C.M., Lauterbach J., J. Catal. 243, 2006, 368–375.
[165] Vijay R., Hendershot R.J., Rivera-Jiménez S.M., Rogers W.B., Feist B.J., Snively C.M., Lauterbach J., Catal. Commun. 6, 2005, 167–171.
[166] DiGiulio C.D., Komvokis V.G., Amiridis M.D., Catal. Today 184, 2012, 8–19.
[167] Milt V.G., Querini C.A., Miró E.E., Ulla M.A., J. Catal. 220, 2003, 424–432.
[168] Liu Z., Ihl Woo S., Catal. Rev. 48, 2006, 43–89.
[169] Nova I., Ciardelli C., Tronconi E., D. Chatterjee, Weibel M., Top. Catal. 42, 2007, 43–46.
[170] Irfan M.F., Goo J.H., Kim S.D., Hong S.C., Chemosphere 66, 2007, 54–59.
[171] Kandylas I.P., Koltsakis G.C., Ind. Eng. Chem. Res. 41 2002, 2115–2123.

Qingling Liu and Mingyu Guo

5 Selective catalytic reduction of nitrogen oxides by NH$_3$

5.1 Introduction

5.1.1 Overview of NO$_x$

Nitrogen oxides (NO$_x$) are a general term for various compounds of nitrogen and oxygen, which are one of the main harmful substances produced by the burning of fossil fuels. NO$_x$ include NO, NO$_2$, N$_2$O, N$_2$O$_3$, and N$_2$O$_5$. NO and NO$_2$ are two major sources of air pollutants, which occupy over 90% and 5–10% of total NO$_x$, respectively.

NO$_x$ can be divided into three kinds: thermal NO$_x$, transient NO$_x$, and fuel NO$_x$ with regard to the formation path. The formation of thermal NO$_x$ is mainly affected by the reaction temperature and is generated at a high temperature above 1,500 °C. Transient nitrogen mainly refers to the part of NO$_x$ formed by the rapid reaction of hydrocarbons and N$_2$ in the flame when the fuel is burned. This formation process does not require a very high temperature. The fuel NO$_x$ is derived from the oxidation of nitrogen-containing organic matter in the fuel component, which is mainly related to the nitrogen content of the substance in the fuel, and the combustion temperature has no significant influence on its production. Usually, most of the NO$_x$ is thermal NO$_x$.

5.1.2 Emission sources of NO$_x$

The NO$_x$ in the air mainly come from two sources: one is natural factors through some natural phenomena such as lightning, burning caused by volcanic eruptions, forest fires, and decomposition of nitrogenous organic matter. The other is human factors, which include stationary and mobile sources. The stationary sources mainly use fossil fuels, such as industrial boilers and thermal power plants. The mobile sources are mainly gasoline and diesel fuel vehicles, such as motor vehicles and ship exhausts. Of the human factors, stationary sources and mobile sources constitute 60% and 40%, respectively.

Qingling Liu and Mingyu Guo, Tianjin Key Laboratory of Indoor Air Environmental Quality Control, School of Environmental Science and Engineering, Tianjin University, Tianjin, China

https://doi.org/10.1515/9783110544183-005

5.1.3 Hazard of NO_x

Harm to human, animal, and plant health

Similar to human CO poisoning, NO_x can also bind to hemoglobin in human blood, but its affinity with hemoglobin is stronger. In addition, NO_x can cause damage to the lungs, causing respiratory diseases such as bronchitis. Therefore, prolonged exposure to NO_x can cause human poisoning and affect normal human life activities. NO_x affects the growth of animals and plants because it can cause leaf necrosis of plants, thereby inhibiting photosynthesis of plants and affecting growth.

Harm to the stability of the overall ecosystem

NO_x are the major source of air pollutants, which will lead to many ecosystem problems such as ozone depletion, acid rain, and photochemical smog.

Photochemical smog: When NO_x and hydrocarbons such as olefins are present together, a series of complex chemical reactions will occur under the sunlight, producing photochemical smog, which not only reduces visibility, but also has strong oxidative properties, which can affect the respiratory tract and cause strong itching effect to eyes.

Acid rain: NO_x and rain contact form acid rain. When acid rain falls to the ground level, it will cause corrosion of vegetation and buildings. When nitrate particles fall into the soil or water, it will cause acidification and the pollution of soil and groundwater.

Ozone depletion and the greenhouse effect: The entering of NO_x into the upper atmosphere may destroy the ozone layer and increases the concentration of ultraviolet light reaching the ground, which in turn leads to a series of environmental pollution and ecological changes, such as aggravating the global greenhouse effect.

5.1.4 Removal of NO_x

Selective catalytic reduction (SCR) of NO_x

Currently, selective catalytic reduction (SCR) is the most efficient and common way to remove NO_x. The catalyst is the core of the SCR system, and the most common used catalyst is V_2O_5–$WO_3(MoO_3)/TiO_2$. SCR is a process in the presence of O_2 and catalysts, in which a reductant reacts with exhausted NO_x to selectively form H_2O and N_2. The following are the basic chemical equations of SCR reaction.

Standard SCR reaction:

$$4NH_3 + 4NO + O_2 \rightarrow 4N_2 + 6H_2O \tag{5.1}$$

$$4NH_3 + 2NO_2 + O_2 \rightarrow 3N_2 + 6H_2O \tag{5.2}$$

Rapid SCR reaction:

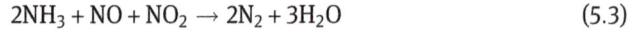

$$2NH_3 + NO + NO_2 \rightarrow 2N_2 + 3H_2O \tag{5.3}$$

The SCR reaction mechanism

The SCR of NO_x has been studied for many years, and three different reaction mechanisms are proposed for NO_x removal, which are Langmuir–Hinshelwood (L-H) mechanism, Eley–Rideal (E-R) mechanism Figures 5.1–5.3, and Mars–Van–Krevelen (M-V-K) mechanism.

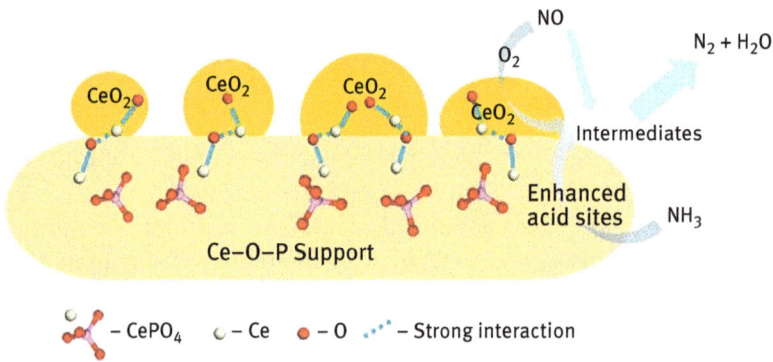

Figure 5.1: L-H reaction pathway on $CeO_2/Ce–O–P$ catalyst [3].

L-H mechanism is the reaction among the adsorbed species. For the SCR reaction, this means the reaction between the adsorbed NO_x species with the adsorbed NH_3 or NH_4^+. Guo et al. [1] studied the SCR mechanism on the Cr_2O_3 catalyst, and the results showed the SCR reaction on this kind of catalyst followed the L-H mechanism and the active intermediate compounds are nitrites. Han et al. [2] held the view that the L-H mechanism dominated the low-temperature NH_3-SCR reaction on the Ce-Cu-SAPO-18 catalysts. The large amounts of acid sites and isolated Cu^{2+} species in Ce-Cu-SAPO-18 promoted the adsorption and activation of NO, thus facilitated the reaction between the adsorbed NH_3 molecules and NO_x species. Yao et al. [3] confirmed that the NO_x-adsorbed species on CeO_2/TiO_2 catalyst were very stable and could hardly react with adsorbed NH_3, following L-H reaction pathway at low temperature. Zhang et al. [4] synthesized $Cu_{0.01}Ce_{0.09}TiO_x$ catalyst through a sol–gel

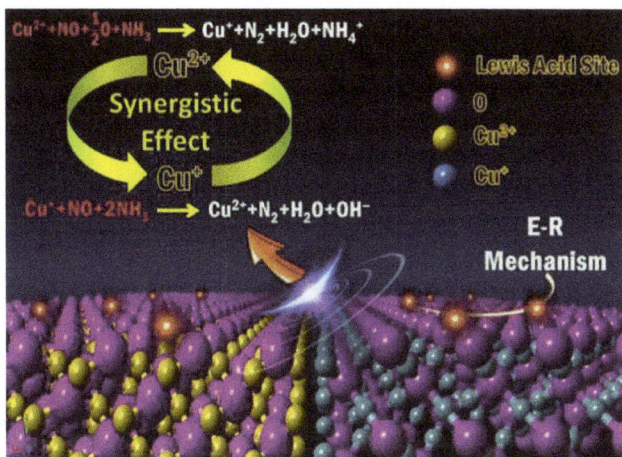

Figure 5.2: E-R mechanism of NH$_3$-SCR-NO over CuO$_x$ [5].

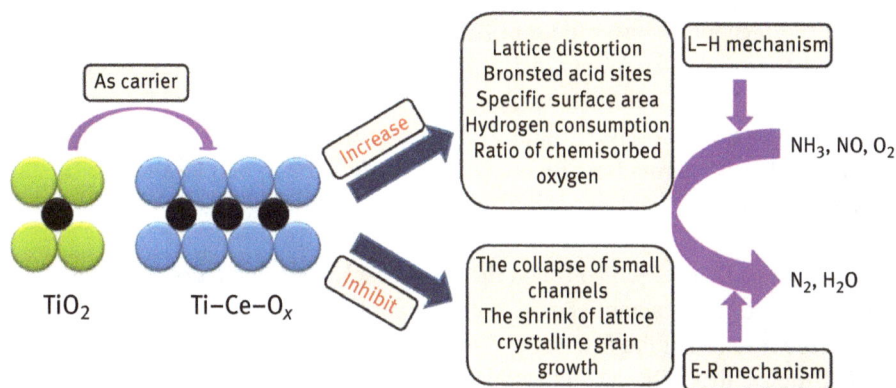

Figure 5.3: L-H and E-R mechanisms over Ti–Ce–O$_x$ catalyst carrier surface.

method, and L-H mechanism was predominant in the NH$_3$-SCR reaction at 150 °C for this Cu$_{0.01}$Ce$_{0.09}$TiO$_x$ catalyst.

E-R mechanism is the reaction between the gas molecules and adsorbed species. As the adsorption of NH$_3$ is a necessary process of SCR reaction, thus the E-R mechanism of SCR is the reaction of gaseous NO$_x$ species with adsorbed NH$_3$ or NH$_4^+$. Wang et al. [5] explained that the NH$_3$-SCR of NO$_x$ over CuO/Cu$_2$O heterostructure followed E-R mechanism and the typical E-R mechanism equations are as follows:

$$NH_3(g) \rightarrow NH_3(\text{Lewis acid sites}) \tag{5.4}$$

$$O_2(g) \rightarrow O_2(a) \tag{5.5}$$

$$NH_3(a) + O_2(a) \rightarrow NH_2 + OH \tag{5.6}$$

$$NO(a) + NH_2(a) \rightarrow NH_2NO(a) \tag{5.7}$$

$$NH_2NO(a) \rightarrow N_2 + H_2O \tag{5.8}$$

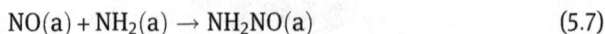

Qi et al. [6] described the reaction process of the SCR reaction on Ce-3Mn catalyst. NH_3 mainly occupied the Lewis acid sites and formed coordinated NH_3 on the catalyst surface, and then generated highly active intermediate (NH_2NO, amide species) by reacting with gaseous NO, and ultimately converted into N_2 and H_2O. This process was in accordance with the E-R mechanism. Wang et al. [7] synthesized Fe/ZSM-5 catalyst by impregnation method and only E-R mechanism existed during the NH_3-SCR reaction process at 150 °C.

Some researchers also reported that both of the L-H and E-R mechanism existed on the catalyst surface. Based on the in situ DRIFT spectroscopy, both the reaction of adsorbed monodentate NO_2^- with adsorbed NH_4^+ and that of gas-phase NO with adsorbed NH_4^+ existed over the Ti–Ce–O_x-500 catalyst carrier surface, which confirmed the reaction followed both of the L-H and E-R mechanism [8]. Fei et al. [9] revealed that the mechanisms of L-H and E-R existed synchronously during the NH_3-SCR reaction, and the L-H mechanism was predominant due to its rapid reaction rate. The results by Wang et al. [10] showed that Sb species on CuSb/TiO_2-0.4 contributed to the formation of more surface adsorbed NH_3 and adsorbed NO_x species, which effectively enhanced the NH_3-SCR process through both the L-H and E-R routes.

M-V-K mechanism is also called the *redox mechanism*, which refers to the reaction process of the reactant species with the catalyst lattice oxygen ions. The first step is the reaction of the catalyst and reactants, creating oxygen vacancies and reducing the catalyst. The second step is that the catalyst is reoxidized and regenerated by dissociation of the adsorbed oxygen to supplement the oxygen vacancies. Previous researches have confirmed that SCR followed M-V-K mechanism on V-based catalysts, and the oxygen is directly involved in the reoxidation of the V-site [11, 12]. Beretta et al. [13] through kinetic analysis on the V_2O_5/MoO_3/TiO_2 catalyst revealed that the observed experimental trends can be described by a M-V-K rate expression assuming the rate of reoxidation depends on NO concentration as well as the rate of reduction.

Li et al. [14] showed that the denitration mechanism of the Fe–Mn/SBA-15 catalyst in the SCR reaction followed L-H, E-R, and M-V-K mechanisms. The Fe–Mn/SBA-15 catalyst has good oxygen storage ability, high oxidation, and abundant amount of chemisorbed oxygen, which promoted the M-V-K route of the SCR reaction.

5.2 The categories of the SCR catalysts

5.2.1 Noble metal catalysts

Noble metal catalysts are the earliest researched SCR catalysts. The most valuable noble metals are Pt, Rh, Pd, Au, and Ag. These catalysts are usually loaded onto the zeolite by ion exchange or directly loaded onto the zeolite or the oxide carrier by other methods.

Yang et al. [15] compared the promoting SCR performances effect of noble metals (Rh, Ru, Pt, Pd) doping of MnO_x–CeO_2/graphene catalysts, as shown in Figure 5.4, and the results showed that Pd-promoted catalyst exhibited the excellent catalytic performance among the tested samples. Moreover, the addition of Pd into MnO_x–CeO_2/graphene enhanced the SO_2-poisoning resistance.

Pt is a typical noble metal catalyst for SCR reaction. Hong et al. [16] used hydrothermal method with tetrapropylammonium hydroxide, and successfully encapsulated Pt nanoparticles in hollow ZSM-5 single crystals with the "dissolution–recrystallization" process. The synthesized Pt/hollow-ZSM-5re had good low-temperature performance with 84% NO removal efficiency and 92% N_2 selectivity at 90 °C, as displayed in Figure 5.5. It also showed good H_2O- and SO_2-poisoning resistance, which was related to the Pt-active sites in the hollow structure of the catalyst and was relatively stable during the SCR reaction.

More et al. [17] studied the Au–Ag/Al_2O_3 catalysts synthesized by successive impregnation method, which was used for hydrocarbon SCR (HC-SCR) of NO_x. The SCR performance of these catalysts depended on the introduction order of Au and Ag. Au–Ag/Al_2O_3 obtained by first introducing Ag and then Au processed a much better SCR performance in a wide temperature range and it was proved to be a potential SCR catalyst for NO_x removal from lean burn engine exhaust gases. Yan et al. [18] studied the SCR catalyst modified by RuO_2, which displayed good SCR performance with an excellent tolerance to SO_2. Meanwhile, the Ru-promoted SCR catalyst also showed a promising performance for coreduction of mercury emission.

Although the noble metal catalyst has certain catalytic performance in the SCR reaction, its N_2 selectivity is not high. Thus, there will be more N_2O formation during the catalytic process, resulting in secondary pollution. In addition, the higher cost and lower poisoning resistance of the noble metal catalyst limit its application in practice.

5.2.2 Zeolite molecular sieve

Zeolite molecular sieve, a typical porous crystalline material, is a highly efficient SCR catalyst. It has a three-dimensional tetrahedral framework formed by sharing vertices between TO_4 tetrahedrons. The framework T atoms can be Si, Al, or P. The aluminosilicate molecular sieve framework is composed of SiO_4 and AlO_4 tetrahedrons. For

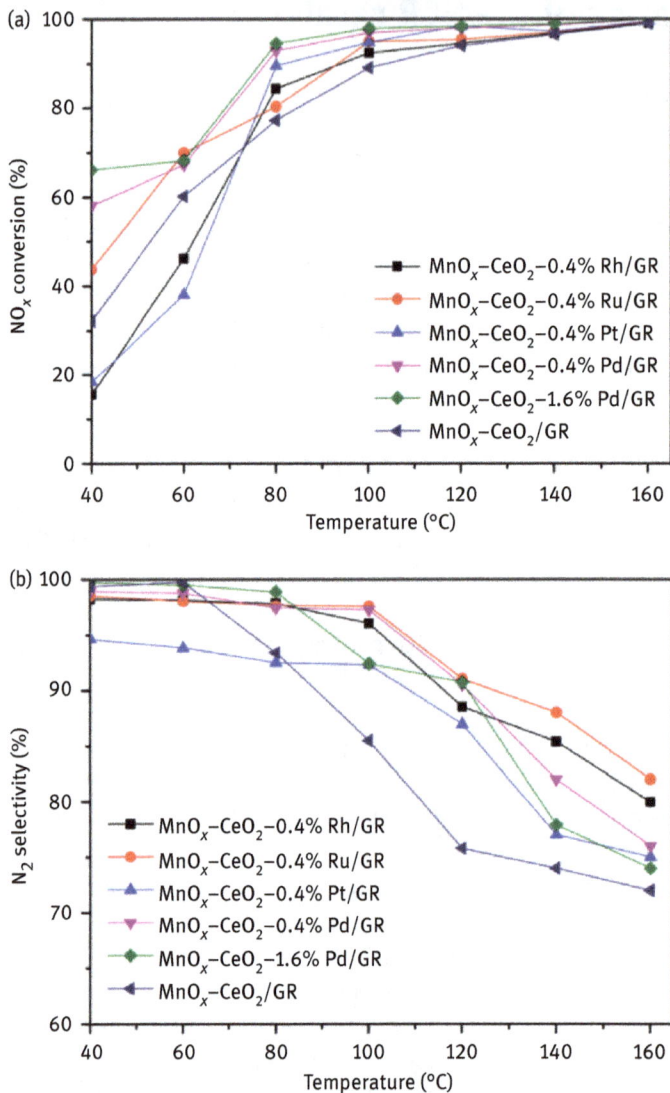

Figure 5.4: (a) NH$_3$-SCR performances and (b) N$_2$ selectivity over the MnO$_x$–CeO$_2$/GR and MnO$_x$–CeO$_2$–M/GR catalysts [15].

aluminum phosphate molecular sieve, the AlO$_4$ and PO$_4$ tetrahedrons are strictly alternated in the framework. The zeolite pore size is between 0.3 and 2.0 nm. According to the size of the pore, the molecular sieve materials can be divided into: 12-membered ring large pore zeolite, 10-membered ring medium pore zeolite, and 8-membered ring small pore zeolite, in addition to an ultralarge pore molecular sieve larger than the 12-membered ring.

(a)   (b)

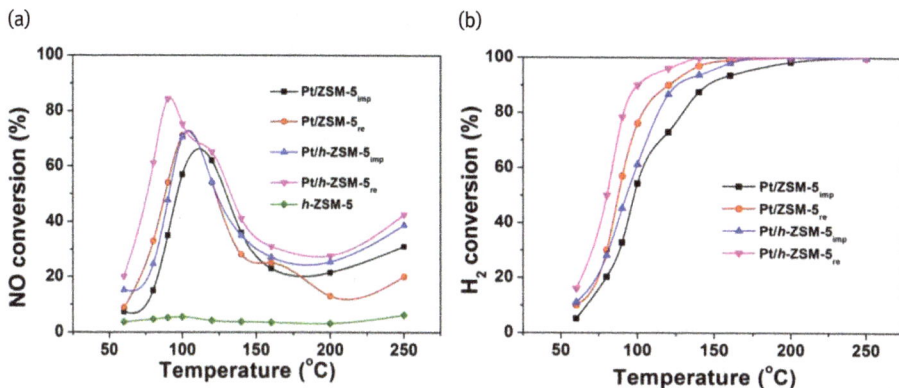

Figure 5.5: Catalytic activity of different catalysts in H_2-SCR (a) NO conversion (b) H_2 conversion [16].

Copper ion-exchanged zeolite catalysts are promising catalysts for SCR of NO_x with NH_3 (NH_3-SCR). Previous studies have indicated that fresh Cu-ZSM-5 catalysts possessed high NH_3-SCR performance. But it was easily dealuminated under hydrothermal condition, which would reduce its SCR activity. Small pore molecular sieve catalysts, such as Cu-SSZ-13 and Cu-SAPO-34, had comparable to or even surpass SCR performance of Cu-ZSM-5. The copper ion-exchanged small pore molecular sieves had higher SCR performance in the temperature range of 150–500 °C and better hydrothermal stability compared with the larger pore Cu-ZSM-5. The substitution of heteroatoms, the dimensionality of the framework, and the degree of copper exchange all affected the SCR performance as well as the hydrothermal stability of the catalysts. Molecular sieve carrier materials with CHA structure have attracted much more attention for SCR reaction.

CHA, also known as *chabazite*, is a small-pore molecular sieve with a cage structure consisting of an eight-membered ring and a double six-membered ring. The size of the eight-membered ring is 3.8 × 3.8 Å, and the size of the elliptical cavity is 8.35 Å (Figure 5.6).

It is the unique pore structure that has the advantages of high N_2 selectivity, excellent catalytic activity, and large specific surface area; thus, CHA has become the most important molecular sieve carrier material for NH_3-SCR catalyst. Studies have shown that Cu-active component supported on the CHA carrier exhibited high catalytic activity [19]. The Cu-CHA catalysts being widely studied now are Cu-SSZ-13 and Cu-SAPO-34.

In 1985, Chevron's Zones first synthesized SSZ-13 molecular sieves (high silicon to aluminum ratio 8–50) by hydrothermal synthesis and improved the synthesis in 1991 [20]. Bull found that Cu-SSZ-13 performed super NH_3-SCR activity and hydrothermal stability after Cu ion exchange. Kwak et al. [21, 22] found that Cu-SSZ-13 showed an excellent SCR performance and N_2 selectivity in the temperature range

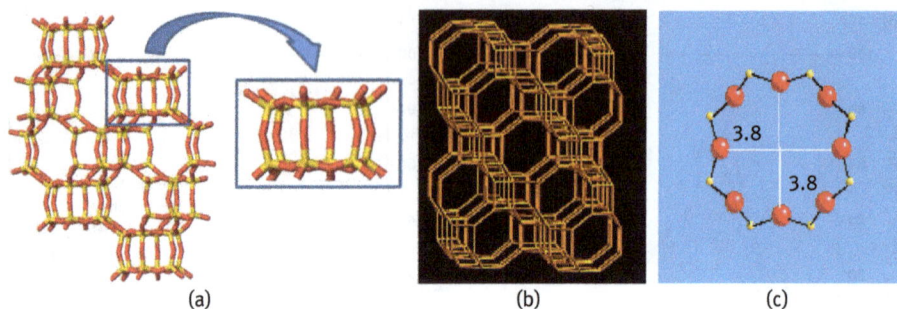

Figure 5.6: (a) CHA type molecular sieve structure, (b) double six-membered ring (D6R) and CHA cage structure, and (c) crystal plane aperture size map.

of 160–550 °C, which was much better than Cu-ZSM-5 and Cu-Beta. Different Cu loading also affected the SCR activity. With the increase of Cu loading, both the reduction of NO and oxidation of NH_3 increased at low temperatures. But at high temperatures (>500 °C), the reduction of NO decreased, and the oxidation of NH_3 increased. With 40–60% Cu loading, the excellent hydrothermal stability of the zeolite was contributed to its unique small pore structure [23]. There are two kinds of isolated Cu^{2+} sites, which is also of benefit to the hydrothermal stability. By determining the specific position of copper ions in the SSZ-13 structure, Fickel et al. [24, 25] found that appropriate Cu loading and the unique small pore structure of CHA made great contribution to the excellent catalytic performance and hydrothermal stability of Cu-SSZ-13. The produced $Al(OH)_3$ by the high-temperature dealumination could reconstitute the CHA structure after the temperature is lowered, thereby stabilizing the framework of the molecular sieve and ensuring hydrothermal stability. Ishihara et al. [26] reported the Cu-SAPO-34 catalyst, which had good hydrothermal stability and SO_2 resistance. It could be recovered after the calcination at 500 °C. Fickel et al. [25] synthesized Cu-SAPO-34 by ion-exchanged method, which had lower NO conversion before calcination. But after the calcination at 550 °C, the SCR activity and stability was improved. Wang et al. [27] compared the hydrothermal stability between the Cu-SSZ-13 and Cu-SAPO-34. It was shown that the active sites were increased after hydrothermal aging, which contributed to the increase of the NO conversion at high temperatures. Although both of the two kinds of zeolites showed excellent low-temperature SCR performance, Cu-SAPO-34 showed higher hydrothermal stability compared with Cu-SSZ-13.

Cu-SSZ-13 catalyst possessed excellent SCR catalytic performance and high N_2 selectivity. It maintained a good SCR catalytic activity at 800,000 h^{-1} ultrahigh space velocity, and showed promising resistance to H_2O, CO_2, and C_3H_6 poisoning. In addition, TMAdaOH (N,N,N-trimethyl-1-adamantamonium hydroxide), the expensive template used for SSZ-13 synthesis, leads to the increase the production costs and its toxicity would pollute the environment, which significantly limited the application of Cu-SSZ-13. Hence, the main issues for the application of Cu-SSZ-13 are SO$_2$ and H$_2$O resistance and the cost.

5.2.3 Metal oxides catalysts

V-based catalysts

The typical V-based catalyst is V$_2$O$_5$–WO$_3$/TiO$_2$. The supported V$_2$O$_5$–WO$_3$/TiO$_2$ catalysts have become the most widely used for SCR applications in industries since early 1970s.

Lai et al. [28] provided a detailed review of the current fundamental understanding and recent progress of the supported V$_2$O$_5$–WO$_3$/TiO$_2$ catalyst system, including the synthesis process, molecular structures of active mental oxides, active sites, surface acidity, reaction intermediates, reaction kinetics, and mechanism of the reaction. They confirmed that the surface V^{+5} oxidation state vanadia species are the active sites for the SCR reaction and the surface tungsta species are the promoter for the SCR reaction The results showed that the structure and dispersion of active compounds on the catalyst were significantly affected by synthesis method. The surface VO$_x$ and WO$_x$ sites on TiO$_2$ form both isolated and oligomeric surface species with incipient wetness impregnation synthesizing method. For the acid sites, the surface tungsta sites possess comparable amounts of Brønsted and Lewis acidities, but the surface vanadia sites mostly exhibit Brønsted acidity and the most abundant surface intermediate is surface NH$_4^{+*}$ species on Brønsted acid sites. The SCR reaction on V$_2$O$_5$–WO$_3$/TiO$_2$ catalyst performed via M-V-K mechanism and the rate-determining step involves decomposition of the surface NO-NH$_x$ reaction intermediate or breaking of an N–H bond during the course of formation (Figure 5.7).

Mn-based catalysts

Mn-based catalysts have attracted much more attention these years for its excellent low-temperature SCR performance. There are two reasons for the excellent low-temperature SCR activity of MnO$_x$ catalyst: (1) Mn has multiple valence states, and the mutual conversion between different valence states can promote the SCR reaction and (2) the surface of MnO$_x$ contains a variety of surface adsorbed oxygen (O$_\alpha$), which can oxidize NO into NO$_2$ and promote fast SCR reaction.

Kapteijin et al. [29] found MnO$_2$ had the highest SCR activity and M$_2$O$_3$ had the highest N$_2$ selectivity at low temperatures among the MnO$_x$ catalysts. The order of the SCR activity of the MnO$_x$ are MnO$_2$>Mn$_2$O$_3$>Mn$_3$O$_4$> MnO. Tang et al. [30] synthesized MnO$_x$ by four different methods; these are low-temperature solid phase reaction

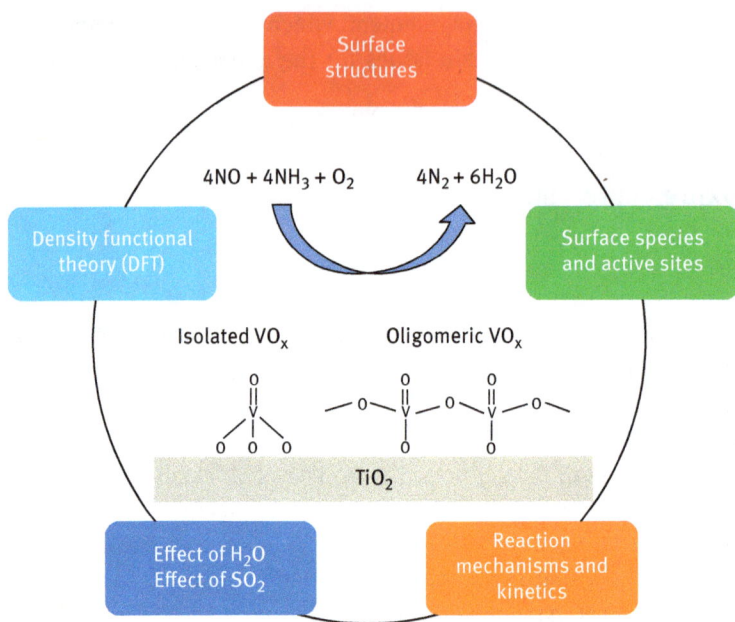

Figure 5.7: Overview of the V_2O_5–WO_3/TiO_2 catalyst system [28].

method (SP), rheological phase reaction method (RP), coprecipitation method (CP), and citric acid method (CA). The results showed that Mn_2O_3 synthesized by citric acid method had the highest crystallinity, and MnO_x synthesized by CP method with the lowest crystallinity had the highest SCR performance at low temperatures.

Supported Mn-based catalysts also had excellent NH_3-SCR performance. Peña et al. [31] tested the Co, Cr, Cu, Fe, Mn, Ni, V oxides loaded on TiO_2, and MnO_x/TiO_2 processed the highest low-temperature SCR performance among the tested catalysts (as shown in Figure 5.8). Jiang et al. [32] synthesized a series of MnO_x/TiO_2 catalysts through different methods. The NO removal efficiency of $MnO_x(0.4)$/TiO_2 made by sol–gel method reached 90% at 200 °C and the catalyst had a good SO_2 resistance.

Although pure MnO_x or supported MnO_x have excellent SCR performance of low temperatures, the SO_2 and H_2O resistance is poor. Many researchers modified the Mn-based catalysts to improve the resistance at low temperatures by adding other metal oxides into the catalysts. Kang et al. [33] found that the novel MnO_x catalyst prepared by precipitation method with sodium carbonate as precipitant exhibited outstanding catalytic activity in the temperature range of 373–473 K, as shown in Figure 5.9. However, the single MnO_x catalyst is sensitive to H_2O. More researches are focused on the modification of manganese-based catalysts with transition metals. Thirupathi and Smirniotis [34] reported Mn–Ni(0.4)/TiO_2 exhibited 100% NO conversion and N_2 selectivity at 200 °C. The hydrothermal stability is excellent with 10 vol% H_2O and 240 h

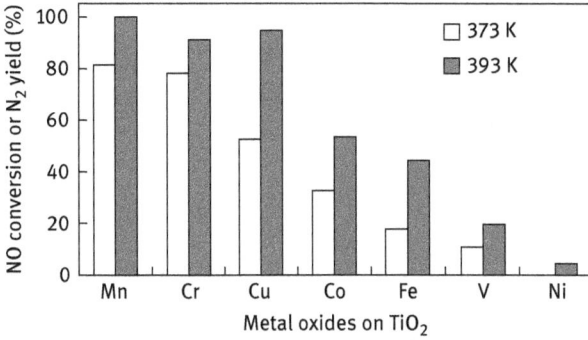

Figure 5.8: Performance of 20 wt% transition metal/TiO_2 catalysts at 373 and 393 K (GHSV = 8,000 h^{-1}, NO = NH_3 = 2,000 ppm, 2.0 vol% O_2, He balance) [31].

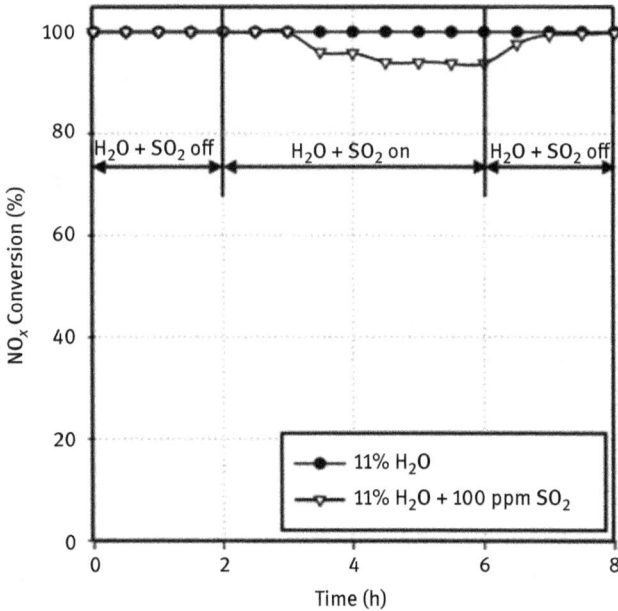

Figure 5.9: Effects of H_2O and SO_2 on NO_x conversions over MnO_x-SC catalyst at 398 K [33].

running time (as shown in Figure 5.10). The excellent SCR performance and stability is related to the Mn^{4+} on the catalyst surface and Mn^{4+} limited the formation of Mn_2O_3. Mn–La–Ce–Ni–O_x also showed good SCR performance and SO_2 resistance [35]. At 200 °C, NO removal efficiencies were 95% and 85% with (300 ppm) and without SO_2. Currently, Mn-based catalyst has good SCR performance at low temperatures, but its poisoning resistance still needs to be improved.

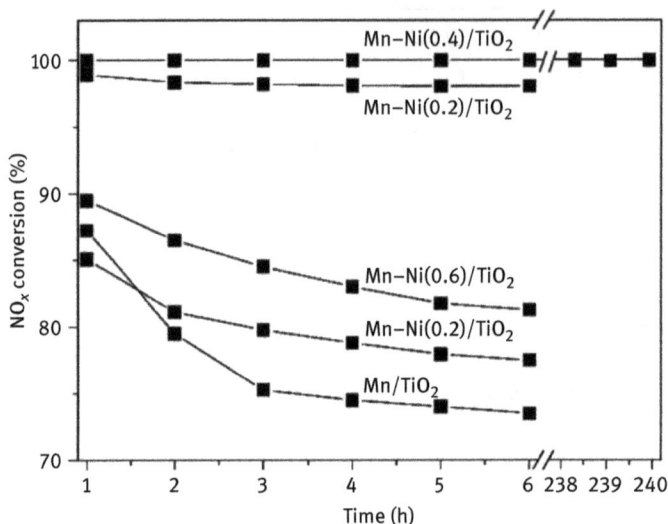

Figure 5.10: SCR of NO with NH_3 at 200 °C over Mn/TiO_2 and $Mn-Ni/TiO_2$ catalysts [34].

Cr-based catalysts

Cr has been studied in SCR reactions recently, and it showed good NO conversion and sulfur resistance. Yang et al. and Huang et al. studied the SCR catalysts with the addition of a certain amount of Cr into the catalyst. The proper addition of Cr weakened the V reduction peak, increased the proportion of oxygen vacancies, and finally promoted the SCR reaction [36, 37]. Li et al. [38] reported that the Cr addition of CeO_2-ZrO_2 mixed oxides exhibited better SCR performance and sulfur and H_2O resistance (as shown in Figure 5.11). The reason was that Cr oxides could increase the surface area, the amount and reactivity of the acid sites, and surface active oxygen sites. Li et al. [39] reported that the incorporated Cr ions into the sargassum-based activated carbon (SAC) catalyst could enhance conversion between Cr^{6+} and lower oxidation states (Cr^{5+}, Cr^{3+}, or Cr^{2+}) in the materials and the reactivity of the acid sites. It was concluded that Cr-doped SAC possessed excellent water and sulfur resistance. Zhou et al. [40] studied the effect of transition metals (Cr, Zr, Mo) on the MnO_x-FeO_x catalysts for the SCR reaction at low temperatures, as shown in Figure 5.12. The results revealed that the addition of Cr obviously reduced the active temperature of MnO_x-FeO_x catalysts on account of better dispersion of active components, enhanced surface specific area, and increased the amount of Lewis acid sites. Therefore, Cr oxide could be a promising additive for SCR catalysts but its toxicity and lower temperature SCR activity need to be considered for further application.

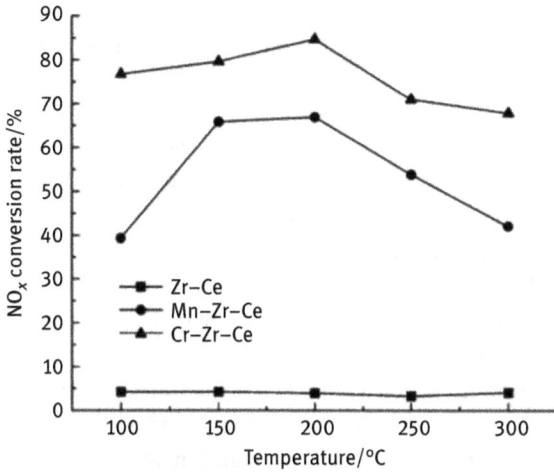

Figure 5.11: SCR activity of different Zr–Ce-based catalysts [38].

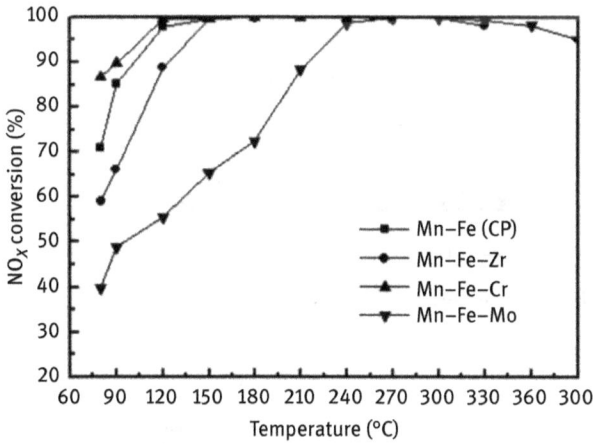

Figure 5.12: SCR activity of different Mn–Fe-based catalysts [40].

Fe-based catalysts

Fe-based catalysts have been studied in these years for its nontoxic ability, high SCR performance, and N$_2$ selectivity at low and medium temperatures. But broadening its high-temperature window and improving its SO$_2$ poisoning capacity is still a technical problem for iron-based catalysts used for SCR unit in coal-fired power plants and industrial furnaces.

Liu et al. [41] compared the SCR performance of γ-Fe$_2$O$_3$ and α-Fe$_2$O$_3$, as shown in Figure 5.13. γ-Fe$_2$O$_3$ catalyst performed higher catalytic activity than α-Fe$_2$O$_3$ in the temperature range of 150–300 °C, because these two catalysts were functioned by different SCR mechanisms. Both NH$_3$ and NO$_x$ species could be easily adsorbed and reacted on the γ-Fe$_2$O$_3$ surface. Compared with gaseous NH$_3$, gaseous NO$_x$ was more easily to be adsorbed on α-Fe$_2$O$_3$; consequently, the formed stable nitrates blocked the active sites, which adversely affected the SCR reaction. Wang et al. [42] studied Mn-based SCR catalyst by doping Fe. The results showed that sulfation of the Mn-active component was significantly reduced by doping the Mn/γ-Al$_2$O$_3$ catalyst with Fe; thus, its SO$_2$ resistance was improved. Jan et al. [43] synthesized a series of Fe$_2$O$_3$ and found the Fe$_2$O$_3$/WO$_3$/Al$_2$O$_3$ catalyst possessed better thermal stability and N$_2$ selectivity. Fabrizioli et al. [44] synthesized Fe$_2$O$_3$/SiO$_2$ catalyst by sol–gel method, and it is proven that Fe$_2$O$_3$ could dispersed well on the SiO$_2$. The NO conversion was over 98% when the temperature was lower than 460 °C, and over 97% of N$_2$ selectivity was achieved below 500 °C.

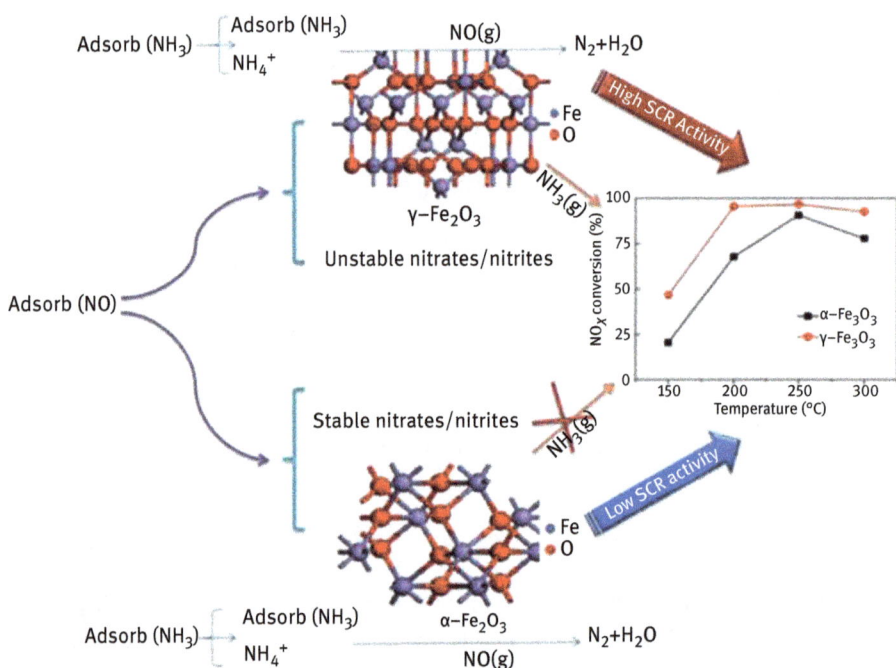

Figure 5.13: Comparison between α-Fe$_2$O$_3$ and γ-Fe$_2$O$_3$ for SCR reaction [41].

5.3 Major issues for catalyst applications

5.3.1 Low-temperature SCR performance

As the SCR unit is placed behind the desulfurizer and electric precipitation in power plant or used for cold start-up process of diesel engine, its working temperature should be around 250 °C or even below. Many studies have been focused on the low-temperature SCR performance. As discussed above, zeolites-, Mn-, Cr-, or Fe-based catalysts all exhibit low-temperature SCR activities. Usually, the actual situation is more complicated for the low temperature with some other poisoning sources, which will be discussed in the following.

Some researchers studied the improvement of the low-temperature SCR activity through adding different transition metal oxides, modifying the supports, or combining several kinds of materials. Liu et al. [45] synthesized Mn–Ce/Cu-SSZ-13 catalyst and the catalyst performed higher activities than Cu-SSZ-13 at 80 °C to 175 °C and Mn–Ce at above 200 °C. The NO_x conversions of this catalyst were above 90% from 125 °C to 450 °C, which extended the active low-temperature range. The mechanism of the promoted low temperature SCR is shown in Figure 5.14. Chen et al. [46] compared the addition of Ti^{4+} and Sn^{4+} into the CeO_2–MnO_x catalyst for SCR reaction at low temperatures; the results showed that Ti^{4+} significantly enhanced the SCR activity below 200 °C.

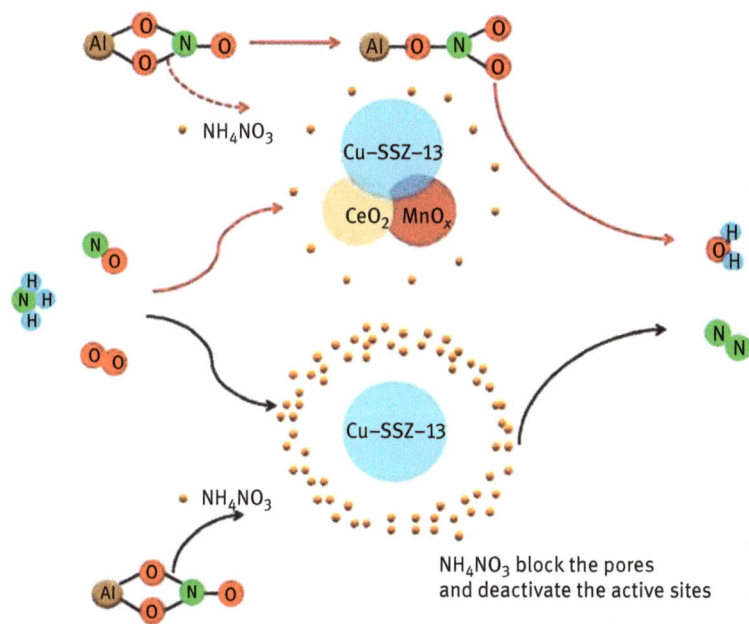

Figure 5.14: Mechanism study of Mn–Ce/Cu-SSZ-13 catalyst for low-temperature SCR [45].

5.3.2 SO$_2$ and H$_2$O resistance

SO$_2$ resistance

When high sulfur fuel is used by the engines, such as coal or heavy fuel oil, there will be a lot of SO$_x$ in the exhaust gases. Sulfur dioxide (SO$_2$) is initially formed by the oxidation of sulfur contained in fuels, and is easily oxidized further to sulfur trioxide (SO$_3$). However, the oxidation of SO$_2$ to SO$_3$ during SCR is undesirable owing to the formation of sulfates that poison the SCR catalysts. Ammonium sulfates are formed when the ammonia combines with SO$_3$ at temperatures below 250 °C, which can block the catalyst's pores, active sites, and foul downstream heat exchangers [47]. Thus, it is necessary to improve the SO$_2$ resistance of the SCR catalysts.

Phil et al. [48] tested some V$_2$O$_5$/TiO$_2$ catalysts modified by several transition metal oxides. It was found that the addition of Sb could reduce the stability of surface sulfates, slow down its deposition rate, and accelerate the decomposition of the sulfates, thus reduced the SO$_2$ poisoning. Peralta et al. [49] modified the catalyst by adding Ce, because Ce could easily combine with SO$_2$ to form cerium sulfate, which had no effect on the catalyst. This could avoid the reaction of the other components with SO$_2$ to protect the active sites. Qi et al. [50] synthesized MnO$_x$–CeO$_2$ by the citric acid method. The SCR activity and SO$_2$ resistance was significantly improved and the SO$_2$ poisoning was reversible, as shown in Figure 5.15. Wu et al. [51] added CeO$_2$ into the Mn/TiO$_2$ catalyst and found that the addition of CeO$_2$ blocked the formation of sulfates and ammonium sulfates on the catalyst surface. Thus, the addition of CeO$_2$ significantly improved the SO$_2$ resistance of Mn-based catalyst.

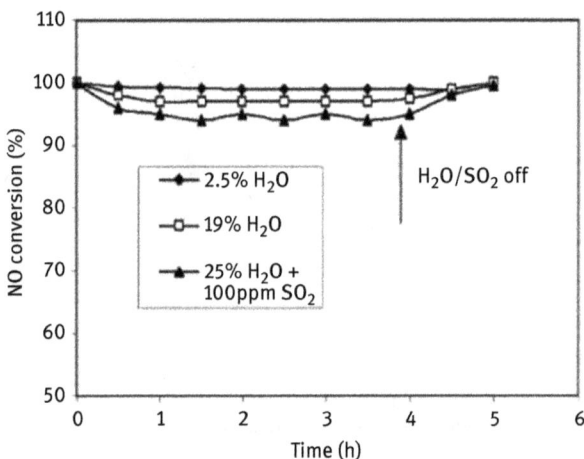

Figure 5.15: Effects of time-on-stream on SCR activity with H$_2$O + SO$_2$ and without H$_2$O + SO$_2$ [50].

With 100 ppm SO$_2$ at 150 °C for 6.5 h, the NO conversion of Mn–Ce/TiO$_2$ still remained at about 84%.

Yao et al. [52] synthesized Ce–O–P catalyst and found that the active oxygen species could capture the SO$_2$ and attenuate its interference with NO adsorption and oxidation, thereby improving sulfur resistance.

Liu et al. [53] synthesized Ce$_3$W$_2$SbO$_x$ catalyst by the coprecipitation method. It has been proved that the catalyst redox performance and surface acidity was significantly improved by the strong interaction between Sb, W, and Ce species, which promoted the adsorption and activation of NH$_3$ substances and improved the SO$_2$ and H$_2$O resistance of the catalyst. The results of SO$_2$ and H$_2$O resistance are shown in Figure 5.16.

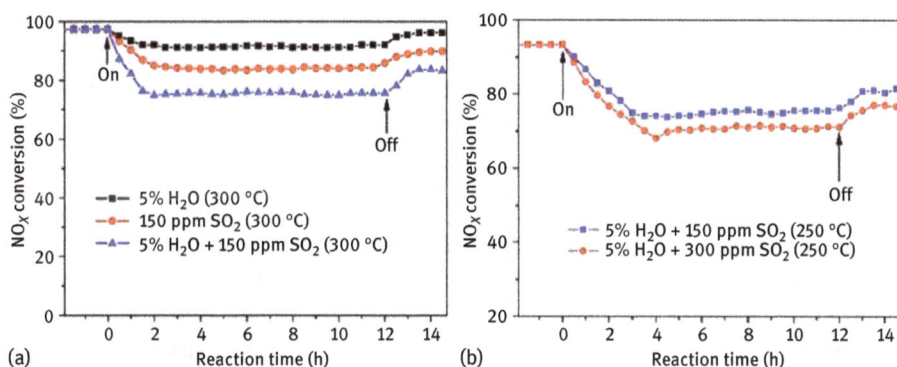

Figure 5.16: NO$_x$ conversions over Ce$_3$W$_2$SbO$_x$ catalyst in the presence of SO$_2$ and SO$_2$ + H$_2$O at (a) 300 °C and (b) 250 °C [53].

Zhu et al. [54] synthesized novel highly efficient SCR catalysts for low temperatures by modifying Fe–Mn/TiO$_2$ catalyst with holmium (Ho). When the temperature was 120 °C and SO$_2$ was 400 ppm, Fe$_{0.3}$Ho$_{0.1}$Mn$_{0.4}$/TiO$_2$ showed excellent SO$_2$ resistance. After stopping the introduction of SO$_2$, the catalytic activity was obviously recovered. The characterization results proved that the specific surface area was significantly enlarged after the addition of Ho$_2$O$_3$; thus, it can effectively promote the dispersion of Fe$_2$O$_3$ on the catalyst and improve its performance.

Only a few researches have proven SO$_2$ can have a promoting effect on SCR reaction. For example, SO$_2$ improved the SCR performance of Co$_2$O$_3$ and Cr$_2$O$_3$ catalyst at 300 °C [55] and 280 °C [1], respectively. Sulfation of the catalysts with SO$_2$ before the SCR reaction could form SO$_4^{2-}$ and increase the Brønsted acid sites on the surface of the catalyst [56–58]. Brønsted acid sites are essential to SCR reaction, because NH$_4^+$ adsorbed on Brønsted acid sites can react with gaseous NO and promote the SCR reaction through E-R mechanism. Zhang et al. [56] showed that sulfation also improved

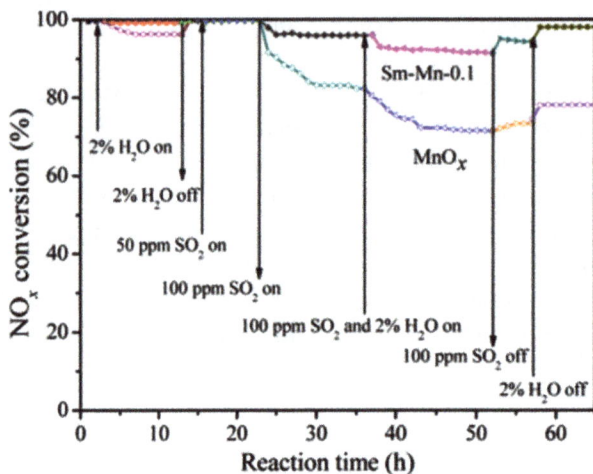

Figure 5.17: Effect of H_2O and SO_2 on catalytic activities of MnO_x and Sm-Mn-0.1 catalysts for the SCR reaction at 100 °C [61].

the mobility of oxygen and increased chemisorbed oxygen. Xue et al. [58] showed that SO_2 destroyed the Lewis acid sites but strengthened the Brønsted acid sites on the $Cr-CeO_x$ catalysts. Wei et al. [57] reported that SO_2 preferentially adsorbed on Ce atom on the $Mn-Ce/TiO_2$ catalysts, generating cerium sulfate with Brønsted acid sites, which reduced the sulfation of Lewis acid sites and Brønsted acid sites and protected the active sites. Liu et al. [59] suggested the acidity was greatly enhanced after sulfation on the catalyst surface, leading to the increase of NH_3 adsorption capability especially in the high-temperature range, which is beneficial to the SCR reaction.

H_2O resistance

For gas-fired boilers, the prominent characteristic of its exhaust gas is that H_2O content is high, up to 15–20 vol%. Such high H_2O content levels are mainly produced by combustion of methane. As we know, H_2O will restrain the removal of NO_x, mainly due to the competitive adsorption of H_2O and NH_3/NO on the active sites of catalysts. Furthermore, the exhaust gas of gas-fired boilers usually needs to pass through waste heat boilers to reduce the energy loss. Compared with coal-fired boiler, the exhaust compositions and content of gas-fired boiler have a great difference, the content of H_2O that is higher. H_2O in the exhaust gases can inhibit the removal efficiency of NO_x; besides, the exhaust temperature of gas-fired boiler is relatively low. Therefore, the catalysts of NH_3-SCR for gas-fired boiler must have outstanding low-temperature activity and H_2O tolerance. Because of the gradual implementation of the "Coal to Gas" project in China, it is of great practical and economic significance to find low-temperature NH$_3$-SCR catalysts for gas-fired boiler exhaust.

Li et al. [39] studied the SCR performance at low temperatures and the effect of H$_2$O on the Cr/AC catalyst. It was shown that the NO$_x$ conversion could reach over 90% at 125 °C, and the catalyst also had a better H$_2$O resistance. After addition of 5% H$_2$O into the system, NO conversion rapidly dropped from 91.9% to 83.1%. But increasing the H$_2$O to 10% into the system, NO still maintained conversion around 80%. The main reasons for the good H$_2$O resistance of the catalyst are the increased acid sites on the catalyst surface and the mutual transformation between Cr^{6+} and its low-valence state after the addition of Cr into the catalyst.

SO$_2$ and H$_2$O resistance

Usually both SO$_2$ and H$_2$O can have adverse effects on the SCR catalyst, especially when they coexist. Many researchers focused on the effect of coexistence of SO$_2$ and H$_2$O. Qi et al. [60] investigated the modification of MnO$_x$–CeO$_2$ catalysts. It was found that the addition of Fe or Zr oxides improved the SCR activity and N$_2$ selectivity at low temperatures, and the addition of Pr enhanced the N$_2$ selectivity and H$_2$O and SO$_2$ resistance. Meng et al. [61] designed a highly effective Sm-MnO$_x$ catalyst, which showed excellent SO$_2$ and H$_2$O resistance at 100 °C, as shown in Figure 5.17.

5.3.3 Hydrothermal aging and coking resistance

Hydrocarbons and particulate matters could result from fuel combustion in the diesel engines, which could adhere to and poison the surface of catalysts. For the actual diesel engine exhaust systems, the characteristic of the gas stream is H$_2$O content generated from the combustion of diesel fuel with a high carbon number. And the rapid burning of collected particulate by DPF (diesel particulate filter) to be regenerated is essential for diesel engines [62]. Thus, the deactivation of SCR catalysts by hydrothermal aging and coking is also a critic challenge for their actual application. During operation process of diesel engines, the SCR catalyst is gradually deactivated by hydrothermal aging and hydrocarbon coking poisoning [62].

Hydrothermal aging resistance

Li et al. [63] compared hydrothermal stability of two kinds of Cu-SAPO-34 catalysts synthesized by different methods. The migration of Cu species during hydrothermal treatment affected the SCR reaction. As the aging time prolonged to 24 h, partial isolated Cu^{2+} transforms into surface CuO particles, consistent with its aggregation during the aging treatment over the one-pot prepared sample. The increase of the amount of surface CuO particles in the aged samples promotes the nonselective

NH_3 oxidation reaction in the high-temperature range and results in lowering of the SCR activity by limiting NH_3 supply. Cao et al. [64] studied a ceria-modified WO_3–TiO_2–SiO_2 monolithic catalyst, which showed higher hydrothermal stability at 550 °C compared with unmodified catalyst. The results showed that no phase transformation of anatase TiO_2 to rutile occurred for all the aged catalysts. The addition of ceria prevented crystallization of WO_3 species and effectively supplemented the surface acidity and redox property of catalyst. Marberger et al. [65] studied the thermal activation and aging of a V_2O_5/WO_3–TiO_2 catalyst and concluded that the onset of catalyst deactivation, observed at lower aging temperature for the hydrothermally aged catalyst compared to the thermally treated one, is possibly due to a larger amount of mobile V and W species and the concurrent loss of specific surface area.

Coking resistance

Li et al. [66] investigated the effect of propene on the SCR performance on Fe-ZSM-5 catalyst and proposed that the deactivation was caused by the blockage of Fe^{3+}-active site by propene residue mechanism. The reaction of NO oxidation to NO_2 mainly occurred on Fe^{3+} sites, which were easy to be blocked by the propene. Thus, the conversion of NO oxidation into NO_2 was significantly inhibited on a poisoned catalyst by propene below 400 °C, which significantly inhibited the fast SCR reaction. Liu et al. [67] synthesized a few-layered micro@meso-porous Fe/Beta@SBA-15 CSCs with SBA-15 film as the shell and Beta supporting FeO_x monometallic NPs as the core using an ultradilute liquid-phase coating method. It was found that Fe/Beta@SBA-15 CSCs exhibited high propene resistance and SO_2 tolerance. The thickness of SBA-15 sheaths could be controllably adjusted by altering the mass proportion of Si/Beta, and different NH_3-SCR performances were achieved.

Hydrothermal aging and coking resistance

As shown in Figure 5.18, Wang et al. [68] indicated that the presence of Fe in the catalyst indeed improved NO removal efficiency and N_2 selectivity in the presence of propene for hydrothermally aged Cu, Fe/SSZ-13 catalyst.

5.3.4 Alkali metal resistance

Usually, alkaline metals exist in the exhaust gases of power plants and industrial boilers. More seriously, the alternative of fossil fuels with biomass in some industrial boilers and power plants could even cause three to four times faster deactivation of the SCR catalysts. There are much higher amounts of poisonous potassium

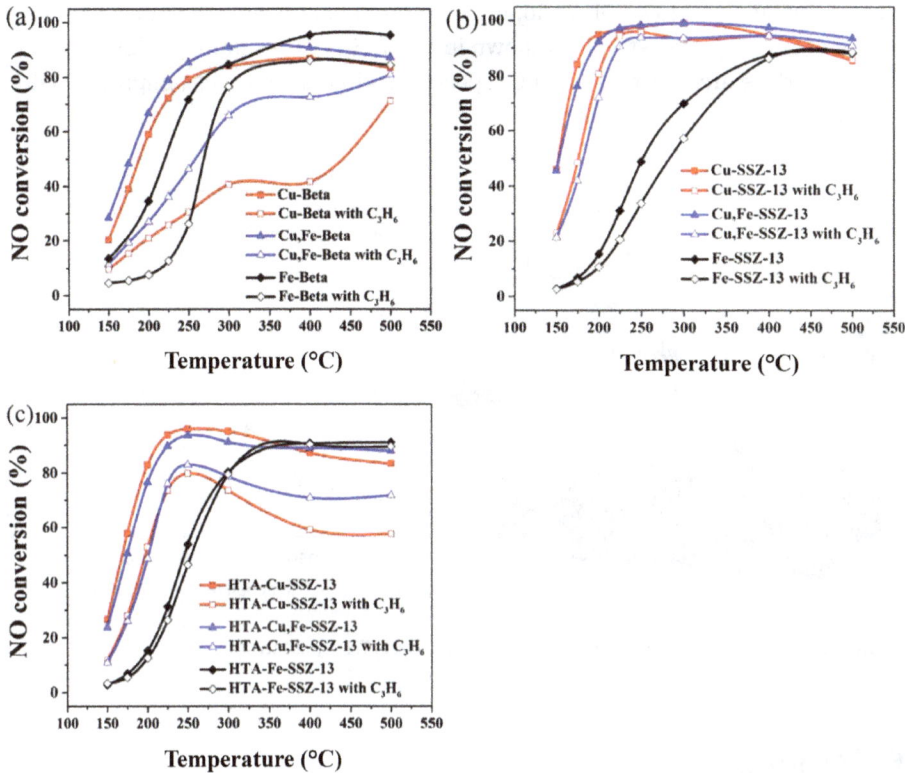

Figure 5.18: NO conversion as a function of temperature during standard SCR with or without the presence of C$_3$H$_6$ for: (a) fresh Beta samples; (b) fresh SSZ-13 samples; and (c) Hydrothermally aged SSZ-13 catalysts [68].

and sodium in the biomass-burned flue gases (almost up to 2 wt%) [69], which may cause the alkali metal poisoning of SCR catalysts. In the SCR actual applications of glass furnaces, refuse incinerators, steel furnaces, and cement plants, alkaline poisoning is still a key issue because of the significant amounts of alkali metals and alkaline earth metals in the original materials. Hence, the study and development of SCR catalysts with tolerance to alkali metal poisoning still need to be further researched.

Wang et al. [69] studied CeO$_2$ loaded on modified titanate nanotubes, and the results showed that the titanated nanotubes treated with ethanol had much better alkaline metal-poisoning resistance. The enhanced alkaline metal poisoning resistance could be attributed to more structural ion-exchangeable OH groups. These OH groups enhanced the surface acid strength, increased the amount of acid sites, and maintained a large proportion of oxygen vacancy and Ce^{3+} on the ceria surface. The active compounds could well performed by the shell protection effect when alkaline metal exists [69, 70]. Zr-doping was found to efficiently enhance the K tolerance of Ce/TiO$_2$ catalyst [71], as shown in Figure 5.19. The addition of Zr increased specific surface area, dispersion of Ce species, acid sites, as well as improved redox property.

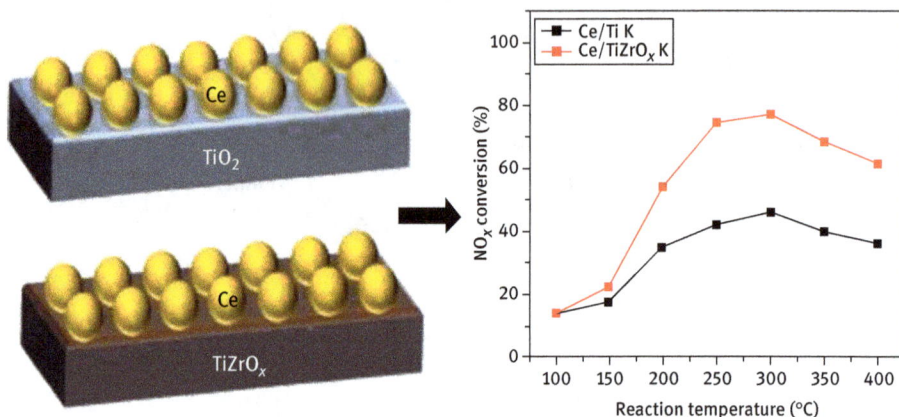

Figure 5.19: K tolerance comparison between Ce/Ti and Ce/TiZrO$_x$ catalysts [71].

References

[1]   Guo M., Liu Q., Zhao P., et al. Promotional Effect of SO$_2$ On Cr$_2$O$_3$ Catalysts for the Marine NH$_3$–SCR Reaction. Chem. Eng. J., 2019, 361:830–838.

[2]   Han S., Cheng J., Ye Q., et al. Ce Doping to Cu-SAPO-18: Enhanced Catalytic Performance for the NH3-SCR of NO in Simulated Diesel Exhaust. Micropor Mesopor Mat., 2019, 276:133–146.

[3]   Yao W., Wang X., Liu Y., et al. Ce O P Material Supported CeO$_2$ Catalysts: A Novel Catalyst for Selective Catalytic Reduction of NO with NH$_3$ at Low Temperature. Appl. Surf. Sci., 2019, 467–468:439–445.

[4]   Zhang T., Ma S., Chen L., et al. Effect of Cu Doping On the SCR Activity Over the Cu$_m$Ce$_{0.1-m}$TiO$_x$ (M = 0.01, 0.02 and 0.03) Catalysts. Appl. Catal. A: Gen., 2019, 570:251–261.

[5]   Wang Q., Xu H., Huang W., et al. Metal Organic Frameworks-Assisted Fabrication of CuO/Cu$_2$O for Enhanced Selective Catalytic Reduction of NO$_x$ by NH$_3$ at Low Temperatures. J. Hazard. Mater., 2019, 364:499–508.

[6]   Qi K., Xie J., Zhang Z., et al. Facile Large-Scale Synthesis of Ce Mn Composites by Redox-Precipitation and its Superior Low-Temperature Performance for NO Removal. Powder Technol., 2018, 338:774–782.

[7]   Wang X., Hu H., Zhang X., et al. Effect of Iron Loading On the Performance and Structure of Fe/ZSM-5 Catalyst for the Selective Catalytic Reduction of NO with NH$_3$. Environ. Sci. Pollut R., 2019, 26(2):1706–1715.

[8]   Jin Q., Shen Y., Ma L., et al. Novel TiO$_2$ Catalyst Carriers with High Thermostability for Selective Catalytic Reduction of NO by NH$_3$. Catal. Today, 2019, 327:279–287.

[9]   Fei Z., Yang Y., Wang M., et al. Precisely Fabricating Ce-O-Ti Structure to Enhance Performance of Ce-Ti Based Catalysts for Selective Catalytic Reduction of NO with NH$_3$. Chem. Eng. J., 2018, 353:930–939.

[10] Wang Z., Guo R., Shi X., et al. The Enhanced Performance of Sb-modified Cu/TiO$_2$ Catalyst for Selective Catalytic Reduction of NOx with NH$_3$. Appl. Surf. Sci., 2019, 475:334–341.

[11] Zhu M., Lai J., Tumuluri U., et al. Reaction Pathways and Kinetics for Selective Catalytic Reduction (SCR) of Acidic NO$_x$ Emissions from Power Plants with NH$_3$. ACS Catal., 2017, 7:8358–8361.

[12] Inomata M., Miyamoto A., Murakami Y. Mechanism of the Reaction of NO and NH$_3$ On Vanadium Oxide Catalyst in the Presence of Oxygen Under the Dilute Gas Condition. J. Catal., 1980, 1(62):140–148.

[13] Beretta A., Lanza A., Lietti L., et al. An Investigation On the Redox Kinetics of NH$_3$-SCR Over a V/Mo/Ti Catalyst: Evidence of a Direct Role of NO in the Re-Oxidation Step. Chem. Eng. J., 2019, 359:88–98.

[14] Li G., Wang B., Wang Z., et al. Reaction Mechanism of Low-Temperature Selective Catalytic Reduction of NO$_x$ over Fe-Mn Oxides Supported on Fly-Ash-Derived SBA-15 Molecular Sieves: Structure-Activity Relationships and in situ DRIFT Analysis. J. Phys. Chem. C, 2018, 122(35):20210–20231.

[15] Yang L., You X., Sheng Z., et al. The Promoting Effect of Noble Metal (Rh, Ru, Pt, Pd) Doping On the Performances of MnO$_x$-CeO$_2$/graphene Catalysts for the Selective Catalytic Reduction of NO with NH3 at Low Temperatures. New J. Chem., 2018, 42(14):11673–11681.

[16] Hong Z., Wang Z., Chen D., et al. Hollow ZSM-5 Encapsulated Pt Nanoparticles for Selective Catalytic Reduction of NO by Hydrogen. Appl. Surf. Sci., 2018, 440:1037–1046.

[17] More P. M., Dongare M. K., Umbarkar S. B., et al. Bimetallic Au-Ag/Al$_2$O$_3$ as Efficient Catalysts for the Hydrocarbon Selective Reduction of NOxfrom Lean Burn Engine Exhaust. Catal. Today, 2018, 306:23–31.

[18] Yan N., Chen W., Chen J., et al. Significance of RuO$_2$ Modified SCR Catalyst for Elemental Mercury Oxidation in Coal-Fired Flue Gas. *Environ. Sci. Technol.*, 2011, 45(13):5725–5730.

[19] Kwak J. H., Tran D., Burton S. D., et al. Effects of Hydrothermal Aging On NH$_3$-SCR Reaction Over Cu/zeolites. J. Catal., 2012, 287:203–209.

[20] Zones S. Conversion of Faujasites to High-Silica Chabazite SSZ-13 in the Presence of N,N,N-Trimethyl-l -adamantammonium Iodide. J. Chem. Soc. Faraday Trans., 1991, 22(87): 3709–3716.

[21] Kwak J. H., Tonkyn R. G., Kim D. H., et al. Excellent Activity and Selectivity of Cu-SSZ-13 in the Selective Catalytic Reduction of NO$_x$ with NH$_3$. J. Catal., 2010, 275(2):187–190.

[22] Kwak J. H., Tran D., Szanyi J., et al. The Effect of Copper Loading on the Selective Catalytic Reduction of Nitric Oxide by Ammonia Over Cu-SSZ-13. Catal. Lett., 2012, 142(3): 295–301.

[23] Gao F., Kwak J. H., Szanyi J., et al. Current Understanding of Cu-Exchanged Chabazite Molecular Sieves for Use as Commercial Diesel Engine DeNO$_x$ Catalysts. Top. Catal., 2013, 56(15–17):1441–1459.

[24] Fickel D. W., Lobo R. F. Copper Coordination in Cu-SSZ-13 and Cu-SSZ-16 Investigated by Variable-Temperature XRD. J. Phys. Chem. C, 2010, 114(3):1633–1640.

[25] Fickel D. W., D Addio E., Lauterbach J. A., et al. The Ammonia Selective Catalytic Reduction Activity of Copper-Exchanged Small-Pore Zeolites. Appl. Catal. B: Environ., 2011, 102(3–4):441–448.

[26] Ishihara T., Kagawa M., Hadama F., et al. Thermostable Molecular Sieves, Silicoaluminophosphate (SAPO)-34, for the Removal of NO$_x$ with C$_3$H$_6$ in the Coexistence of O$_2$, H$_2$O, and SO$_2$. Ind. Eng. Chem. Res., 1997, 36(1):17–22.

[27] Wang D., Jangjou Y., Liu Y., et al. A Comparison of Hydrothermal Aging Effects On NH$_3$-SCR of NO$_x$ Over Cu-SSZ-13 and Cu-SAPO-34 Catalysts. Appl. Catal. B: Environ., 2015, 165: 438–445.

[28] Lai J., Wachs I. E. A Perspective on the Selective Catalytic Reduction (SCR) of NO with NH$_3$ by Supported V$_2$O$_5$-WO$_3$/TiO$_2$ Catalysts. ACS Catal., 2018, 8(7):6537–6551.

[29] Kapteijn F., Singoredjo L., Andreini A., et al. ChemInform Abstract: Activity and Selectivity of Pure Manganese Oxides in the Selective Catalytic Reduction of Nitric Oxide with Ammonia. Chem. Inform., 1994, 3(2–3):173–189.

[30] Tang X., Hao J., Xu W., et al. Low Temperature Selective Catalytic Reduction of NO$_x$ with NH$_3$ Over Amorphous MnO$_x$ Catalysts Prepared by Three Methods. Catal. Commun., 2007, 8(3):329–334.

[31] Peña D. A., Uphade B. S., Smirniotis P. G. TiO$_2$-supported Metal Oxide Catalysts for Low-Temperature Selective Catalytic Reduction of NO with NH$_3$. I. Evaluation and Characterization of First Row Transition Metals. J. Catal., 2004, 221(2):421–431.

[32] Jiang B., Liu Y., Wu Z. Low-Temperature Selective Catalytic Reduction of NO On MnO$_x$/TiO$_2$ Prepared by Different Methods. J. Hazard. Mater., 2009, 162(2–3):1249–1254.

[33] Kang M., Yeon T. H., Park E. D., et al. Novel MnO$_x$ Catalysts for NO Reduction at Low Temperature with Ammonia. Catal. Lett., 2006, 106(1–2):77–80.

[34] Thirupathi B., Smirniotis P. G. Nickel-Doped Mn/TiO$_2$ as an Efficient Catalyst for the Low-Temperature SCR of NO with NH$_3$: Catalytic Evaluation and Characterizations . J. Catal., 2012, 288:74–83.

[35] Yang B., Zheng D., Shen Y., et al. Influencing Factors On Low-Temperature deNO$_x$ Performance of Mn-La-Ce-Ni-O$_x$/PPS Catalytic Filters Applied for Cement Kiln. J. Ind. Eng. Chem., 2015, 24:148–152.

[36] Yang R., Huang H., Chen Y., et al. Performance of Cr-doped Vanadia/Titania Catalysts for Low-Temperature Selective Catalytic Reduction of NO$_x$ with NH$_3$. Chinese J. Catal., 2015, 36(8):1256–1262.

[37] Huang H., Jin L., Lu H., et al. Monolithic Cr-V/TiO$_2$/cordierite Catalysts Prepared by In-Situ Precipitation and Impregnation for Low-Temperature NH$_3$-SCR Reactions. Catal. Commun., 2013, 34:1–4.

[38] Li W., Liu H., Chen Y. Promotion of Transition Metal Oxides On the NH$_3$-SCR Performance of ZrO$_2$-CeO$_2$ Catalyst. Front. Env. Sci. Eng., 2017, 11(2):6.

[39] Li S., Wang X., Tan S., et al. CrO$_3$ Supported On Sargassum-Based Activated Carbon as Low Temperature Catalysts for the Selective Catalytic Reduction of NO with NH$_3$. Fuel, 2017, 191:511–517.

[40] Zhou C., Zhang Y., Wang X., et al. Influence of the Addition of Transition Metals (Cr, Zr, Mo) On the Properties of MnO$_x$-FeO$_x$ Catalysts for Low-Temperature Selective Catalytic Reduction of NO$_x$ by Ammonia. J.Colloid Interf. Sci., 2013, 392:319–324.

[41] Liu C., Yang S., Ma L., et al. Comparison on the Performance of α-Fe$_2$O$_3$ and γ-Fe$_2$O$_3$ for Selective Catalytic Reduction of Nitrogen Oxides with Ammonia. Catal. Lett., 2013, 143(7):697–704.

[42] Wang J., Nie Z., An Z., et al. Improvement of SO2 Resistance of Low-Temperature Mn-Based Denitration Catalysts by Fe Doping. ACS Omega, 2019, 4(2):3755–3760.

[43] Jan M. T., Kureti S., Hizbullah K., et al. Reduction of Nitrogen Oxides by Ammonia Over Iron-Containing Catalysts. Chem. Eng. Technol., 2007, 30(10):1440–1444.

[44] Fabrizioli P., Bürgi T., Baiker A. Environmental Catalysis on Iron Oxide–Silica Aerogels: Selective Oxidation of NH3 and Reduction of NO by NH$_3$. J. Catal., 2002, 206(1):143–154.

[45] Liu Q., Fu Z., Ma L., et al. MnO$_x$-CeO$_2$ Supported On Cu-SSZ-13: A Novel SCR Catalyst in a Wide Temperature Range. Appl. Catal. A: Gen., 2017, 547:146–154.

[46] Chen L., Yao X., Cao J., et al. Effect of Ti^{4+} and Sn^{4+} Co-Incorporation On the Catalytic Performance of CeO$_2$-MnO$_x$ Catalyst for Low Temperature NH$_3$-SCR. Appl. Surf. Sci., 2019, 476:283–292.

[47] Dunn J. P., Stenger H. G., Wachs I. E. Molecular Structure-Reactivity Relationships for the Oxidation of Sulfur Dioxide Over Supported Metal Oxide Catalysts. Catal. Today, 1999, 53(4):543–556.

[48] Phil H. H., Reddy M. P., Kumar P. A., et al. SO$_2$ Resistant Antimony Promoted V$_2$O$_5$/TiO$_2$ Catalyst for NH$_3$-SCR of NO$_x$ at Low Temperatures. Appl. Catal. B: Environ., 2008, 78(3–4):301–308.

[49] Peralta M., Milt V., Cornaglia L., et al. Stability of Ba, K/CeO$_2$ Catalyst During Diesel Soot Combustion: Effect of Temperature, Water, and Sulfur Dioxide. J. Catal., 2006, 242(1):118–130.

[50] Qi G., Yang R. T. Performance and Kinetics Study for Low-Temperature SCR of NO with NH$_3$ Over MnO$_x$–CeO$_2$ Catalyst. J. Catal., 2003, 217(2):434–441.

[51] Wu Z., Jin R., Wang H., et al. Effect of Ceria Doping On SO$_2$ Resistance of Mn/TiO$_2$ for Selective Catalytic Reduction of NO with NH$_3$ at Low Temperature. Catal. Commun., 2009, 10(6):935–939.

[52] Yao W., Liu Y., Wang X., et al. The Superior Performance of Sol-Gel Made Ce-O-P Catalyst for Selective Catalytic Reduction of NO with NH$_3$. J. Phys. Chem. C, 2016, 120(1):221–229.

[53] Liu J., Li G., Zhang Y., et al. Novel Ce-W-Sb Mixed Oxide Catalyst for Selective Catalytic Reduction of NO$_x$ with NH$_3$. Appl. Surf. Sci., 2017, 401:7–16.

[54] Zhu Y., Zhang Y., Xiao R., et al. Novel Holmium-Modified Fe-Mn/TiO$_2$ Catalysts with a Broad Temperature Window and High Sulfur Dioxide Tolerance for Low-Temperature SCR. Catal. Commun., 2017, 88:64–67.

[55] Ke R., Li J., Liang X., et al. Novel Promoting Effect of SO$_2$ On the Selective Catalytic Reduction of NOx by Ammonia Over Co$_3$O$_4$ Catalyst. Catal. Commun., 2007, 8(12):2096–2099.

[56] Zhang L., Qu H., Du T., et al. H$_2$O and SO$_2$ Tolerance, Activity and Reaction Mechanism of Sulfated Ni-Ce-La Composite Oxide Nanocrystals in NH$_3$-SCR. Chem. Eng. J., 2016, 296: 122–131.

[57] Wei L., Cui S., Guo H., et al. DRIFT and DFT Study of Cerium Addition On SO$_2$ of Manganese-based Catalysts for Low Temperature SCR. J. Mol. Catal. A: Chem., 2016, 421:102–108.

[58] Xue L. Cr-CeO$_x$ mixed-oxide catalysts for the selective catalytic reduction (SCR) of NO$_x$ with NH$_3$ at medium-low temperature. Xiangtan University, 2015.

[59] Liu F., Asakura K., He H., et al. Influence of Sulfation On Iron Titanate Catalyst for the Selective Catalytic Reduction of NO$_x$ with NH$_3$. Appl. Catal. B: Environ., 2011, 103(3–4):369–377.

[60] Qi G., Yang R. T., Chang R. MnO$_x$-CeO$_2$ Mixed Oxides Prepared by Co-Precipitation for Selective Catalytic Reduction of NO with NH$_3$ at Low Temperatures. Appl. Catal. B: Environ., 2004, 51(2):93–106.

[61] Meng D., Zhan W., Guo Y., et al. A Highly Effective Catalyst of Sm-MnO$_x$ for the NH$_3$-SCR of NO$_x$ at Low Temperature: Promotional Role of Sm and Its Catalytic Performance. ACS Catal., 2015, 5(10):5973–5983.

[62] Li J., Chang H., Ma L., et al. Low-Temperature Selective Catalytic Reduction of NO$_x$ with NH$_3$ Over Metal Oxide and Zeolite Catalysts-a Review. Catal. Today, 2011, 175(1): 147–156.

[63] Li X., Zhao Y., Zhao H., et al. The Cu Migration of Cu-SAPO-34 Catalyst for Ammonia Selective Catalytic Reduction of NO$_x$ During High Temperature Hydrothermal Aging Treatment. Catal. Today, 2019, 327:126–133.

[64] Cao L., Wu X., Xu Y., et al. Ceria-Modified WO_3-TiO_2-SiO_2 Monolithic Catalyst for High-Temperature NH_3-SCR. Catal. Commun., 2019, 120:55–58.

[65] Marberger A., Elsener M., Nuguid R. J. G., et al. Thermal Activation and Aging of a V_2O_5/WO_3-TiO_2 Catalyst for the Selective Catalytic Reduction of NO with NH_3. Appl. Catal. A: Gen., 2019, 573:64–72.

[66] Li J., Zhu R., Cheng Y., et al. Mechanism of Propene Poisoning on Fe-ZSM-5 for Selective Catalytic Reduction of NO_x with Ammonia. Environ. Sci. Technol., 2010, 44(5):1799–1805.

[67] Liu J., Liu J., Zhao Z., et al. Fe/Beta@SBA-15 Core-Shell Catalyst: Interface Stable Effect and Propene Poisoning Resistance for No Abatement. AIChE J., 2018, 64(11):3967–3978.

[68] Wang A., Wang Y., Walter E. D., et al. NH_3-SCR On Cu, Fe and Cu+Fe Exchanged Beta and SSZ-13 Catalysts: Hydrothermal Aging and Propylene Poisoning Effects. Catal. Today, 2019, 320:91–99.

[69] Wang P., Chen H., Xiongbo W., et al. Novel SCR Catalyst with Superior Alkaline Resistance Performance: Enhanced Self-Protection Originated From Modifying Protonated Titanate Nanotubes. J. Mater. Chem. A, 2015, 3:680–690.

[70] Chen X., Wang H., Wu Z., et al. Novel $H_2Ti_{12}O_{25}$-Confined CeO_2 Catalyst with Remarkable Resistance to Alkali Poisoning Based on the "Shell Protection Effect". J. Phys. Chem. C, 2011, 115(35):17479–17484.

[71] Liu S., Guo R., Sun X., et al. Selective Catalytic Reduction of NO_x Over Ce/$TiZrO_x$ Catalyst: The Promoted K Resistance by $TiZrO_x$ Support, Mol. Catal., 2019, 462:19–27.

Jiang Wu and Fengguo Tian

6 Photocatalytic removal of gaseous elemental mercury

Hg from firing off-gas occurs in three modes: elemental mercury (Hg^0), oxidized mercury (Hg^{2+}), and particulate-bound mercury (Hg^P). Hg^P could be lightly trapped in fabric filter or electrostatic precipitator, while Hg^{2+} could be deprived in wet flue-gas desulfurization for water solubility. Nevertheless, Hg^0 is the hardest to get rid of by current air pollutants-controlling apparatus in thermal power station. The current Hg^0 eliminating methodologies include physical adsorption, oxidant injection, and catalytic oxidation. Compared with current methodologies, the booming photocatalysis possesses great oxidation potency and content recyclable features any minor contamination. Hence, photocatalytic oxidation of Hg^0 elimination from flue gas has huge feasibility.

Titanium dioxide (TiO_2), one of the versatility semiconductor photocatalysts, has been discovered to solve environmental problem and energy dilemma due to its steel oxidizability, avirulence, and senior chemical firmness. However, its broad band gap (3.02 eV as to rutile, 3.20 eV as to anatase) inhibits the adhibition in photocatalyzing, only being irradiated by UV radiation. Moreover, photoexcited electrons and holes lightly recombine and hardly isolate, which causes a finite photo-oxidation coefficient. To build up the photocatalytic capability of TiO_2, many researches have been implemented to improve its visual qualities and enlarge its response limit to visible ray. These researches comprise modifying surface, metal or nonmetal ions incorporation, and combination of metal oxides.

Different morphologies and different exposed faces have huge effect on photocatalytic reaction. Incorporating metal or nonmetal ions is a efficient way to increase the production of defects to facilitate photo-induced electron–hole pairs separating. Combination of metal oxides and TiO_2 support is a remarkable way to boost the isolation of photoexcited electron–hole pairs and minimize the band gap of TiO_2. Generating TiO_2 hollow spheres by hydrothermal approach could endow the as-prepared samples with different photocatalytic characteristics in removing Hg^0 with different amounts of hollow sphere. CuO/TiO_2 heterostructures display enhanced photoactivity in oxidizing Hg^0 under visible light when compared with P25.

Jiang Wu, College of Energy and Mechanical Engineering, Shanghai University of Electric Power, Shanghai, China
Fengguo Tian, College of Environmental Science and Engineering, Donghua University, Shanghai, China

https://doi.org/10.1515/9783110544183-006

6.1 Introduction

Mercury (Hg), because of its volatileness, acute toxicity, and serious diffusibility, has been paid consideration around the world in these decades [1]. The biggest artificial exhaust source of Hg is coal-combustion power plants, and so lots of regulations have been issued to reduce its release. The first global and legally binding instrument, Minamata Convention on Hg, was agreed by members of 147 countries in January 2013 [2], which presented a very precise release regulation for coal-combustion power stations, and it has come into force since August 16, 2017. Therefore, developing efficient technologies has become a critical task to regulate Hg release from thermal power stations.

Mercury from thermal power station off-gas is in three modes: elemental mercury (Hg^0), oxidized mercury (Hg^{2+}), and particulate-bound mercury (Hg^P) [3, 4]. Hg^P can be lightly trapped by bag-house or fabric filter (FF) or dust collector or electrostatic precipitator (ESP), while Hg^{2+} could be absorbed in wet flue-gas desulfurization (WFGD) as a result of water solubility (particularly in acidic circumstances), but it is very tough to remove Hg^0. The general Hg^0 emission control methods include physical adsorption, oxidants/chemicals emission, and catalysis oxidation [5–7]. In comparison with other methods, photocatalytic oxidation has better removal efficiency, without secondary pollution, and the catalyst is reusable. So, photocatalytic oxidation has a significant potential.

Photocatalysis based on semiconductive nanoparticles is a cutting-edge method. The exploitation of photocatalytic method has been thriving for its purity,cleanliness and larger potency figuring out the environmental issue and energy peril [8–11]. Titanium dioxide (TiO_2), one of the intensively used photocatalysts, has been widely studied for several decades on environmental contamination purification and energy peril [12]. However, the conventional TiO_2 nanoparticles can solely be activated under extraviolet exposure. Its big band gap (3.20 eV as to anatase, 3.02 eV as to rutile) hinders the utilization of TiO_2 nanophotocatalysts [13]. Besides, the motivated electron–hole pairs of TiO_2 photocatalysts readily recombine and they are difficult to transfer, leading to small quantum yield potency, which limits the oxidation potency [14]. Numerous strategies have been tried to bedizen TiO_2 or other semiconductors to improve electron–holes splitting and motion, of which fabricating structure junction in a photocatalyst is a very effective technique. Homojunction can endow photogenerated electrons jump more easily over the boundary surface by improving the charge separation, because of the junction built with materials possessing identical compositions and/or crystal structures [15, 16]. It has been demonstrated that homojunctions were generated between anatase and rutile TiO_2, which could promote the detachment of photogenerated electron–hole pairs, thus intensifying the function of TiO_2 without introducing any other photocatalyst [17]. Besides, to enhance photocatalytic capability of TiO_2 ulteriorly, many researches have demonstrated doping it to attain utility under the condition of visible light, including surface modification engineering [18], merging metal and/or metalloid ions [19–20], and associating with other metal oxides [21].

The morphologies and crystal face-exposure have significant influence on photocatalytic reaction. Incorporating metal/nonmetal ion is a useful method to produce more defects for muscling the elimination of photoexcited electrons and holes. Combining TiO_2 with metal oxides is a recommendable approach to boost the migration of photoinduced electron–hole pairs and to narrow its band gap. It was noted that different amounts of TiO_2 hollow spheres compounded by hydrothermal method have different photocatalytic properties in removing Hg^0. Compared with commercial TiO_2 (Degussa P25), CuO/TiO_2 heterostructure exhibits improved photoactivity to oxidize Hg^0 under visible light.

6.2 Photocatalytic eliminating of elemental mercury

6.2.1 The influence of surface heterojunction on photocatalytic removal of elemental mercury

In the three polymorphs of TiO_2, including anatase, rutile, and brookite, the anatase displays the most prominent photocatalytic capability [22]. In the exposed planes of anatase, the three low-index planes, including (001), (010), and (101) planes, are with the surface energy of 0.90, 0.53, and 0.44 J/m^2, respectively [23–25]. Commonly, when the surface energy is higher, its photocatalytic reactivity is better [26], and so the exposure of (001) facet with high energy becomes a target. Lu et al. proved the fabrication of anatase TiO_2 crystals with 47% exposure of (001) facets [27]. Chen et al. prepared the anatase microspheres with high-surface-area nanosheets of almost 100% (001) facets [28]. Nevertheless, the photoactivity is not successively enhanced with the increasing amount of (001) facets [29]; for example, Cheng et al. reported that pure (001) planes display lower photoactivity than (101) planes in hydrogen evolution process [30].

Fabrication of anatase TiO₂ with coexposed (001) and (101) facets

In conventional fabrication process, 5 mL of tetrabutyl titanate and x mL (x = 0, 0.25, 0.5, 0.75, and 1) HF (hydrofluoric acid) liquor with a density of 40 wt% were dripped into 40 ml di water (D.I. water). The obtained liquor was milled consecutively for 30 min and then put into 50 mL Teflon-lined autoclave at 160 °C for 24 h. After hydrothermal procedure, the formed white precipitation was collected and washed with pure ethanol and D.I. water for three times. After that, the cleaned material was dried in a ventilating device all night at 80 °C. The names of the as-prepared chemicals were termed as A-HF-0, A-HF-0.25, A-HF-0.5, A-HF-0.75, and A-HF-1 severally.

Characterization of anatase TiO$_2$ with coexposed (001) and (101) facets

The patterns of XRD are shown in Figure 6.1. It demonstrates that the peaks at 25.28, 37.80, 48.05, 53.89, and 55.06 ° are attained, corresponding to (101), (004), (200), (105), and (211) facets of anatase TiO$_2$, respectively, which agree well with JCPDS card (NO. 21–1272) [31]. It is notable that only A-HF-0 was brookite, while other as-prepared nanosamples are anatase TiO$_2$. It can be seen that the diffraction peak of (004) plane of the as-prepared samples is widened, indicating an increase in the percentage of (001) planes [32]. Meanwhile, the peak intensity of the (200) plane is improved and the diffraction peak of (200) plane is straitened, which demonstrates that the side length of (001) direction increases with the addition of HF.

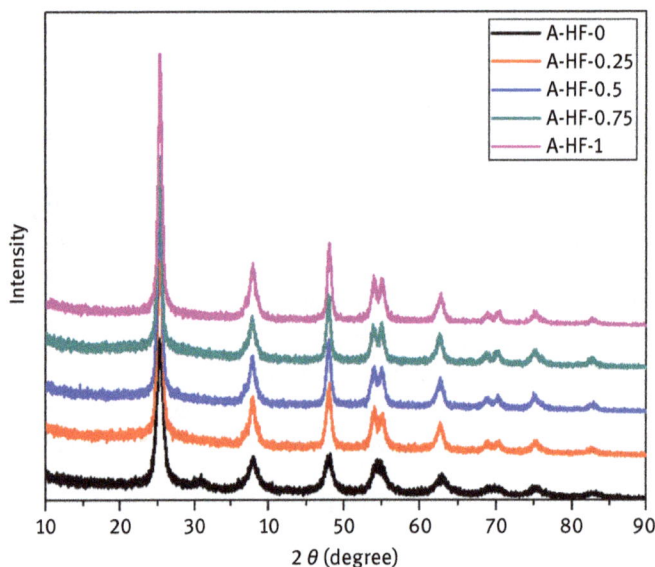

Figure 6.1: XRD graphs of as-prepared samples. (Reprinted from Ref. [33], Copyright 2017, with permission from Elsevier.).

To further analyze the morphological and structural characteristics of the as-prepared nanosamples, the transmission electron microscope (TEM) and high-resolution transmission electron microscope (HRTEM) are adopted for analysis, and the consequences are depicted in Figure 6.2. Figure 6.2(a) and (c) describes the TEM

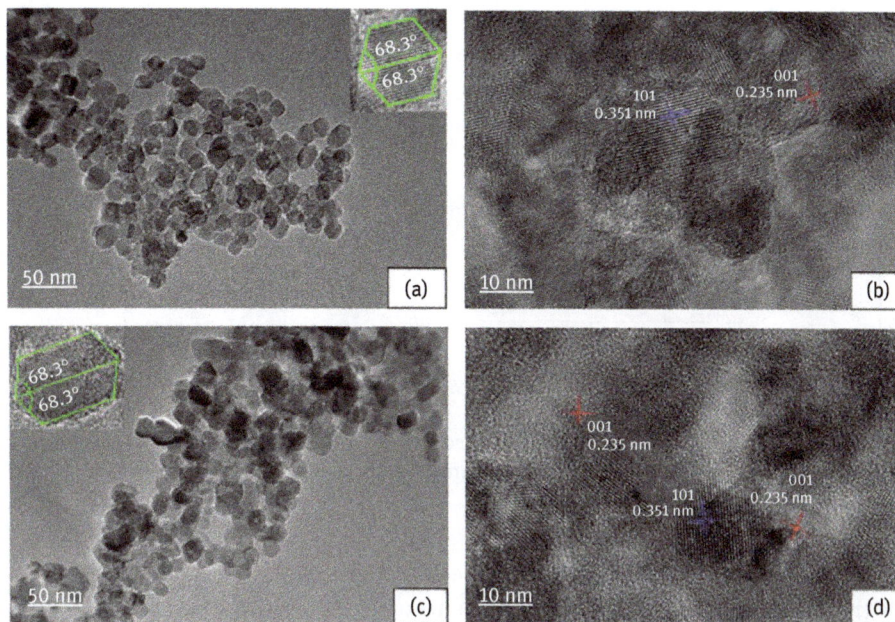

Figure 6.2: TEM images of (a) A-HF-0.25; (c) A-HF-0.75; HRTEM images of (b) A-HF-0.25; (d) A-HF-0.75. (Reprinted from Ref. [33], Copyright 2017, with permission from Elsevier.).

graphs of A-HF-0.25 and A-HF-0.75, displaying the truncated octahedral nanoparticles, and the interfacial angle is 68.3 °, which corresponds to the value between the (001) plane and (101) plane of the anatase TiO$_2$ [34]. In the mean time, the amount of (001) facets increases. The matching HRTEM results are shown in Figure 6.2(b) and (d). It can be observed that the interplane spaces of 0.351 nm and 0.235 nm are due to the exposure of (101) and (001) facets of anatase TiO$_2$ [35].

The UV–vis diffuse reflectance spectroscopy (DRS) of the as-prepared nanomaterials is shown in Figure 6.3(a) and the consequences are listed in Table 6.1. The absorption threshold of A-HF-0 is 389 nm, which agrees well with the computational results of band gap for anatase TiO$_2$ (3.20 eV). Furthermore, in comparison with A-HF-0, A-HF-x (x = 0.25, 0.5, 0.75, and 1) has an impressive improvement for ray assimilation capability.

The fluorescence spectrum is generally adopted to characterize the peril of the photogenerated electron–hole pairs. The photocatalytic liveness is largely dependent on the isolation potency of the photoinduced electron–hole pairs. The smaller emission signals of photoluminescence (PL) are related to better separation capability of photogenerated electron–hole pairs. It can be seen from Figure 6.3(b) that the fluorescence level of the samples fell off gradually from A-HF-0 to A-HF-0.75. Nevertheless, when compared with A-HF-0.75, the fluorescence level of the sample

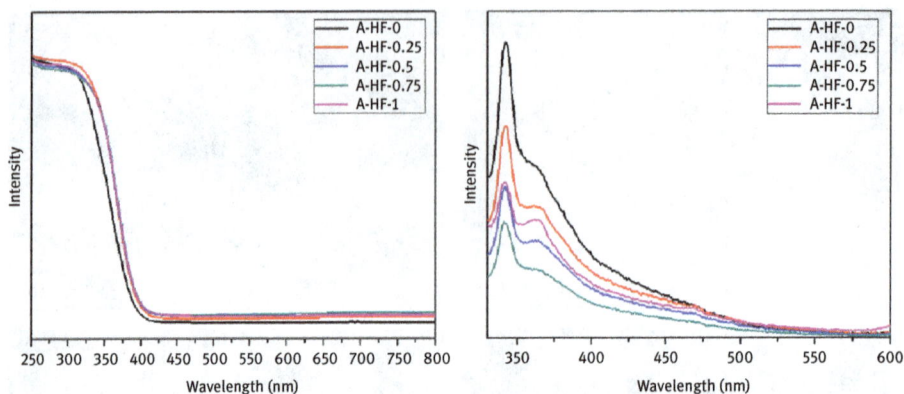

Figure 6.3: (a) UV–vis diffuse reflectance spectroscopy of as-prepared samples; (b) photoluminescence emission spectra of as-prepared samples. (Reprinted from Ref. [33], Copyright 2017, with permission from Elsevier.).

A-HF-1 displays an increasing trend, which indexes that a suitable ratio of (001) to (101) planes can impressively manage the recombination of photoexcited electrons and holes.

Performance of the as-prepared anatase TiO$_2$

The photocatalytic ability of the as-prepared samples was tested under the ultraviolet light. Each experiment was carried out for about 2 h (120 min). Shown in Figure 6.4, the elimination rate of P25 is better than that of A-HF-0, and the elimination efficiencies of P25 and A-HF-0 are 55% and 49% severally, which is displayed in Table 6.1. At the same time, the elimination rate increases with the increase of (001) planes. However, when compared with sample A-HF-0.75, the removal rate of the sample A-HF-1 displays a decreasing trend. It indicates that a suitable ratio of (001) and (101) planes can quite boost the photocatalytic capability of TiO$_2$.

Photocatalytic reaction mechanism

Based on the description data and experimental consequences, the feasible chemical behavior theory is proposed to explore the improved photoactivity, which is shown in Figure 6.5. With the irradiation of the UV lamp, photogenerated electrons were delivered from conduction band (CB) to valence band (VB). The Fermi value of the (001) planes enters its VB and that of (101) planes is at the roof of VB. When (001) and (101) are linked, the Fermi levels would be equal [35]. Therefore, the surficial heterojunction is obtained, which improved peril and transfer of photogenerated electron–hole

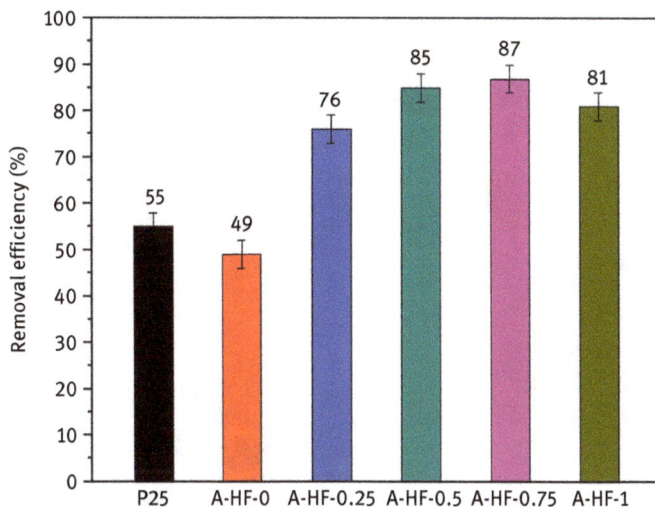

Figure 6.4: The removal efficiency of Hg^0 by the as-prepared samples. (Reprinted from Ref. [33], Copyright 2017, with permission from Elsevier.).

Table 6.1: The percentage of (001) facets, physical, chemical, and photocatalytic efficiency of the as-prepared samples. (Reprinted from Ref. [33], Copyright 2017, with permission from Elsevier).

Samples	Percentage of (001) facets	Absorbing boundary(nm)	Bandgap(eV)	Hg^0 removal efficiency (UV lamp) η （%）
P25	–	–	–	55
A-HF-0	8%	389	3.19	49
A-HF-0.25	17%	393	3.16	76
A-HF-0.5	25%	394	3.15	85
A-HF-0.75	32%	395	3.14	87
A-HF-1	38%	396	3.13	81

pairs. After surface heterojunction is formed, the photogenerated electrons on (001) planes are transferred to (101) planes, and photogenerated holes upon the (101) planes are transferred to (001) planes, making photogenerated holes and electrons assembled on (101) and (001) facets, respectively. The water molecules (H$_2$O) and hydroxyl (OH$^-$) adsorbed on the surface can be react with hydroxyl radicals (•OH) by photogenerated holes. The O$_2$ adsorbed on the surface of the samples can react with

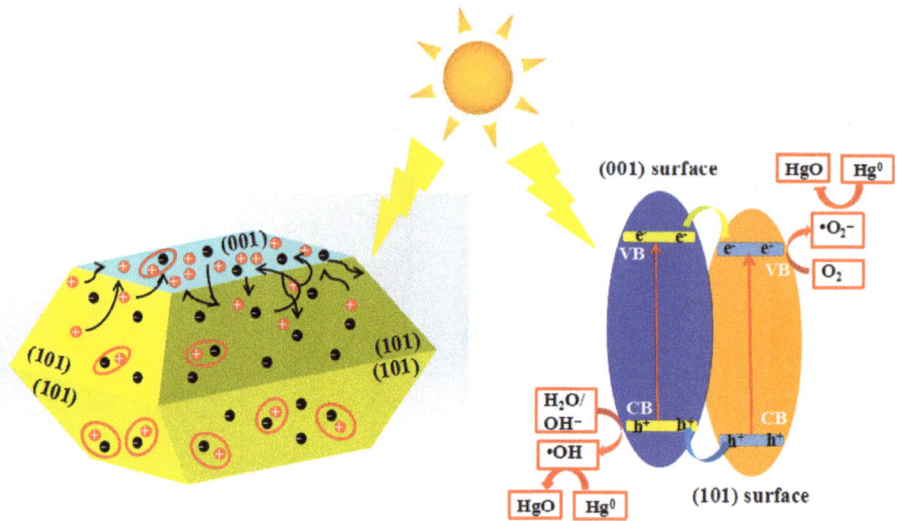

Figure 6.5: The proposed feasible photo induced electrons and holes conveyance theory and the possible reaction mechanism between different facets and Hg^0. (Reprinted from Ref. [33], Copyright 2017, with permission from Elsevier.).

the photoinduced electrons, producing superoxide radicals ($\bullet O_2^-$). With the yielded active species, $\bullet OH$ and $\bullet O_2^-$, Hg^0 could be oxidized to HgO.

6.2.2 The influence of morphology on photocatalytic removal of elemental mercury

A number of TiO_2 nanomaterials have been developed, including spheres [37, 38], nanorods [39–41], nanofibers [42–44], nanotubes [45–46], nanosheets [47–49], and interconnected architectures [50]. Hollow structured TiO_2 nanomaterials have been paid increscent attention since they possess low density, impressive specific surface area (SSA), and attractive photocatalytic liveness [51]. From the literature, the hollow structured TiO_2 nanomaterials have been developed smoothly and implemented to a lot of application areas such as the decomposition of organic contamination [52]. But as far as we know, there is a few literature reporting on the applications of the hollow-sphere TiO_2 in the field of the Hg^0 elimination from coal-fire-derived flue gas through photocatalytic oxidation. In our research, hollow-sphere TiO_2 was synthesized with nontemplate method by hydrothermal technique adopting trifluoroacetic acid (TFA) and $Ti(SO_4)_2$ as the feed stocks [53]. The photocatalytic ability of hollow-sphere TiO_2 for the Hg^0 peril from coal-combustion flue gas with photocatalytic oxidation was studied.

Preparing TiO₂ hollow microspheres

All materials adopted in the research were in analytical reagent grade with no more distillation before being used. $Ti(SO_4)_2$, which was bought from Sinopharm Chemical Reagent Co., Ltd, was applied as titanium source. TFA was attained from Aladdin Reagent Co. Ltd. In a representative process, 1.5 g of $Ti(SO_4)_2$ and some specific quantity of TFA were dropped into 125 ml of D.I. water with magnetic milling. The pH of the liquor was changed to about 1.3 with 1.0 M H_2SO_4 liquor. With continuous magnetic stirring for 60 min, the attained lucid liquor was separated into two equal fractions and put into 100 ml Teflon-lined autoclave. Accordingly, the hydrothermal approach was carried out at 180 °C for 12 h. Thereafter, the Teflon-lined autoclave was refrigerated at room temperature. The resulted sediment was baptized with D.I. water and ethyl alcohol for three times and then desiccated at 80 °C for 12 h. Samples were prepared, with the molarity of TFA to $Ti(SO_4)_2$ (M) was changed from 1–4.

Characterization of TiO₂ hollow microspheres

The XRD modes of the as-prepared TiO_2 nanomaterial photocatalyst are shown in Figure 6.6. The size of the crystal was analyzed by using Scherrer's equation into the whole width at half-maximum of the (101) peak of anatase TiO_2:

$$D = K\lambda/B\cos\theta \tag{6.1}$$

D is the crystallite magnitude to be computed, B is the half-height-width of the diffraction peak of anatase TiO_2, $K = 0.89$ is a parameter, θ is the characteristic X-ray

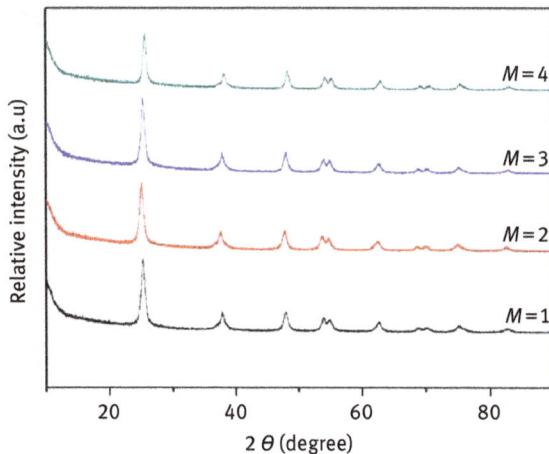

Figure 6.6: XRD patterns of TiO_2 samples prepared with $M = 1, 2, 3, 4$ at 180 °C for 12 h. (Reprinted from Ref. [56], Copyright 2015, with permission from Elsevier.).

irradiation, and λ is the X-ray wavelength that corresponds to the Cu Ka irradiation (0.15406 nm), with R-silicon (99.9999%) as a criterion for the mechanical line widening [54]. The XRD patterns of the hollow-sphere TiO_2 show that all the peaks are suited to the anatase phase of TiO_2 (JCPDS No. 21–1272, space group I41/amd (141)). Based on the Scherrer equation, the mean sizes of crystallites given in Table 6.2 demonstrate an increase in size with an increase in M. It was due to the fact that at high temperature, a fraction of TFA was resolved in liquor, producing a specific quantity of F^-, which can improve crystallization of anatase phase and enhance crystallite growth by adsorbing superficially the as-prepared TiO_2 particles [55].

Table 6.2: Effects of M on specific surface areas, average pore parameters, crystalline size, and Hg^0 removal efficiency of titania powders. (Reprinted from Ref. [56], Copyright 2015, with permission from Elsevier).

M	Specific surface area (m²/g)	Porevolume (cm³/g)	Porediameter (nm)	Crystalline size(nm)	Hg⁰ removalefficiency (visiblelight) η (%)
P25	55.1	0.06	3.9	20	77.81
1	166.73	0.198	7.48	13.33	74.43
2	164.77	0.274	10.49	13.67	78.64
3	152.41	0.252	10.2	14.05	76.36
4	148.63	0.274	11.23	14.50	72.70

The hierarchically nanoporous structuredTiO_2 hollow spheres were verified by the SEM and TEM images. The characteristic SEM and TEM patterns of the hollow spheres are shown in Figure 6.7. It illustrates the formation of hollow-sphere TiO_2 with diminutive size distribution, with ca. 800 nm average size. These results agree well with the layered and vesicular structure of the TiO_2 hollow spheres, which were proved by nitrogen adsorption–desorption analysis. The thickness of hollow-sphere TiO_2 wall ranges from 200–400 nm. It also indicates that TiO_2 surface gradually eroded with the increase in TFA capacity, making the microspheres founder owing to F-decomposed from TFA. To attain more data about the hollow structure TiO_2, the as-prepared samples were depicted by TEM. Figure 6.7d shows the TEM pattern of the hollow-sphere TiO_2 samples fabricated at $M = 2$. The concentration diversity of electrons from the dark edge to pale center additionally verifies the hollow inner obviously. It demonstrates that fluoride in an acidic surroundings could enhance self-transformation of solid sphere TiO_2 and could yield hollow-sphere TiO_2. The outcome agrees well with the former researches [57].

For investigating the optical characteristics of the hollow microsphere TiO_2, the materials were characterized by UV–vis DRS. The outcome is depicted in Figure 6.8.

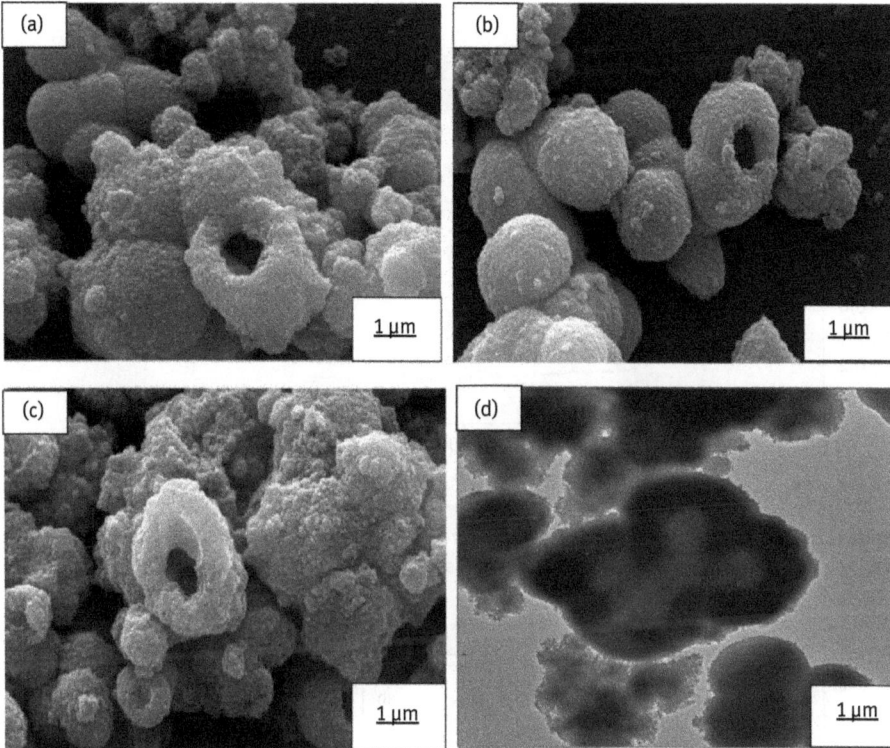

Figure 6.7: SEM (a, b, c) and TEM (d) images of TiO$_2$ samples prepared at $M = 1$ (a), 2 (b, d), and 4 (c). (Reprinted from Ref. [56], Copyright 2015, with permission from Elsevier.).

It shows that the as-prepared hollow microsphere TiO$_2$ possesses wider absorption range than P25, reaching the wide region of the UV–vis from 240–700 nm. The E_{bg} ranging from 396 to 404 nm is useful to promote the photocatalytic capability under the UV radiation, making it possess higher elimination potency of Hg than P25 under identical reaction conditions. Nevertheless, the accurate theory of the structure of hollow sphere to improve the photocatalytic liveness needs more research in detail.

Performance of TiO$_2$ hollow microspheres

The photocatalytic peril efficiency of the as-prepared TiO$_2$ hollow spheres on Hg^0 was investigated. As shown in Figure 6.9(a), each experiment was conducted for about 60 min; the results demonstrate that when molar ratio (M) increased from 1 to 4, elimination efficiencies η of Hg^0 were 74.43%, 78.64%, 76.36%, and 72.70% severally. Hg^0 photocatalytic elimination rate of the P25 was 77.81%.

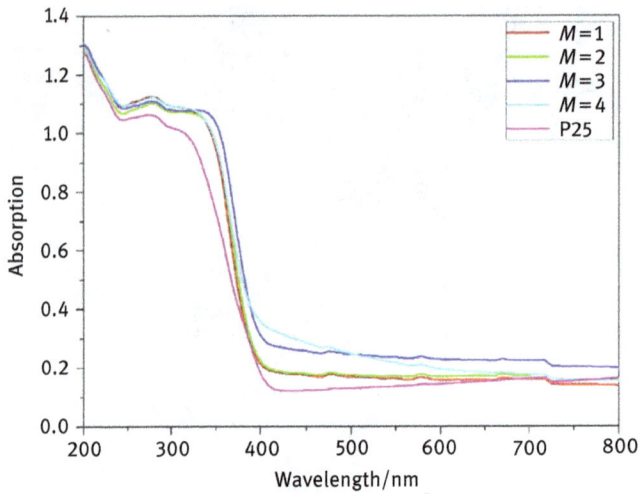

Figure 6.8: UV–vis spectra of TiO_2 samples prepared at M (M = 1, 2, 3, 4) and P25. (Reprinted from Ref. [56], Copyright 2015, with permission from Elsevier.).

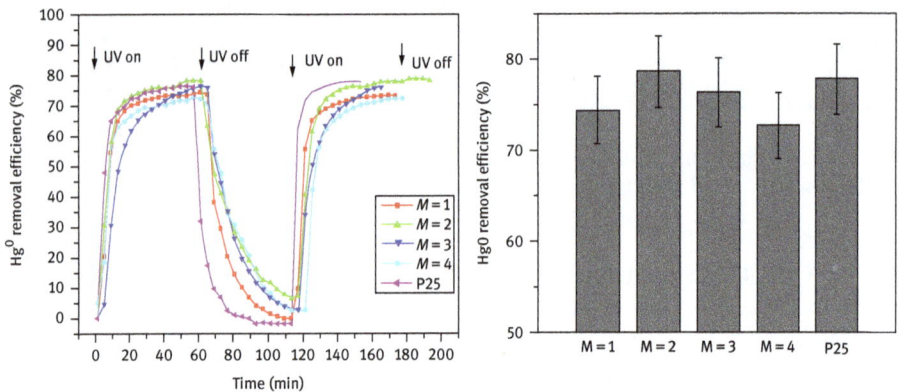

Figure 6.9: (a) The experiment procedure diagram; (b) effect of M on Hg^0 peril rate. (Reprinted from Ref. [56], Copyright 2015, with permission from Elsevier.).

The graph of the experimental process is shown in Figure 6.9(a). It demonstrates that the removal efficiency of mercury gradually enhanced after the light was turned on. After about 60 min, the mercury concentration in the simulated flue gas remained unchanged, reaching a still state. When UV light was turned off, mercury concentration enhanced sharply, meaning that the mercury peril rate weakened. Then UV light was turned on again repeatedly until the mercury recovered to the primary content remained unchanged.

When M was less than 2, the photocatalytic liveness of the hollow-sphere TiO_2 was distinctly increased owing to the improvement of TiO_2 crystallization, TiO_2 hollow structure development, and facial fluorination of TiO_2 via F provided by TFA. From Figure 6.9(b), it could be seen that with the greater increase of M, photocatalytic liveness of the hollow-sphere TiO_2 became lower and its mercury peril rate reduced, which manifested that the photocatalytic capability of the as-prepared hollow-sphere TiO_2 was relative to F. When the addition of TFA was higher than a specific score, photocatalytic liveness of the as-prepared hollow-sphere TiO_2 would begin to decrease. This is consistent with previous studies [58], that is, too high F concentration restrained the photocatalytic liveness.

The elimination rate of mercury in this research group was over 70%. When M was 2, it possessed nearly identical potency as P25, while the latter comprises Fe^{3+}, which helped to improve the photocatalytic capability of TiO_2 nanomaterial. The as-prepared hollow-sphere TiO_2 with no Fe-doping attained the efficiency similar to that of P25, owing to the fact that the added TFA degraded to provide F, making TiO_2 surface corrode and form hollow structure. Therefore, the SSA of the as-prepared hollow-sphere TiO_2 was higher than that of P25. The large SSA enhanced its adsorption ability so that it promoted the photocatalytic capability. But too much F would hold active sites on the TiO_2 surface, which inhibits the adsorption of Hg^0, thereby decreasing its photocatalytic potency.

The influence of chemical reaction temperature on the peril potency of mercury was carried out under the condition of 303.45 μWcm^{-2} UV intensity, and the hollow microsphere in TiO_2 materials was fabricated when $M = 2$. Each experiment was conducted for about 60 min, and the experimental data are shown in Figure 6.10. When reaction temperatures were set as 14, 55, 90, 120, and 150 °C, the removal efficiencies of Hg^0 were 78.64%, 82.75%, 82.15%, 76.96%, and 67.76% severally. It showed that the photocatalytic removal potency became higher first and then lower following the increase of the reaction temperature. When chemical reaction temperature was above 100 °C, the mercury oxidation elimination potency reduced sharply. This was owing to the fact that there would be lots of collisions among the reactive species of gas-phase and the as-prepared TiO_2 nanomaterial samples, and the amount of Hg^0 sedimentation and adsorption on the surface of the photocatalyst would decrease, resulting in a feasible reduction of both quantity of reactive species and chemical reaction potency [59].

Mechanism of the photocatalysis

The band gap of TiO_2 nanomaterial is 3.2 eV, corresponding to the photon energy at 387.5 nm. Under the condition of the UV radiation where the wavelength is lower than 387.5 nm, the electrons at the VB of the nano-TiO_2 would move to the CB to form the electrons and holes, which are of strong chemical liveness.

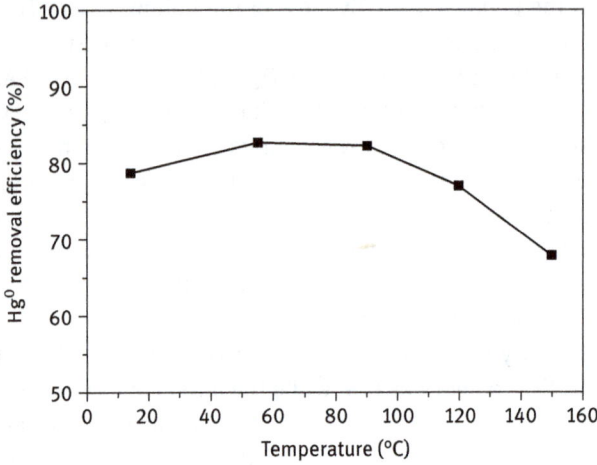

Figure 6.10: Effect of reaction temperature on Hg^0 removal efficiency η. (Reprinted from Ref. [56], Copyright 2015, with permission from Elsevier.).

As shown in Figure 6.11, at the starting stage of the reaction, a little of mercury was adsorbed on the surface of the hollow-microsphere TiO_2. At the time of the photocatalytic reaction process, the UV radiation excited the electrons in

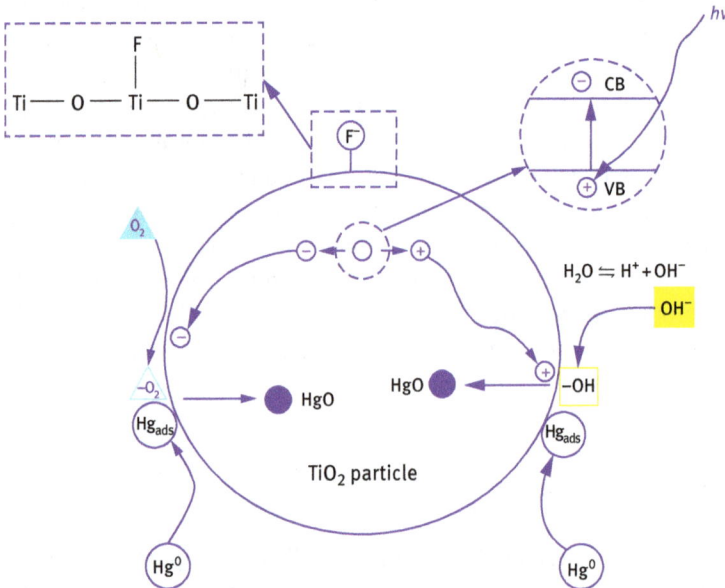

Figure 6.11: A schematic diagram of generation and transfer of charge carriers in F–TiO_2 under UV irradiation. (Reprinted from Ref. [56], Copyright 2015, with permission from Elsevier.).

TiO_2, which produces electron–hole pairs, and afterward the generated electrons and holes were distributed on the surface of the as-prepared hollow-microsphere TiO_2. O_2, H_2O, and OH^- adsorbed on the surface of the as-prepared TiO_2 catalyst reacted with the generated holes to produce super oxygen ions O_2^- and hydroxyl radicals •OH. These free radicals with intensive oxidizing ability would transform Hg^0 to Hg^{2+}. The as-prepared TiO_2 hollow microspheres possessed the band gap from 396–404 nm, being longer than 387.5 nm, and so the photocatalytic ability with UV irradiation of the as-prepared TiO2 photocatalyst was much improved. It also partly explains the reason why TiO_2 hollow microsphere had better elimination potency of mercury than P25 in specific chemical reaction surroundings.

As discussed already, the formation of the hollow microspheres was owing to the existence of F. Because fluorine is the most electronegative element, the Ti–F group on the surface can trap the electrons of the CB by tightly capturing the captured electrons and transfer them to O_2 that are adsorbed on the TiO_2 surface, which enhances the photocatalytic liveness of the as-prepared catalyst [60–61].

The fabrication process of the hollow-sphere TiO_2 was as follows. The spheres that were with glabrous surfaces were obtained at the initial state of the reaction. Then, after the reaction continued for 6 h, the crystalline plates yielded primarily on the surface of the microspheres. At the same time, the hollow interiors start to act on the core of the spheres. With the increase of the reaction time, the hollow spheres were gradually formed. In time of the hollow-sphere-forming process, the center and inner parts of the solid spheres were melted and recrystallized on shell of the solid spheres stage by stage, forming hollow spheres eventually [62].

6.2.3 The influence of metal oxides with TiO_2 on photocatalytic removal of elemental mercury

As is well known, TiO_2 photocatalyst can solely be irradiated by extraviolate radiation, which holds for barely around 3–5% of the heliacal radiation. Meanwhile, the electron transfer rate is low and the recombination between excited electrons and holes is fast, which often makes quantum yield be of very low rate and photo-oxidation rate be limited [63–64]. CuO, one of the inexpensive metal oxides, has been regarded as an excellent active content to reduce water under sacrificial surroundings [65–68], because it can inhibit the recombination of photoexcited electron–hole pairs. Meanwhile, it can restrain the band gap energy. Nevertheless, there are few researches on the application of CuO loaded on TiO_2 nanomaterial as a photocatalyst to photocatalytically oxidize Hg^0. Furthermore, there are few researches dealing with the influence of light irradiation on photocatalytic oxidation ability.

Preparation of CuO/TiO$_2$ photocatalysts

The CuO/TiO$_2$ nanomaterial photocatalysts with 0–15 wt.% of CuO loading were fabricated by the coprecipitation method described by Yoong et al. [69]. In this process, Cu(NO$_3$)$_2$ hemihydrate and glycerol (molarity of 1:2) were put into ultrapure water (100 mL) to yield an aqueous copper(II)–glycerol compound. The mass of the used Cu(NO$_3$)$_2$ hemihydrate and glycerol was dependent on the loaded CuO. Degussa P25(5.0 g) was put into the liquor by continuously milling. The copper–glycerol compound was deposited afterward on TiO$_2$ host by dropping 0.5 M NaOH while milling until pH of 12. The obtained liquid was milled for another 60 min, and then the sediment, which may be presumable Cu(OH)$_2$/TiO$_2$), was gathered by centrifugal migration. After being washed by ultrapure water repeatedly, the Cu(OH)$_2$/TiO$_2$ powder was obtained by desiccating all night at 70 °C in chamber atmosphere. CuO/TiO$_2$ photocatalysts were attained by calcinating the Cu(OH)$_2$/TiO$_2$ nanoparticles at 300 °C for 120 min.

Characterization of CuO/TiO$_2$ photocatalysts

The XRD graphs of CuO/TiO$_2$ nanomaterial catalysts are shown in Figure 6.12 (a). The graphs for all the prepared chemicals are dominantly the peaks owing to anatase and rutile in TiO$_2$. Based on the result of anatase (101) and rutile (110) reflection, the weight ratio of anatase to rutile in the Degussa P25 host was around 6:1, which agrees well with the specifications by the business firm. It is interesting that when the doped CuO reached 5 wt.%, rutile (110) disappeared, and then reappeared when the weight of CuO increased (Figure 6.12 (a)).

Figure 6.12 (b) demonstrates an extensive view of XRD graphs in the range of $2\theta = 30–50$, where the CuO reflections were most intense. It can be seen that no characteristic peak of CuO existed in the as-prepared catalysts when the loaded CuO was no more than 5 wt.%, which shows that the CuO crystalline phase was so tiny that it cannot be observed in the TiO$_2$ phase. Well-distributed active contents in the photocatalyst can improve its catalytic liveness [61]. Nevertheless, when CuO doping was 7.5 wt.% or more, the monoclinic CuO peaks heightened. The XRD data verified that the diffusive CuO on TiO$_2$ had an impressive shift when the CuO loading ranges from 5–7.5 wt.%.

As mentioned above, the prepared 0–15 wt.% CuO/TiO$_2$ nanomaterial photocatalysts are shown in Figure 6.13. When the CuO loading was low (0.31–1.25 wt.%), the color of the as-prepared CuO/TiO$_2$ nanomaterial samples ranged from light blue to green, which is characteristic of octahedrally coordinated Cu(II) ions [72]. When CuO density was 2.5 wt.% or more, the as-prepared CuO/TiO$_2$ nanomaterial photocatalysts were grey, with intensity increasing in a proportionate trend with the increase of CuO loading. Since the color of CuO is black and that of TiO$_2$ is white, the

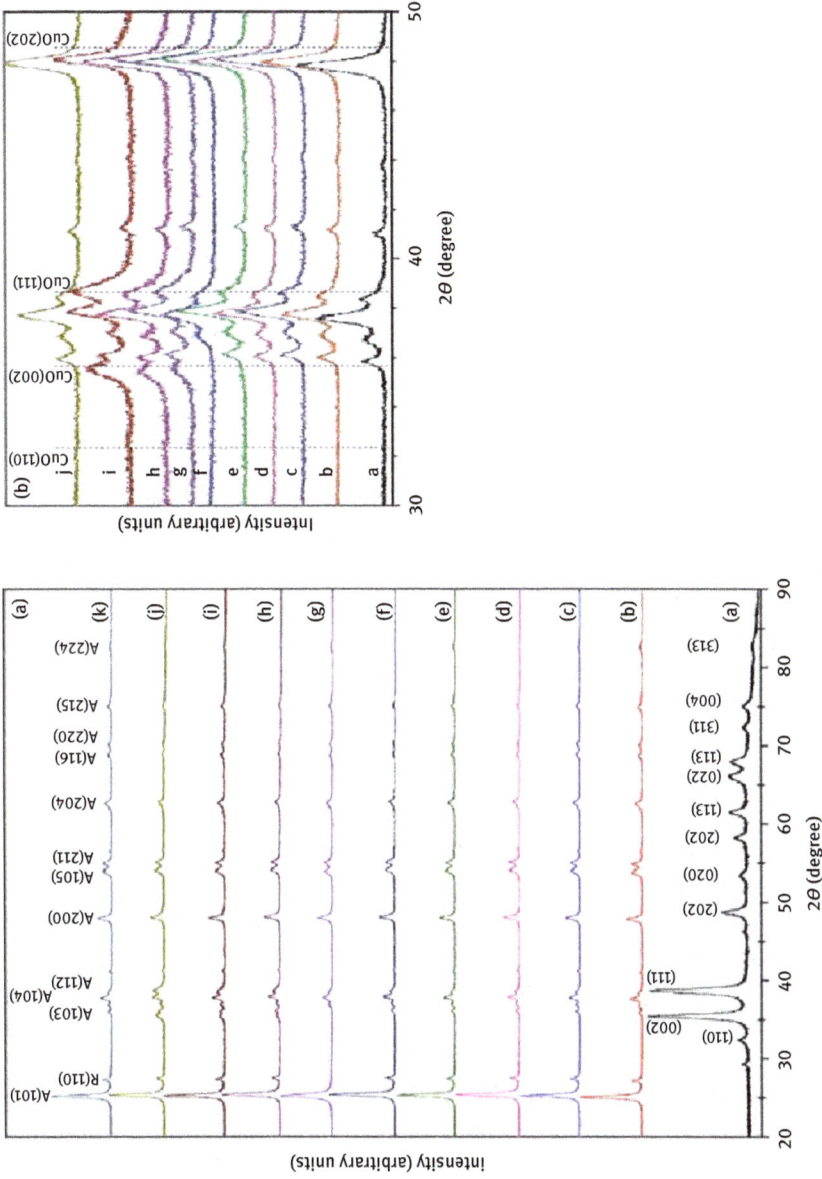

Figure 6.12: (a) Powder XRD patterns of the 0–15 wt.% CuO/TiO$_2$ photocatalysts after calcinating at 300 °C for 2 h; (b) the expanded view of the XRD patterns. (Reprinted from Ref. [70], Copyright 2015, with permission from Elsevier.).

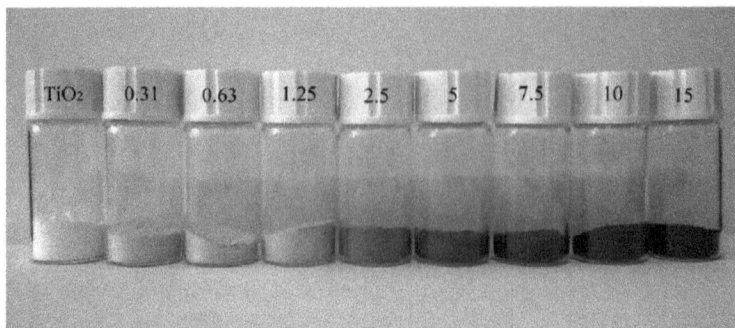

Figure 6.13: Photograph of the 0–15 wt.% CuO/TiO_2 photocatalysts. (Reprinted from Ref. [70], Copyright 2015, with permission from Elsevier.).

reason why 2.5–15 wt.% CuO/TiO_2 nanomaterial chemicals were grey could be because of the fact that CuO nanoparticles were formed on the TiO_2 surface. XRD data are shown in Figure 6.12, which show a single CuO phase appeared in the 2.5–15 wt.% CuO/TiO_2 nanomaterials. The morphology diversity in the CuO/TiO_2 photocatalysts fabricated at tiny intensity CuO (<2.5 wt.%) and strong intensity CuO (2.5–15 wt.%) strongly indicates that the electronic state of CuO(II) among these samples was varying.

To obtain CuO density in the as-prepared CuO/TiO_2 nanomaterial photocatalysts and their nanoporous structures, the FESEM/EDAX analysis was conducted and the micrographs of CuO/TiO_2 samples are depicted in Figure 6.14, which demonstrate no remarkable diversity in the morphology of the as-prepared catalyst with varying scale of doped CuO compared with that of P25. When the doped density of CuO was over 5 wt.%, the as-prepared nanomaterial catalyst started to slightly aggregate. As shown in Figure 6.14, Cu, Ti, and O are present in the whole as-prepared samples. Although when doped CuO content was small, some Cu^{2+} was probed on the surface of as-prepared TiO_2. With the increase in CuO content, CuO concentration in the as-prepared samples also gradually increased. Through the analyzed results of EDAX listed in Table 6.3, CuO level in the as-prepared CuO/TiO_2 nanomaterial photocatalysts was nearly the same as that of the actually loaded CuO.

TEM was adopted to detect the distribution of CuO on the basis of TiO_2 as a role of loading CuO. TEM graphs of the chosen as-prepared CuO/TiO_2 photocatalysts are shown in Figure 6.15. As shown in Figure 6.15(a), the TEM images of the 1.25 wt.% CuO/TiO_2 nanomaterial photocatalyst possesses the characteristics of P25 TiO_2 host, consisting of fine spherical anatase TiO_2 crystallites and larger angular rutile TiO_2 crystallites, with diameters 20–30 nm and 40–60 nm, respectively. Although copper was probed by EDAX analysis, no single CuO particles were observed, which demonstrate that when low CuO was loaded, Cu^{2+} was largely distributed in the TiO_2 support particles. When more CuO was loaded, especially 10 wt.% or more CuO,

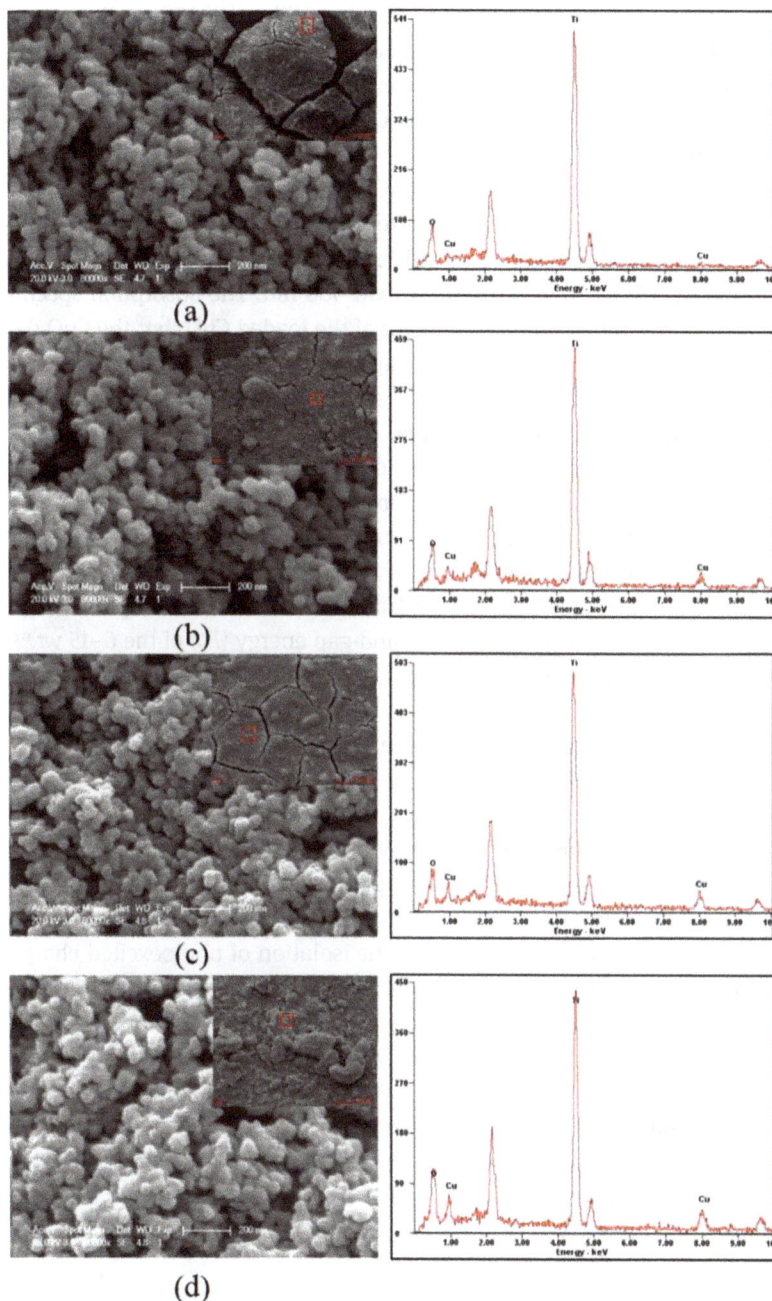

Figure 6.14: FESEM images (left) and EDAX spectra (right) for chosen CuO/TiO$_2$ photocatalysts. (Reprinted from Ref. [70], Copyright 2015, with permission from Elsevier.).

scattered CuO would be detected on the nano-TiO$_2$ host (Figure 6.15(b)). The size of the CuO nanoparticles loaded on the TiO$_2$ support was around 2–4 nm in diameter. Then those nanoparticles grew priorly on anatase TiO$_2$ particle host. The TEM values shown in Figure 6.15 demonstrate that the CuO nanoparticle was formed only when the loaded CuO surpassed certain amount.

The optical characteristics of 0–15 wt.% CuO/TiO$_2$ nanomaterial photocatalysts were investigated with absorption spectroscopy from 250 to 800 nm at room temperature, and the data are shown in Figure 6.16. All the fabricated samples presented impressive absorption of light at wavelengths less than 400 nm. The absorption spectra of visible light heavily depended upon the content of the loaded CuO, but the CuO deposition made it easier for absorption of visible light by all the as-prepared samples. When the loaded CuO was lower than 2.5 wt.%, absorption performance at wavelength of 450 nm or longer than 800 nm was significant, which is in compliance with the as-prepared samples' blue-green color. When the loaded CuO was 2.5 wt.% or more, strong absorption over the whole visual spectrum was detected, which agrees well with the formation of CuO nanoparticle, that is, bulk CuO absorbs intensely in the extent of 300–800 nm. According to the UV–vis result shown in Figure 6.16, it can be observed that the electrical state of CuO(II) was different across the samples, which had tiny and large loaded CuO. Eventually, the band-gap energy (E_g) of the 0–15 wt.% CuO/TiO$_2$ nanomaterial photocatalysts was computed, corresponding to the absorption spectrum of the samples based on the equation of $E_g = 1,240/\lambda_{\text{Absorp.Edge}}$[73]. The results are listed in Table 6.3, which demonstrate that the absorption edge of the as-prepared CuO/TiO$_2$ nanomaterial photocatalysts significantly motioned to less energy when the loaded CuO increased.

The PL emission spectra have been diffusely applied to detect the transferring potency of the charge carriers and the lifetime of photoexcited electron–hole pairs in semiconductors. As shown in Figure 6.17, the study adopting PL spectroscopy also indicates that there was a great enhancement in the isolation of photoexcited charge carriers [74]. The intense characteristics of the PL signal detected for the P25 TiO$_2$ demonstrate that with the light excitation, rapid recombination of generated electrons and holes occurred. After TiO$_2$ was doped with CuO, the PL strength of the as-prepared photocatalysts was lower than that of P25 TiO$_2$, indicating CuO effactually suppressed the recombination of electron–hole pairs in TiO$_2$, regardless of mild CuO loading. This is ascribed to CuO doping, because the loaded CuO suppressed the recombination between the generated electrons and holes, which heightened the photocatalytic ability of TiO$_2$ for the reaction of Hg^0.

Performance of CuO/TiO$_2$ photocatalysts

The photocatalytic oxidation of Hg^0 in the gas phase was investigated. Initially, the oxidation potency of the as-prepared CuO/TiO$_2$ photocatalyst with varying content

Table 6.3: List of the physical, chemical, and photocatalytic properties of the 0–15 wt.% CuO/TiO_2 photocatalysts. (Reprinted from Ref. [70], Copyright 2015, with permission from Elsevier).

CuO loading (wt.%)	BET surface area (m²/g)	Pore volume (cm³/g)	Pore diameter (nm)	CuO content by EDAX (wt.%)	Absorbing boundary (nm)	Bandgap (eV)	Hg^0 removal efficiency (visible light) η (%)
TiO_2	47.65	0.176	16.02	–	394	3.15	19.9
0.31	43.83	0.460	43.07	–	464	2.67	54
0.63	46.04	0.457	39.92	–	512	2.42	47
1.25	47.26	0.315	31.01	1.1	540	2.29	57.9
2.5	46.98	0.388	36.41	–	600	2.07	42.7
5	46.84	0.308	31.46	5.5	612	2.03	45.3
7.5	46.49	0.347	37.30	–	628	1.97	37
10	42.18	0.386	43.83	9	645	1.98	20.2
15	42.41	0.351	40.68	14.7	718	1.73	7.3
CuO	31.37	0.268	27.36	–	–	1.34	3

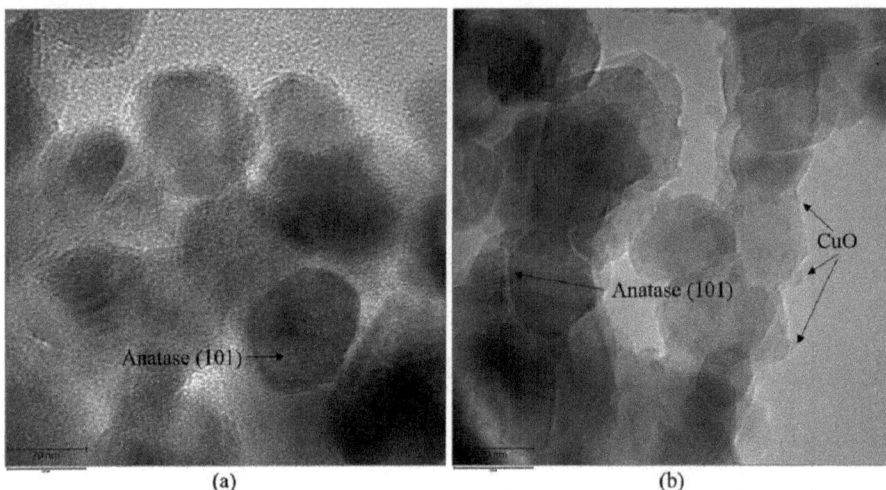

Figure 6.15: TEM images of (a) 1.25 wt.% CuO/TiO_2; (b) 10 wt.% CuO/TiO_2. (Reprinted from Ref. [70], Copyright 2015, with permission from Elsevier.).

of loaded CuO under extraviolet light and xenon light circumstances was observed. The experimental data are shown in Figure 6.18. The CuO itself had no ability to oxidize Hg^0 photocatalytically with the UV irradiation or xenon lamp circumstances because the edge energy of CuO (E_g = 1.34 eV) was so narrow that when it was irradiated by UV light, the energy was insufficient to yield •OH. With the increasing loading of CuO, the elimination potency of mercury reduced but remained higher than 70%, which being much better than P25 TiO_2 with the UV irradiation. This can be mainly attributed to the few Cu ions inserted into the TiO_2 lattices, which enhanced the photocatalytic performance. CuO effactually restrained the recombination of the generated electron–hole pairs in TiO_2 even when CuO amount was moderate [75]. But excess loading of CuO induced a decreased photocatalytic potency. The critical theory is that excess CuO covered the TiO_2 surface, blocking the active sites [76]. In the range of loaded CuO more than 1.25 wt.%, when the loading CuO content was 5 wt.%, the efficiency reached the highest level. As is well known, rutile TiO_2 is with higher band than anatase TiO_2, so it is easier for anatase TiO_2 to be excited by photons. So decreasing rutile TiO_2 amount in the catalyst can be a promising prospect to raise the photocatalytic potency for future studies.

For the experimental results of Hg^0 oxidation efficiency with catalytic process under sunlight, that is, xenon lamp, the excitation is demonstrated in Figure 6.18. When the amount of doped CuO reached 1.25 wt.%, the removal efficiency of mercury increased to 57.8%, which being much better than that of P25 with day light irradiation. Raw TiO_2 had photocatalytic ability below daylight. Nevertheless, a little of Fe ions in P25 TiO_2 improved its photocatalytic performance [77], which increased the

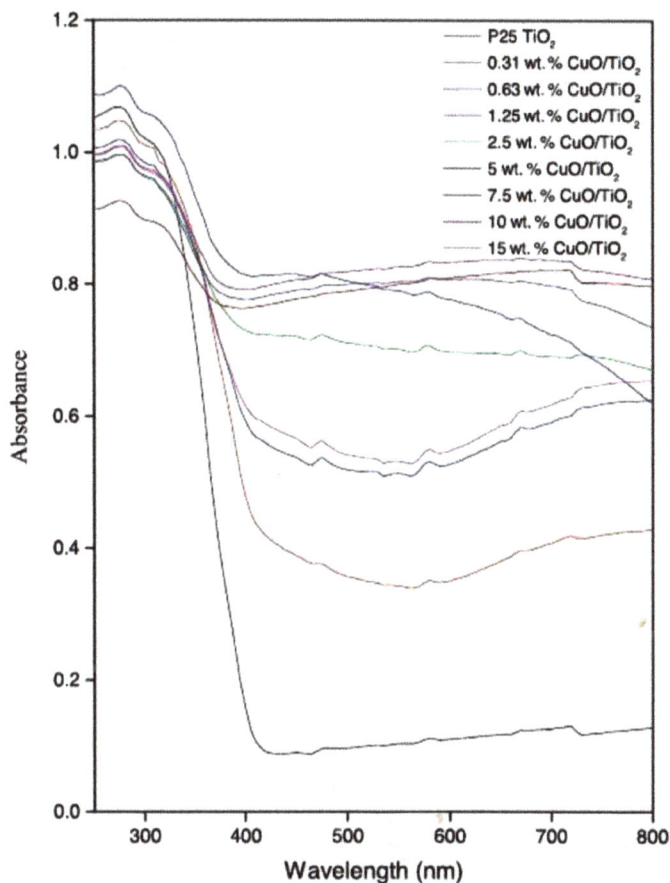

Figure 6.16: UV–vis absorption spectra of the 0–15 wt.% CuO/TiO$_2$ photocatalysts. (Reprinted from Ref. [70], Copyright 2015, with permission from Elsevier.).

peril potency of mercury by the P25 TiO$_2$ with sunlight irradiation to about 20%. When the loaded CuO was low (1.25 wt.%), the copper existed as a submonolayer Cu (II) or CuO species on the TiO$_2$ support. The Fermi energy of the adsorbed species was positive corresponding to the CB of TiO$_2$, facilitating the migration of photogenerated electrons from TiO$_2$ to Cu(II) or CuO centers on the surface. When the loaded CuO was above 1.25 wt.%, the oxidation potency of Hg^0 reduced keenly. Raw CuO was with no active ability of photocatalytic oxidation for Hg^0 in our study, as shown in Figure 6.18. Therefore, the reason why the performance of the as-prepared CuO/TiO$_2$ nanomaterial photocatalysts decreased when the loaded CuO was higher than 1.25%, the optimal value, was due to the fact that the TiO$_2$ sites were partially blocked by the CuO nanoparticles. This result demonstrates that the highest oxidation potency of Hg^0 was found for 0.31–1.25 wt.% CuO/TiO$_2$ photocatalyst.

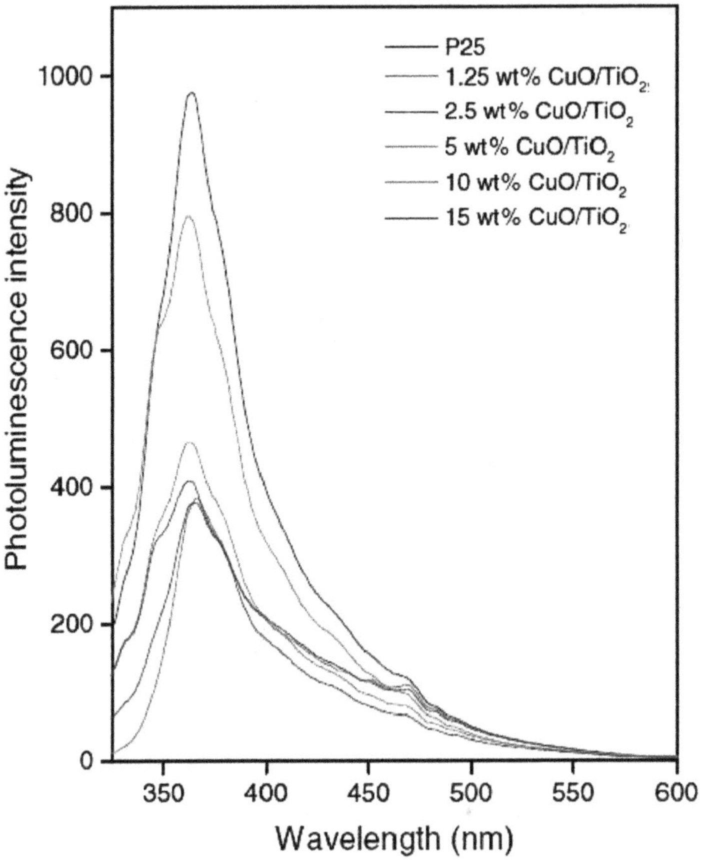

Figure 6.17: Photoluminescence spectra for 0–15 wt.% CuO/P25 TiO$_2$ photocatalysts. (Reprinted from Ref. [70], Copyright 2015, with permission from Elsevier.).

Photocatalytic reaction mechanism

Figure 6.19 demonstrates the schematic diagram for the separation of current carriers of the CuO/TiO$_2$ nanomaterial catalysts with irradiation of the visual light. In our work, the as-prepared CuO/TiO$_2$ nanomaterial photocatalyst displayed remarkable oxidation potency for mercury under the visible light irradiation condition. The mechanism can be as the following: first, the CuO doping suppressed the recombination process of photoexcited electron–hole pairs, enhancing the photocatalytic oxidation efficiency. Secondly, the electron transfer helped the CuO/TiO$_2$ nanomaterial catalyst to be of good photocatalytic performance with the irradiation of visible light. As a fact, the VB value of the as-prepared TiO$_2$ or CuO/TiO$_2$ lies

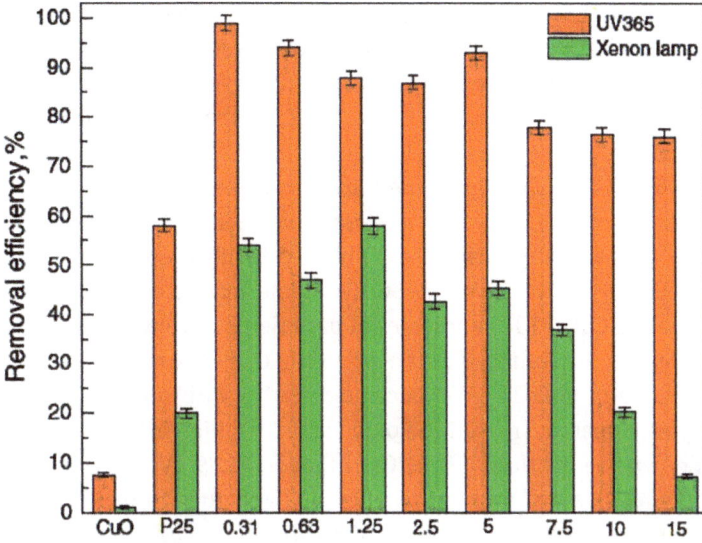

Figure 6.18: Influence of CuO-doping density on the mercury peril potency of CuO/TiO$_2$ under UV365 light and visible light (xenon lamp). (Reprinted from Ref. [70], Copyright 2015, with permission from Elsevier.).

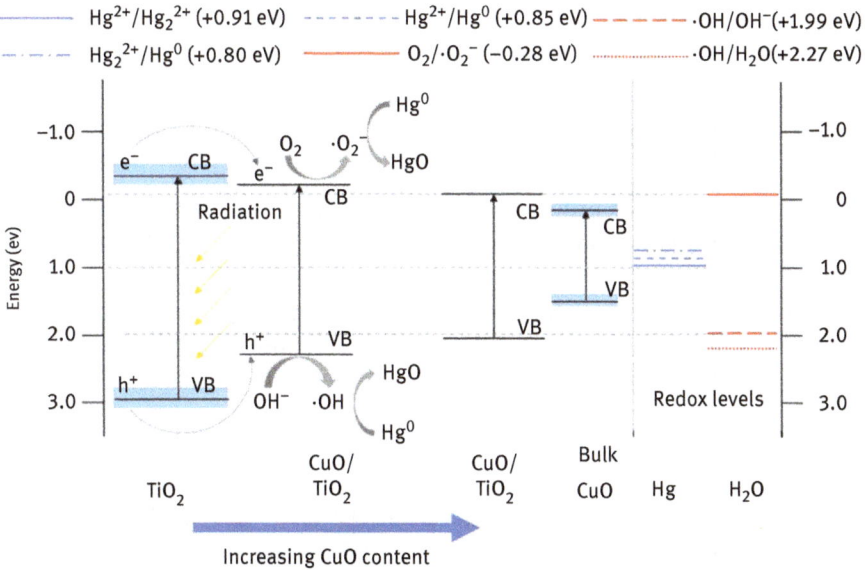

Figure 6.19: Influence of light source on the mercury elimination potency of CuO/TiO$_2$. (Reprinted from Ref. [70], Copyright 2015, with permission from Elsevier.).

under the photoredox reaction, which involves the direction of oxidation of mercury or the snatched of the allodial hydroxyl radicals. When the loaded CuO was low, the existing species of copper was submonolayer Cu(II) or CuO on the TiO$_2$ support. The Fermi energy of the adsorbed species was positive based on the CB of TiO$_2$, which resulted in the migration of photogenerated electrons from TiO$_2$ to the centers of Cu (II) or CuO on the surface as depicted in Figure 6.19. This is the reason that the as-prepared photocatalyst possesses the ability to oxidize elemental mercury with the irradiation of sunlight. Hg^0 was photocatalytically reacted by photoexcited holes or the photogenerated allodial OH radicals [78–80]. With the increasing ratio of the loaded CuO, the CB of the CuO/TiO$_2$ nanomaterial became positive with regard to O$_2$/•O$_2^-$ (−0.28 eV) couples, inhibiting the transportation of electrons from CuO to O$_2$. Meanwhile, the value of CuO/TiO$_2$ VB became negative with regard to •OH/ OH$^-$ (+1.99 eV) couples, which decreased the oxidizing capability. While the level of catalyst VB energy was below that of •OH/OH$^-$, that is, +1.99 eV, the photoexcited holes can barely oxidize Hg^0 and the potency was very low. In reality, with UV irradiating, a small amount of CuO was deoxidized to Cu$_2$O or Cu [81]. Nevertheless, O$_2$ density in the flue gas is much higher than that in liquid, and the existed O$_2$ could inhibit CuO to be deoxidized into Cu$_2$O or Cu. Therefore, the opposite influence in oxidizing Hg^0 is little. Generally, the band gap of the photocatalyst got narrower with the enhancement of loaded CuO, while the oxidation capability is still at a loss. Therefore, the Hg^0 oxidation efficiency decreased when the loaded CuO was high.

6.3 Conclusion

1) Anatase TiO$_2$ with coexpose (001) and (101) facets has a remarkable influence on the photocatalytic liveness, and the appropriate scale of (001) to (101) planes is 32% based on the experimental data.
2) The band gap width of the as-prepared hollow microsphere TiO$_2$ ranged from 396–404 nm, helping enhance photocatalytic performance with the UV excitation, making its higher removal efficiency of mercury than that of P25 under the testing conditions.
3) CuO/TiO$_2$ nanomaterial photocatalyst was fabricated by applying the coprecipitation strategy by adjusting the level of loaded CuO. The as-prepared photocatalysts possessed satisfactory removal efficiency for Hg^0 under UV light irradiation, which reached more than 70% under all the testing conditions. When the content of doped CuO exceeded 1.25 wt %, the peril rate of mercury under the visible light reduced with the increasing doped CuO.

References

[1] Ariya P.A., Amyot M., Dastoor A., Deeds D., Feinberg A., Kos G., Poulain A., Ryjkov A., Semeniuk K., Subir M., Mercury physicochemical and biogeochemical transformation in the atmosphere and at atmospheric interfaces: a review and future directions, Chem. Rev., 115, 2015, 3760–3892.

[2] O. Us Epa, Aieo, Minamata Convention on Mercury, Inter. Legal Mater., 55, 2016, 582–616.

[3] Zhen W. , Jing L. , Zhang B., Yang Y., Zhen Z., Miao S., Mechanism of Heterogeneous Mercury Oxidation by HBr over V_2O_5/TiO_2 Catalyst, Environ. Sci. Technol., 50, 2016, 5398–5404.

[4] Liu Z., Sriram V., Lee J.Y., Heterogeneous oxidation of elemental mercury vapor over $RuO_2/$ rutile TiO_2 catalyst for mercury emissions control, Appl. Catal., B., 207, 2017, 143–152.

[5] Fan L., Ling L., Wang B., Zhang R., The adsorption of mercury species and catalytic oxidation of Hg^0 on the metal-loaded activated carbon, Appl. Catal., A, 520, 2016, 13–23.

[6] Fang P., Cen C.P., Tang Z.J., Experimental study on the oxidative absorption of Hg^0 by $KMnO_4$ solution, Chem. Eng. J., 198–199, 2012, 95–102.

[7] Zhao B., Liu X. , Zhou Z., Shao H. , Xu M., Catalytic oxidation of elemental mercury by Mn-Mo/ CNT at low temperature, Chem. Eng. J., 284, 2015, 1233–1241.

[8] Liu X., Iocozzia J., Wang Y., Cui X., Chen Y., Zhao S., Li Z., Lin Z., Noble Metal Metal Oxide Nanohybrids with Tailored Nanostructures for Efficient Solar Energy Conversion, Photocatalysis and Environmental Remediation, Energy Environ. Sc., 10, 2016, 402–404.

[9] Fujishima A., Honda K., Electrochemical photolysis of water at a semiconductor electrode, Nature, 238, 1972, 37–38.

[10] Asahi R., Morikawa T., Ohwaki T., Aoki K., Taga Y., Visible-light photocatalysis in nitrogen-doped titanium oxides, Science, 293 (2001) 269–271.

[11] Guo Z., Cheng S., Cometto C., Anxolabéhère-Mallart E., Ng S.M., Ko C.C., Liu G., Chen L., Robert M., Lau T.C., Highly Efficient and Selective Photocatalytic CO_2 Reduction by Iron and Cobalt Quaterpyridine Complexes, J. Am. Chem. Soc., 138, 2016, 9413–9414.

[12] Park H., Kim H., Moon G., Choi W., Photoinduced charge transfer processes in solar photocatalysis based on modified TiO_2, Energy Environ. Sc., 9, 2015, 411–433.

[13] Wu Q., Huang F., Zhao M., Xu J., Zhou J., Wang Y., Ultra-small yellow defective TiO_2 nanoparticles for co-catalyst free photocatalytic hydrogen production, Nano Energy., 24, 2016, 63–71.

[14] Tian J., Hao P., Wei N., Cui H., Liu H., 3D Bi_2MoO_6 Nanosheet/TiO_2 Nanobelt Heterostructure: Enhanced Photocatalytic Activities and Photoelectochemistry Performance, Acs Catal., 5, 2015, 4530–4536.

[15] Li P., Zhou Y., Zhao Z., Xu Q., Wang X., Xiao M., Zou Z., Hexahedron Prism-Anchored Octahedronal CeO_2: Crystal Facet-Based Homojunction Promoting Efficient Solar Fuel Synthesis, J. Am. Chem. Soc., 137, 2015, 9547–9550.

[16] Pan L., Wang S., Xie J., Wang L., Zhang X., Zou J.J., Constructing TiO2 p-n homojunction for photoelectrochemical and photocatalytic hydrogen generation, Nano Energy., 28, 2016, 296–303.

[17] Scanlon D.O., Dunnill C.W., Buckeridge J., Shevlin S.A., Logsdail A.J., Woodley S.M., Catlow C.R., Powell M.J., Palgrave R.G., Parkin I.P., Band alignment of rutile and anatase TiO_2, Nature Mater., 12, 2013, 798–801.

[18] Zhou W. , Li W. , Wang J.Q., Qu Y., Yang Y., Xie Y., Zhang K., Wang L., Fu H., Zhao D., Ordered mesoporous black TiO_2 as highly efficient hydrogen evolution photocatalyst, J. Am. Chem. Soc., 136, 2014, 9280–9283.

[19] Lee C.H., Shie J.L., Yang Y.T., Chang C.Y., Photoelectrochemical characteristics, photodegradation and kinetics of metal and non-metal elements co-doped photocatalyst for pollution removal, Chem. Eng. J., 303, 2016, 477–488.

[20] Zhang J., Jin X., Moralesguzman P.I., Yu X., Liu H., Zhang H., Razzari L., Claverie J.P., Engineering the Absorption and Field Enhancement Properties of Au-TiO$_2$ Nanohybrids via Whispering Gallery Mode Resonances for Photocatalytic Water Splitting, Acs Nano., 10, 2016, 4496–4503.

[21] Wu J., Li C., Zhao X., Wu Q., Qi X., Chen X., Hu T., Cao Y., Photocatalytic oxidation of gas-phase Hg^0 by CuO/TiO$_2$, Appl. Catal. B Environ., 176–177, 2015, 559–569.

[22] Kong M., Li Y., Chen X., Tian T., Fang P., Zheng F., Zhao X., Tuning the Relative Concentration Ratio of Bulk Defects to Surface Defects in TiO$_2$ Nanocrystals Leads to High Photocatalytic Efficiency, J. Am. Chem. Soc., 133, 2011 16414–16417.

[23] Liu S. , Yu J. , M. Jaronic, Tunable Photocatalytic Selectivity of Hollow TiO$_2$ Microspheres Composed of Anatase Polyhedra with Exposed {001} Facets, J. Am. Chem. Soc., 132, 2010, 11914–11916.

[24] Etgar L., Zhang W., Gabriel S., Hickey S.G., Nazeeruddin M.K., Eychmüller A., Liu B., Grätzel M., High efficiency quantum dot heterojunction solar cell using anatase (001) TiO$_2$ nanosheets, Adv. Mater., 24, 2012, 2202–2206.

[25] Ong W.J., Tan L.L., Chai S.P., Yong S.T., Mohamed A.R., Highly reactive {001} facets of TiO$_2$-based composites: synthesis, formation mechanism and characterization, Nanoscale., 6, 2014, 1946–2208.

[26] Liu X., Dong G., Li S., Lu G., Bi Y., Direct Observation of Charge Separation on Anatase TiO$_2$ Crystals with Selectively Etched {001} Facets, J. Am. Chem. Soc., 138, 2016, 2917–2920.

[27] Yang H.G., Sun C.H., Qiao S.Z., Zou J. , Liu G., Smith S.C., Cheng H.M., Lu G.Q., Anatase TiO$_2$ single crystals with a large percentage of reactive facets, Nature., 453, 2008, 638.

[28] Chen J.S., Tan Y.L., Li C.M., Yan L.C., Luan D., Madhavi S., Boey F.Y.C., Archer L.A., Lou X.W., Constructing Hierarchical Spheres from Large Ultrathin Anatase TiO$_2$ Nanosheets with Nearly 100% Exposed (001) Facets for Fast Reversible Lithium Storage, J. Am. Chem. Soc., 132, 2010, 6124–6130.

[29] Zheng Z., Huang B., Lu J., Qin X., Zhang X., Dai Y., Hierarchical TiO$_2$ Microspheres: Synergetic Effect of {001} and {101} Facets for Enhanced Photocatalytic Activity, Chem. Eur. J., 17, 2011, 15032–15038.

[30] Liu G., Yang H.G., Pan J. , Yang Y.Q. , Lu G.Q. , Cheng H.M., Titanium Dioxide Crystals with Tailored Facets, Chem. Rev., 114, 2014, 9559–9612.

[31] Liu S.G. , Yu J.G., Jaroniec M., Anatase TiO$_2$ with Dominant High-Energy {001} Facets: Synthesis, Properties, and Applications, Chem. Mater., 23, 2011, 4085–4093.

[32] Zhang J., Wu J., Lu P., Liu Q.Z., Huang T.F., Tian H., Zhou R.X., Ren J.X., Yuan B.X., Sun X.M., Zhang W.B. , The effect of pH on Synthesis of BiOCl and its photocatalytic oxidization performance, Mater. Lett., 186, 2017, 353–356.

[33] Zhou X., Wu J., Zhang J., He P., Ren J., Zhang J., Lu J., Liang P., Xu K., Shui F.,The effect of surface heterojunction between (001) and (101) facets on photocatalytic performance of anatase TiO2,Materials Letters., 205,(2017),173–177.

[34] Wang W., Zhou Y., Lu C., Ni, Y., Rao W., The effect of hydrothermal temperature on the structure and photocatalytic activity of {001} faceted anatase TiO$_2$, Mater. Lett., 160, 2015, 231–234.

[35] Xiang Q., Lv K., Yu J., Pivotal role of fluorine in enhanced photocatalytic activity of anatase TiO$_2$ nanosheets with dominant (001) facets for the photocatalytic degradation of acetone in air, Appl. Catal. B Environ., 96, 2010, 557–564.

[36]	Hao F., Wang X., Zhou C., Jiao X. , Li X., Li J., Lin H., Efficient Light Harvesting and Charge Collection of Dye-Sensitized Solar Cells with (001) Faceted Single Crystalline Anatase Nanoparticles, J. Phys. Chem. C., 116, 2012, 19164–19172.

[37]	Chen J.S., Chen C., Liu J., Xu R., Qiao S.Z., Lou X.W., Ellipsoidal hollow nanostructures assembled from anatase TiO$_2$ nanosheets as a magnetically separable photocatalyst, Chem. Commun., 47, 2011, 2631–2633.

[38]	Chen J.S., Luan D., Li C.M., Boey F.Y.C., Qiao S., Lou X.W., TiO$_2$ and SnO$_2$@TiO$_2$ hollow spheres assembled from anatase TiO$_2$ nanosheets with enhanced lithium storage properties, Chem. Commun., 46, 2010, 8252–8254.

[39]	Joo J., Kwon S.G., Yu T., Cho M., Lee J., Yoon J., Hyeon T., Large-Scale Synthesis of TiO$_2$ Nanorods via Nonhydrolytic Sol–Gel Ester Elimination Reaction and Their Application to Photocatalytic Inactivation of E. coli, J. Phys. Chem. B., 109, 2005, 15297–15302.

[40]	Wu J.J., Yu C.C., Aligned TiO$_2$ Nanorods and Nanowalls, J. Phys. Chem. B 108, 2004, 3377–3379.

[41]	Yun H.J., Lee H., Joo J.B., Kim W., Yi J., Influence of Aspect Ratio of TiO$_2$Nanorods on the Photocatalytic Decomposition of Formic Acid, J. Phys. Chem. C., 113, 2009, 3050–3055.

[42]	Cheng Y., Huang W., Zhang Y., Zhu L., Liu Y., Fan X., Cao X., Preparation of TiO$_2$ hollow nanofibers by electrospining combined with sol–gel process, CrystEngComm., 12, 2010, 2256–2260.

[43]	Sun C., Wang N., Zhou S., Hu X., S. Zhou, P. Chen, Preparation of self-supporting hierarchical nanostructured anatase/rutile composite TiO$_2$ film, Chem. Commun., 2008 3293–3295.

[44]	Zhao T., Liu Z., Nakata K., Nishimoto S., Murakami T., Zhao Y., Jiang L., Fujishima A., Multichannel TiO$_2$ hollow fibers with enhanced photocatalytic activity, J. Mater. Chem., 20, 2010, 5095–5099.

[45]	Z. Liu, X. Zhang, Nishimoto S., Murakami T., Fujishima A., Efficient Photocatalytic Degradation of Gaseous Acetaldehyde by Highly Ordered TiO$_2$ Nanotube Arrays, Environ. Sci. Technol., 42, 2008, 8547–8551.

[46]	Yu J., Dai G., Cheng B., Effect of Crystallization Methods on Morphology and Photocatalytic Activity of Anodized TiO$_2$ Nanotube Array Films, J. Phys. Chem. C., 114, 2010, 19378–19385.

[47]	Aoyama Y., Oaki Y., Ise R., Imai H., Mesocrystal nanosheet of rutile TiO$_2$ and its reaction selectivity as a photocatalyst, CrystEngComm, 14, 2012, 1405–1411.

[48]	Gan X., Gao X., Qiu J., He P., Li X., Xiao X., TiO$_2$ Nanorod-Derived Synthesis of Upstanding Hexagonal Kassite Nanosheet Arrays: An Intermediate Route to Novel Nanoporous TiO$_2$ Nanosheet Arrays, Cryst. Growth Des, 12, 2011, 289–296.

[49]	Sakai N., Fukuda K., Shibata T., Ebina Y., Takada K., Sasaki T., Photoinduced Hydrophilic Conversion Properties of Titania Nanosheets, J. Phys. Chem. B 110, 2006, 6198–6203.

[50]	Hasegawa G., Kanamori K., Nakanishi K., Hanada T., Facile Preparation of Hierarchically Porous TiO$_2$ Monoliths, J. Am. Ceram. Soc. 93, 2010, 3110–3115.

[51]	Syoufian A., Satriya O.H., Nakashima K., Photocatalytic activity of titania hollow spheres: Photodecomposition of methylene blue as a target molecule, Catal. Commun. 8, 2007, 755.

[52]	Wang R., Cai X., Shen F., TiO$_2$ hollow microspheres with mesoporous surface: Superior adsorption performance for dye removal, Appl. Surf. Sci. 305, 2014, 352–358.

[53]	Yu J., Shi L., One-pot hydrothermal synthesis and enhanced photocatalytic activity of trifluoroacetic acid modified TiO$_2$ hollow microspheres, J. Mol. Catal. A: Chem. 326, 2010, 8–14.

[54]	Nasir M., Bagwasi S., Jiao Y., Chen F., Tian B., Zhang J. , Characterization and activity of the Ce and N co-doped TiO$_2$ prepared through hydrothermal method, Chem. Eng. J. 236, 2014, 388–397.

[55] Yu J., Wang W., Cheng B., Su B.-L., Enhancement of Photocatalytic Activity of Mesporous TiO_2 Powders by Hydrothermal Surface Fluorination Treatment, J. Phys. Chem. C 113, (2009), 6743–6750.

[56] J.Wu, X. Li, J. Ren, X. Qi, P. He, B. Ni, C. Zhang, C. Hu, Jun Zhou,Experimental study of TiO2 hollow microspheres removal on elemental mercury in simulated flue gas, Journal of Industrial and Engineering Chemistry, 32, 2015, 49–57.

[57] Liu S.W., Yu J.G., Mann S. , Spontaneous construction of photoactive hollow TiO_2 microspheres and chains, Nanotechnology 20, 2009, 325606–325613.

[58] Yu J.C., Yu J.Q., Ho W., Effects of F-Doping on the Photocatalytic Activity and Microstructures of Nanocrystalline TiO_2 Powders, Chem. Mater. 14 (9), 2002, 3808–3816.

[59] Granite E.J., Pennline H.W. , Photochemical Removal of Mercury from Flue Gas, Ind. Eng. Chem. Res. 41, 2002, 5470–5476.

[60] Yu J.G., Wang W.G., Cheng B., Su B.L., Enhancement of Photocatalytic Activity of Mesporous TiO_2 Powders by Hydrothermal Surface Fluorination Treatment, J. Phys. Chem. C 113, 2009, 6743–6750.

[61] Lv K., Cheng B. , Yu J. , Liu G. , Fluorine ions-mediated morphology control of anatase TiO_2 with enhanced photocatalytic activity, Phys. Chem. Chem. Phys., 14, 2012, 5349–5362.

[62] Wang X., He H., Chen Y., Zhao J. , Zhang X., Anatase TiO_2 hollow microspheres with exposed {001} facets: Facile synthesis and enhanced photocatalysis, Appl. Surf. Sci., 258 (15), 2012 5863–5868.

[63] Chen X., Titanium dioxide nanomaterials and their energy applications, Chin. J. Catal., 30, 2009, 839–851.

[64] Chen X., Mao S.S. , Titanium dioxide nanomaterials: synthesis, properties, modifications, and applications, Chem. Rev., 107 (7), 2007, 2891–2959.

[65] Xu S., Du A.J. , Liu J., Ng J., Sun D.D., Highly efficient CuO incorporated TiO_2 nanotube photocatalyst for hydrogen production from water, Int. J. Hydrogen Energy., 36, 2011, 6560–6568.

[66] Yu J., Hai Y., Jaroniec M., Photocatalytic hydrogen production over CuO-modified titania, J. Colloid Interface Sci., 357, 2011, 223–228.

[67] Bandara J., Udawatta C.P.K., Rajapakse C.S.K., Highly stable CuO incorporated TiO_2 catalyst for photocatalytic hydrogen production from H_2O, Photochem. Photobiol. Sci., 4, 2005, 857–861.

[68] Barreca D., Fornasiero P., Gasparotto A., Gombac, V., Maccato, C., Montini, T., Tondello, E., The potential of supported Cu_2O and CuO nanosystems in photocatalytic H_2 production, ChemSusChem 2, 2009, 230–233.

[69] Yoong L.S., Chong F.K., Dutta B.K., Development of copper-doped TiO_2 photocatalyst for hydrogen production under visible light, Energy 34, 2009, 1652–1661.

[70] J.Wu, C. Li, X. Zhao, Q. Wu, X. Qi, X. Chen, T. Hu, Y. Cao, Photocatalytic oxidation of gas-phase Hg^0 by $CuO/TiO2$,Applied Catalysis B: Environmental,176–177,(2015)559–569

[71] Elzinga E.J., Reeder R.J., X-ray absorption spectroscopy study of Cu^{2+} and Zn^{2+}adsorption complexes at the calcite surface: Implications for site-specific metal incorporation, Geochimica et Cosmochimica preferences during calcite crystal growth, Geochim. Cosmochim. Acta, 66, 2002, 3943–3954.

[72] Wu Z., Jiang B., Liu Y., Effect of transition metals addition on the catalyst of manganese/titania for low-temperature selective catalytic reduction of nitricoxide nitricoxide with ammonia, Appl. Catal. B: Environ. 79, 2008, 347–355.

[73] Hu S., Zhou F., Wang L., Zhang J., Preparation of Cu_2O/CeO_2 heterojunction photocatalyst for the degradation of acid orange 7 under visible light irradiation, Catal. Commun. 12, 2011, 794–797.

[74] Jin Z., Zhang X., Li Y., Li S., Lu G., 5.1% apparent quantum efficiency for stable hydrogen generation over eosin sensitized CuO/TiO_2 photocatalyst under visible light irradiation, Catal. Commun. 8, 2007, 1267–1273.

[75] Chen W.T., Jovic V., Sun-Waterhouse D., Idriss H., Waterhouse G.I.N., The role of CuO in promoting photocatalytic hydrogen production over TiO_2, Int. J. Hydrogen Energy 38, 2013, 15036–15048.

[76] Xu W., Wang H., Zhou X., Zhu T., CuO/TiO_2 catalysts for gas-phase Hg^0 catalytic oxidation, Chem. Eng. J. 243, 2014, 380–385.

[77] Wang B., Li Q., Wang W., Li Y., Zhai, J., Preparation and characterization of Fe^{3+}-doped TiO_2 on fly ash cenospheres for photocatalytic application, Appl. Surf. Sci. 257, 2011, 3473–3479.

[78] Snider G., Ariya P., Photocatalytic oxidation reaction of gaseous mercury over titanium dioxide nanoparticle surfaces, Chem. Phys. Lett. 491, 2010, 23–28.

[79] Botta S.G. , Rodrí´iuez D.J. , Leyva A.G. , Litter M.I. , Features of the transformation of HgII by heterogeneous photocatalysis over TiO_2, Catal. Today 76, 2002, 247–258.

[80] Gustin M.S., Lindberg S.E., Weisberg P.J., An update on the natural sources and sinks of atmospheric mercury, Appl. Geochem. 23, 2008, 482–493.

[81] Montini T., Gombac V., Sordelli L., Delgado J.J., Chen X., Adami G., Fornasiero P., Nanostructured Cu/TiO_2 photocatalysts for H_2 production from ethanol and glycerol aqueous solutions, Chem Cat Chem 3, 2011,574–577.

Wen Xiao and Ziyi Zhong

7 Recovery of valuable metals from e-waste via applications of nanomaterials

7.1 Introduction

The booming of electronic industry and rapid development of semiconductor technologies accelerate the prevalence of electrical and electronic equipment (EEE) in people's daily life, such as mobile phones, tablet, or laptops. Today, electronic products have a decreasing life span, which can be due to the fast introduction of new advanced products in the market and to the rapid change in people's lifestyle with modern facilities [1]. As a result, a huge amount of used EEE is discarded annually, which accelerates the generation of electronic waste (e-waste). The amount of e-waste generated in the United States increased by 65% from 1,900,000 ton in 2000 to 3,140,000 ton in 2013. However, less than 50% of the total e-waste is recycled as shown in Figure 7.1a and b [2]. The majority of e-waste is being trashed and treated by landfilling or incineration as seen in Figure 7.2 [2]. e-Waste contains toxic metals (Cr, Hg, and Pb) and plastics, which represent 70% of the total toxic waste in the United States. Landfilling of e-waste can cause leakage of toxic substances that pollute underground water and incineration of e-waste generates harmful flue gas residues and toxic air emissions [1–3]. Therefore, recycling and treating e-waste are necessary for environmental protection.

e-Wastes such as TV board, mobile phone, PC motherboard, printed circuit boards (PCBs) contain various valuable and even precious metals (PMs), such as copper (Cu), silver (Ag), palladium (Pd), and gold (Au) as listed in Table 7.1 [4]. There is also a high demand for these metals to be used in EEE as shown in Table 7.2 [5]. It is noted that gold has particularly high value in the waste of PCBs [1]. Therefore, e-waste can be a promising secondary resource for metal production, especially for gold production. Recovering the valuable metals from the secondary resource of e-waste will reduce mining of virgin materials [6]. The demand for gold in electronics has increased consistently since the 1980s [5]. However, the gold ore grade declines over time and the trend is generally seen in many metals as well [7]. The mining of low-grade ore requires a large amount of water and energy and causes increased land disruption and pollution [7]. Strikingly, primary Cu and Au ores have Cu and Au contents of 0.5–1% and 1–10 g/ton, respectively, while the Cu and Au contents in e-waste are 20% and

Wen Xiao, Department of Materials Science and Engineering, National University of Singapore, Singapore

Ziyi Zhong, College of Engineering, Guangdong Technion Israel Institute of Technology (GTIIT), Shantou, Guangdong, China; Technion–Israel Institute of Technology (IIT), Haifa, Israel

https://doi.org/10.1515/9783110544183-007

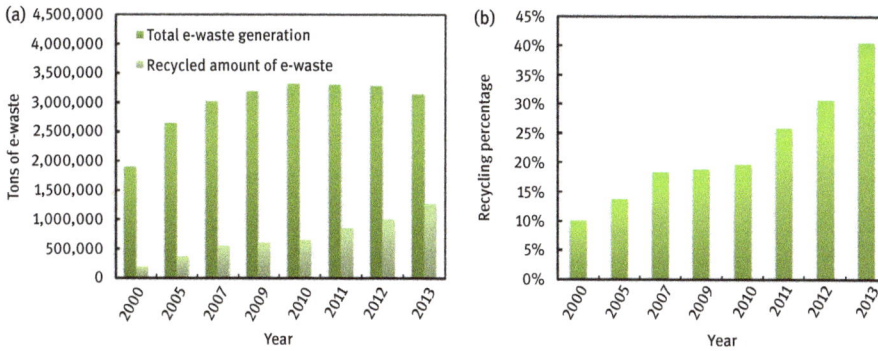

Figure 7.1: (a) Total amount of e-waste generated and recycled in the United States and (b) the corresponding e-waste recycling percentage in selected years from 2000 to 2013. Data retrieved from Ref. [2].

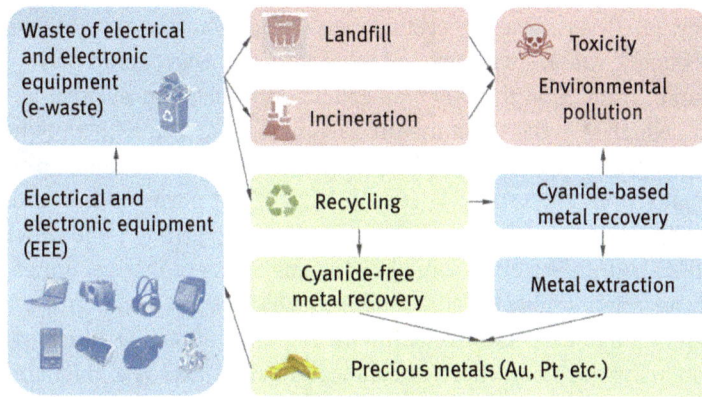

Figure 7.2: Generation and recycling of e-waste.

Table 7.1: Metal content in typical types of e-waste. Reproduced with permission from Ref. [4].

e-Waste	Fe (wt.%)	Cu (wt.%)	Al (wt.%)	Pb (wt.%)	Ni (wt.%)	Ag (ppm)	Au (ppm)	Pd (ppm)
TV board scrap	28	10	10	10	0.3	280	20	10
Mobile phone scrap	5	13	1	0.3	0.1	1,380	350	210
PC mainboard scrap	4.5	14.3	2.8	2.2n	1.1	639	566	124
Printed circuit boards scrap	12	10	7	1.2	0.85	280	110	–
Typical electronic scrap	8	20	2	2	2	2,000	1,000	50

Table 7.2: Demand and value of precious metals in electrical and electronic equipment (EEE). Data retrieved from Ref. [5].

Metals	Demand for EEE (ton/year)	Value in EEE (10^9 US $)	Main applications
Au	300	6.7	Bonding wires, contacts, ICs
Ag	6,000	2.6	Contact switches, solders
Pd	33	0.4	Multilayer capacitors, connectors
Pt	13	0.5	Hard disk, fuel cells
Ru	27	0.5	Hard disk, plasma display

250 g/ton, respectively [4, 8]. As a result, recycling metals from e-waste are highly desirable in terms of economic value, environmental protection, and energy efficiency.

The e-waste market from 2004 to 2009 was estimated to have an average annual growth of 8.8%, which is higher than the production rate [9]. The total amount of wasted computers, phones, TVs, and appliances produced annually doubled to 42 million tons globally from 2009 to 2014 [10, 11]. Most of the e-waste flows into developing countries that lack regulation, such as China (~70%), India, and other eastern Asia and Africa countries [12]. In some low-income towns of these counties, such as Guiyu in China, Sangrampur in India, and Agbogbloshie in Ghana, unregulated and informal recycling of the e-waste is the main source of people's income. As e-waste also contains various hazardous substances, such as lead (Pb), mercury (Hg), chromium (Cr), toxic chemicals in plastics and flame retardants, unsafe and improper handling, dismantling, and disposal of e-waste by recycling workers seriously harm people's health and pollute the environment [6]. e-Waste site in Agbogbloshie was rated to be "among the top ten most toxic sites in the world" [13]. Therefore, developing green technologies to recover and recycle valuable metals from e-waste is urgent.

In this chapter, the processes of e-waste recycling and methods of metal recovery are introduced. Pyrometallurgical and hydrometallurgical processes are described and discussed. Metal extraction from leach liquor is briefed. Methods of gold extraction from solution are particularly focused, including cementation, solvent extraction, carbon adsorption, and coprecipitation. Theory of adsorption isotherms and kinetics and mechanism of selective adsorption of metal ions are reviewed. Moreover, typical examples of recent progress on selective extraction of PMs from e-waste leach liquor are presented. This chapter illustrates a concise and comprehensive picture on the recovery of metals from e-waste covering theories, mechanisms, materials synthesis, and processes.

7.2 Processes of metal recovery from e-waste

Collection, preprocessing, and end processing are the three main steps for e-waste recycling [14]. e-Waste is collected by storing end-of-life components at collection facilities and returning useable components to the supply chain, which depends on government policies, public awareness, and installation of collection facilities. At collection facilities, e-waste is first of all preprocessed as shown by the flowchart in Figure 7.3. The discarded equipment is disassembled manually to isolate individual components from e-waste, liberating wiring boards, housing, drivers, and other components [15]. Then the size of e-waste scrap is reduced by being mechanically shredded into pieces using hammer mills [14]. After that, techniques similar to that used in mineral dressing, such as screening, magnetic, eddy current, and density separation, are applied to separate nonmetals and metals in the shredded e-waste scrap. A new cryomilling process is also proposed recently for the separation of nonmetal and metal components [3]. The nonmetal and metal parts of e-waste are further processed in the end processing. The nonmetal parts of e-waste normally consist of organic substances, such as fiberglass, resin, or plastics. A number of methods are proposed to recycle the nonmetal substances, which will not be the focus of this chapter [16–18]. The metal components consisting of valuable metal resources can be further recycled by two common methods – pyrometallurgical and hydrometallurgical processes, which will be discussed in detail in the following sections. Finally, metals can be extracted by electrometallurgical process or chemical processes such as cementation, solvent extraction, carbon adsorption, or coprecipitation.

7.2.1 Pyrometallurgical processes

Pyrometallurgical process has been utilized for metal recovery from end-of-life PCBs for the last two decades. Pyrometallurgical route typically includes smelting in furnaces, incineration, combustion, and pyrolysis, currently dominating e-waste recycling [19]. Smelting e-waste in a furnace at high temperature liberates valuable metals, which are sorted depending on their chemical and metallurgical properties. PMs, for instance, are segregated into a solvent metal phase of copper and lead. Thus, e-waste scrap can be fed into a furnace of copper or lead smelters to extract Cu, Pb, and PMs, where metals in molten phase and oxides in a slag phase are collected [15]. It is important for smelters to extract valuable metals and meanwhile isolate hazardous substances efficiently, so that recycling facilities can recover valuable metals with minimized environmental impact arising from a large amount of e-waste.

Copper smelters are widely used to recycle and extract PMs from e-waste, which is found to be more environmental friendly than lead smelters that generate toxic fumes [21]. Copper smelters can be installed near cities where e-waste is generated, reducing the transportation cost of e-waste and hence improving the recycling

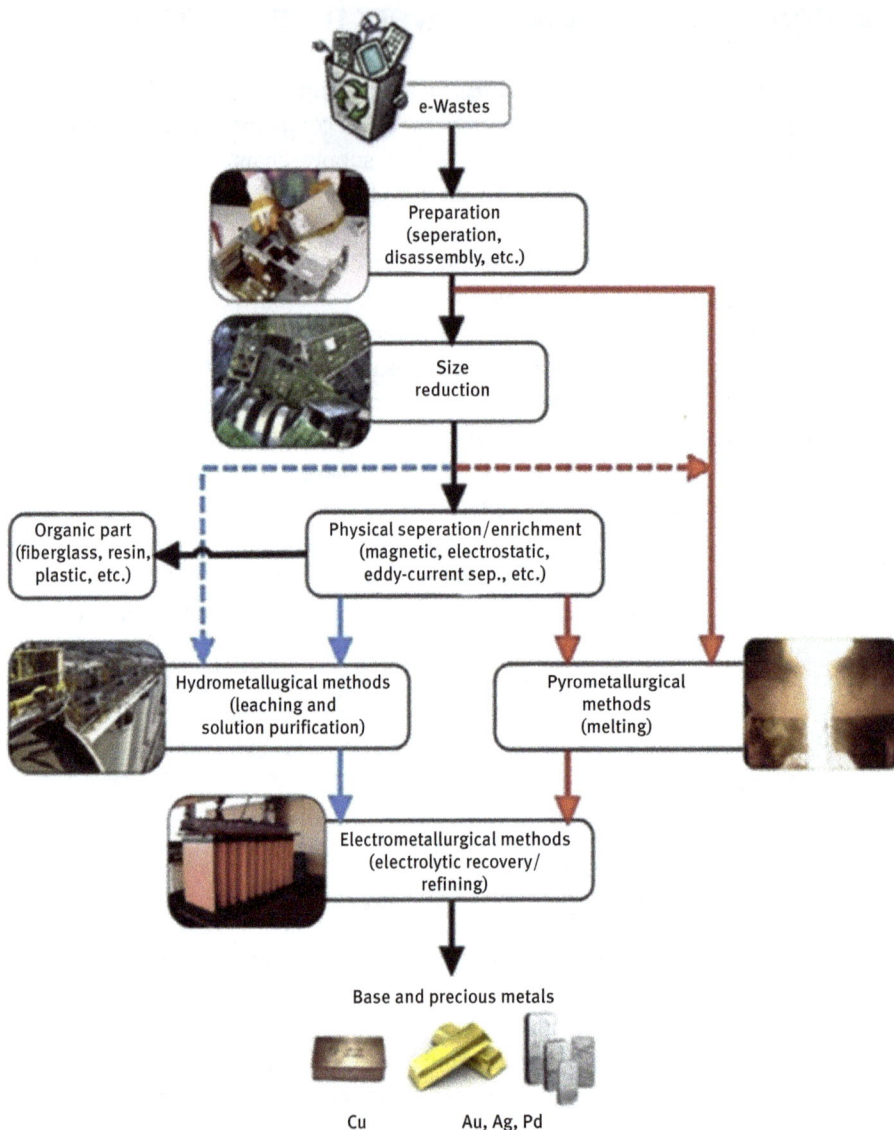

Figure 7.3: A flowchart of the potential metal recovery process from e-wastes including pyrometallurgical and hydrometallurgical methods. Reproduced with permission from Ref. [1].

economy. There are two types of copper smelting routes including primary and secondary copper smelting for e-waste recycling. In the primary copper smelting (sulfur-based), copper matte (40%) and blister copper (98.5%) are produced and pure copper is further produced by fire refining of blister copper. In the secondary copper smelting (black copper route), a reduction process produces crude copper, which is then refined in a converter by oxidation. The black copper route is attractive, because it can consume high levels of impurities, such as Fe, Zn, Pb, and Sn, which are finally removed by oxidation as seen in Figure 7.4. Reduction and oxidation cycles are repeated in the black copper smelting process. Most impurities are segregated into the vapor phase and are discharged in the off-gas [21]. Figure 7.5 shows a schematic process flowchart of Umicore for recovering valuable metals, including several pyrometallurgical, hydrometallurgical, and electrochemical processes. A secondary copper route is adapted to extract PMs from e-waste at Umicore plant, which generates by-products of sulfuric acid (from off-gas purification) and slag (consisting of silicon, aluminum, and iron oxides) that are used as construction material. Waste from the nonferrous industry, PM residues, and PCBs can be fed into the plant. The plant treats on average 250,000 tons of feed material every year and recovers over 50 tons of precious group metals, 100 tons of gold, and 2,400 tons of silver annually [7, 22, 23].

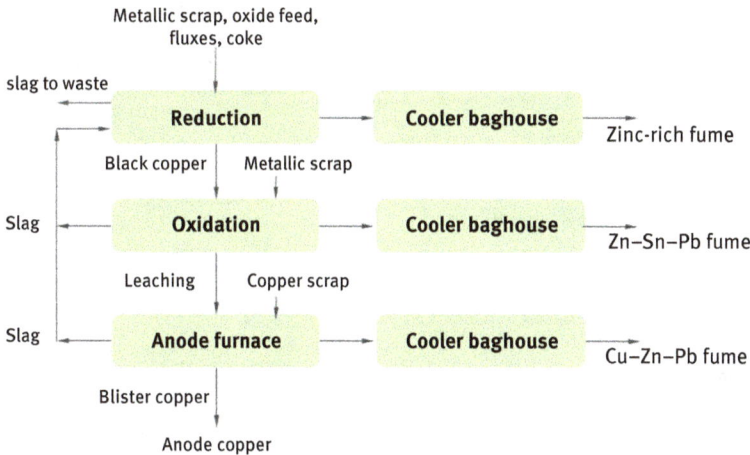

Figure 7.4: A flowchart of black copper smelter process. Reproduced with permission from Ref. [20].

Although the pyrometallurgical process is, in general, economic and has a good recovery rate of PMs, installing start-of-art e-waste recycling plants needs large investment for maximizing recovery of valuable metals and environmental protection. In addition, heating e-waste containing halogenated flame retardants at high temperature leads to the release of hazardous gases such as dioxins and furans, which requires off-gas treatment to reduce environmental pollution [24, 25]. The pyrometallurgical process is also energy intensive and requires high-grade feed materials (high content of copper and PMs). Furthermore, feed materials are complex, posing challenges to handle the smelting and refining processes [4, 22]. Another drawback of the pyrometallurgical

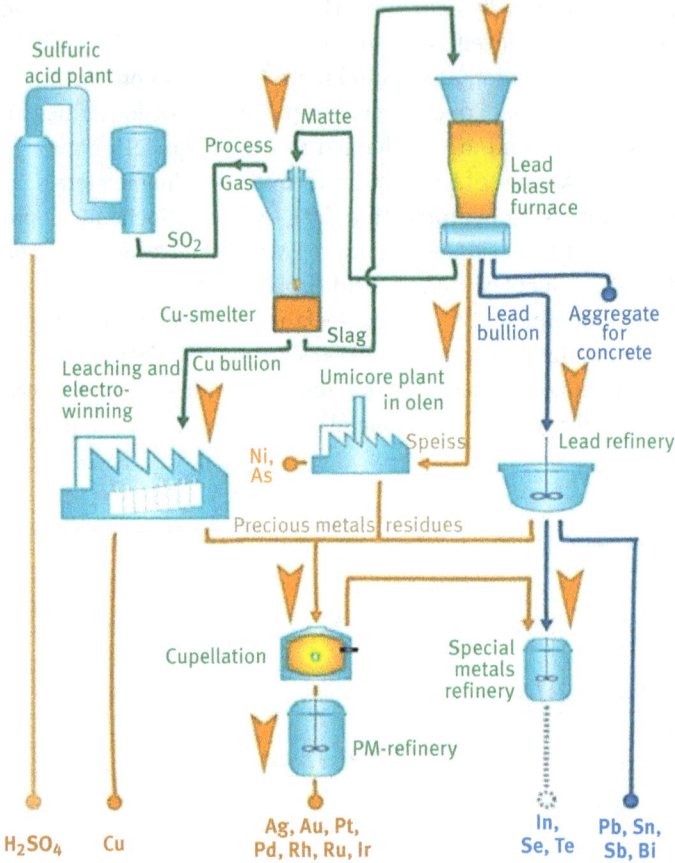

Figure 7.5: A flowchart of integrated metals smelter and refinery process of Umicore. Reproduced with permission from Ref. [22].

process is PMs are segregated into base metals (BMs) (copper, lead, and nickel) in the smelting process. As a result, it is necessary to perform additional hydrometallurgical or electrochemical processes to extract pure metals from the BMs [15].

7.2.2 Hydrometallurgical processes

Hydrometallurgical processes to recycle metals from e-waste originate from the hydro-metallurgical techniques for extracting metals from primary ores. PMs in the primary ores are leached by halides, cyanides, thiourea, or thiosulfates. Similar leaching process can be applied to extract metals from e-waste as well, which however is more complicated than the leaching process of natural ores due to the complex composition of e-waste. A typical example of a hydrometallurgical metal recovery process is shown in Figure 7.6 [26]. Collected e-waste will first undergo sorting, shredding, magnetic and eddy current separation, and aluminum separation. Thereafter, tin (Sn) can be recovered by solder leaching using HBF_4 and copper (Cu) can be recovered by copper leaching using $(NH_4)_2SO_4$. Furthermore, PMs such as gold (Au) and silver (Ag) can be recovered by dissolving in HNO_3, H_2SO_4, or HCl-based solutions like aqua regia solution [26]. Notably, halide leaching, cyanide leaching, thiosulfate leaching, and thiourea leaching are also commonly used [24, 26]. Therefore, leaching is a critical step in the hydrometallurgical process for metal recovery from e-waste like PCBs. Separation and purification of the pregnant solution from a leaching process will be conducted to enrich metal content and remove impurities. Finally, metals are extracted from the leach liquor via electrometallurgical or chemical processes. Chemical processes

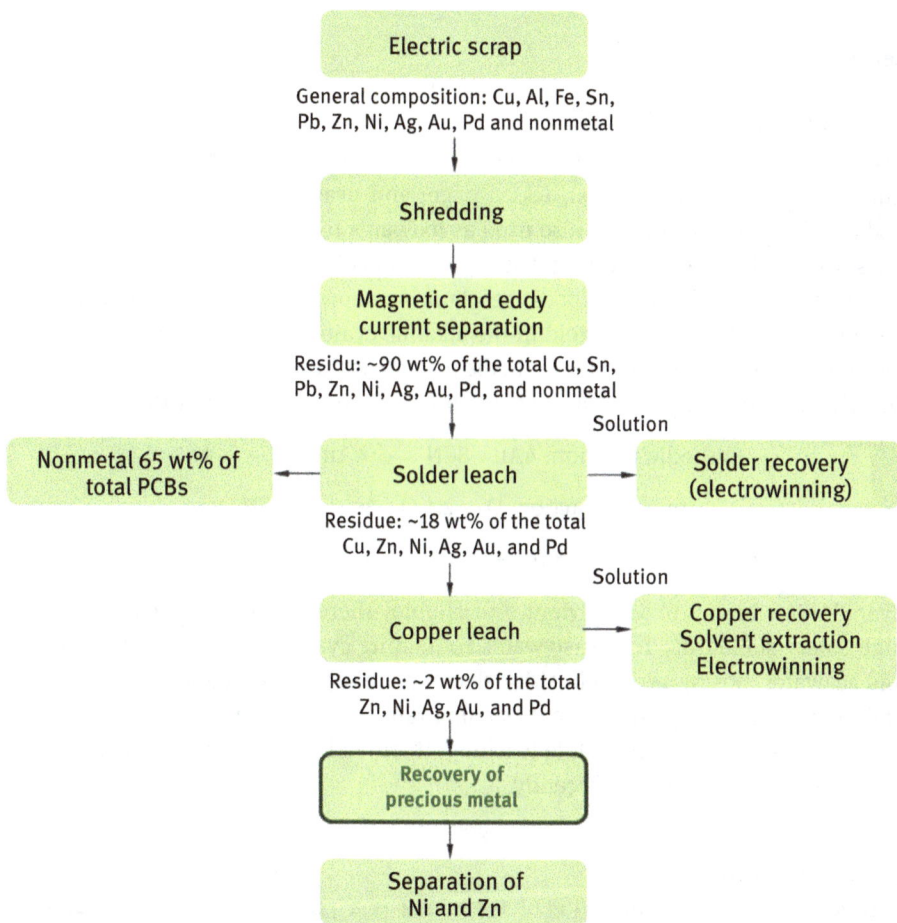

Figure 7.6: A flowchart of hydrometallurgical process for metal recovery from printed circuit boards. Reproduced with permission from Ref. [26].

including cementation, solvent extraction, carbon adsorption, and coprecipitation will be discussed later in this chapter [27–32].

The hydrometallurgical process is a promising alternative of pyrometallurgical process for metal recycling from e-waste due to low initial capital cost and reduced environmental impact [33]. When compared to pyrometallurgical route, the hydrometallurgical route has simple working conditions and low or no emissions of toxic gas and dust [1, 34, 35]. Most importantly, hydrometallurgical process is energy efficient and possesses high recovery rate particularly for PMs such as gold (Au), silver (Ag), and platinum (Pt) [4, 36]. However, it is challenging to find suitable lixiviants that are effective, economical, and safe and easy to implement into the leaching process [37]. In addition, hydrometallurgical routes are generally slow and time consuming, posing difficulties to be applied on industrial scale [4].

Leaching processes

Cyanide leaching

Cyanide has been widely used in gold and silver mining industry for more than a hundred years due to low cost, less dosage, and usage in alkaline condition [38]. Today, alkali metal cyanide is also used as lixiviants to dissolve gold and silver from the surface of PCBs for recovering PMs. Cyanidation of gold is an electrochemical process including anodic and cathodic reaction. As shown by the following reactions, the anodic reaction dissolves gold in the alkaline cyanide solution forming gold cyanide complex ($Au(CN)_2^-$) and the cathodic reaction is the reduction of oxygen forming OH^- [1]. Thus, oxygen (O_2) in the air is required for the cyanidation process.

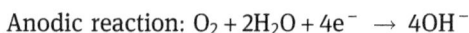

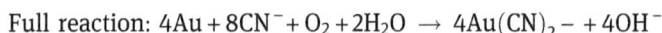

$$\text{Cathodic reaction: } 4Au + 8CN^- \rightarrow 4Au(CN)_2 - + 4e^-$$

$$\text{Anodic reaction: } O_2 + 2H_2O + 4e^- \rightarrow 4OH^-$$

$$\text{Full reaction: } 4Au + 8CN^- + O_2 + 2H_2O \rightarrow 4Au(CN)_2 - + 4OH^-$$

Cyanide is a highly toxic lixiviant and should, therefore, be used carefully with high safety standards. The wastewater containing cyanide can contaminate rivers and seawater, posing serious health risks to the inhabitants. In addition, the cyanidation process is long due to slow leaching rate [35]. Therefore, noncyanide leaching processes, such as thiosulfate leaching, thiourea leaching, and halide leaching, have attracted great interest recently.

Thiosulfate leaching

Noncyanide-based thiosulfate ($S_2O_3^{2-}$) leaching is a potential substitute of cyanide leaching with less interference from foreign cations and causes less environmental pollution [39]. Two different types of thiosulfate are commonly used in gold leaching:

sodium thiosulfate and ammonium thiosulfate. A stable anionic complex forms as per following reaction among gold, thiosulfate, and oxygen.

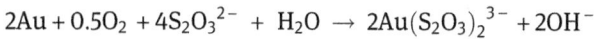

$$2Au + 0.5O_2 + 4S_2O_3^{2-} + H_2O \rightarrow 2Au(S_2O_3)_2^{3-} + 2OH^-$$

Ammonia (NH_3) and copper cations (Cu^{2+}) can act as catalysts to increase the gold leaching rate in thiosulfate [40]. Copper cations (Cu^{2+}) react with thiosulfate ($S_2O_3^{2-}$) based on the following reduction reaction and oxidization reaction in the presence of oxygen (O_2) [41, 42]:

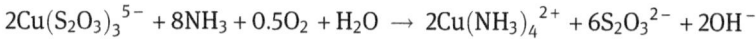

$$2Cu(NH_3)_4^{2+} + 8S_2O_3^{2-} \rightarrow 2Cu(S_2O_3)_3^{5-} + S_4O_6^{2-} + 8NH_3$$

$$2Cu(S_2O_3)_3^{5-} + 8NH_3 + 0.5O_2 + H_2O \rightarrow 2Cu(NH_3)_4^{2+} + 6S_2O_3^{2-} + 2OH^-$$

Gold leaching adopts an electrochemical reaction in the presence of thiosulfate, copper and ammonia as shown:

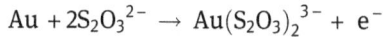

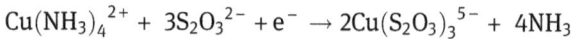

$$Au + 2S_2O_3^{2-} \rightarrow Au(S_2O_3)_2^{3-} + e^-$$

$$Cu(NH_3)_4^{2+} + 3S_2O_3^{2-} + e^- \rightarrow 2Cu(S_2O_3)_3^{5-} + 4NH_3$$

Thiosulfate leaching has high selectivity, nontoxic, and noncorrosive; however, the process is slow and consumes a large amount of reagent (50% thiosulfate loss in thiosulfate solutions containing ammonia and copper), which is uneconomical [35, 39].

Thiourea leaching
Thiourea (($NH_2)_2CS$) leaching is another type of noncyanide leaching having low toxicity, fast kinetics, high efficiency, and low ion interference. It also possesses a high extraction rate for gold leaching. Thiourea is a reducing organic agent being able to form soluble cationic complexes with gold as per following reaction:

$$Au + 2(NH_2)_2CS \rightarrow Au((NH_2)_2CS)_2^+ + e^-$$

Thiourea leaching with the addition of oxidant commonly Fe^{3+} will facilitate the gold and silver recovery as per following reactions [43–45]:

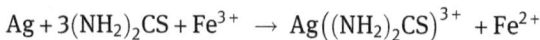

$$Au + 2(NH_2)_2CS + Fe^{3+} \rightarrow Au((NH_2)_2CS)^+ + Fe^{2+}$$

$$Ag + 3(NH_2)_2CS + Fe^{3+} \rightarrow Ag((NH_2)_2CS)^{3+} + Fe^{2+}$$

However, thiourea can be oxidized by ferric iron in acidic condition forming formamidine disulfide, which is unstable in acidic condition and decomposes into elemental sulfur and cyanamide, as shown by the reactions below. The elemental sulfur will hinder gold dissolution by forming a stable passivation layer on the gold surface. The oxidization of thiourea also causes high reagent consumption, which limits the application of thiourea for metal recovery on large scale [46]. Therefore, the selection of suitable oxidant and its concentration is critical to achieving effective gold leaching and limited thiourea oxidation:

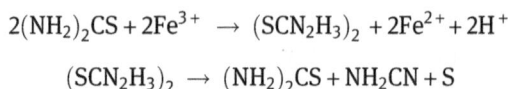

$$2(NH_2)_2CS + 2Fe^{3+} \rightarrow (SCN_2H_3)_2 + 2Fe^{2+} + 2H^+$$

$$(SCN_2H_3)_2 \rightarrow (NH_2)_2CS + NH_2CN + S$$

In alkaline condition, thiourea is unstable and easy to break down into sulfide and cyanamide, further forming urea. Due to the high cost and consumption of thiourea, thiourea lixiviant is currently limited to gold leaching, which also needs further development [15].

Acid and halide leaching
Inorganic acids such as HNO_3, HCl, and H_2SO_4 solutions are commonly used for leaching in hydrometallurgical processes. Figure 7.7 shows the leaching time and effect of HCl, HNO_3, H_2SO_4, $C_2H_4O_2$ (acetic acid), and $C_6H_8O_7$ (citric acid) lixiviants on recovering metal from PCBs pieces [25]. HCl is found to be most effect to recover metals requiring the shortest time to remove 100% of metals from the PCBs piece followed by HNO_3 [25]. All Cu, Zn, Sn, Ni, Pb, Fe, Al, Ag, Au, and Pd are solubilized by HCl in 22 h. H_2SO_4, acetic acid, and citric acid are relatively poor in metal

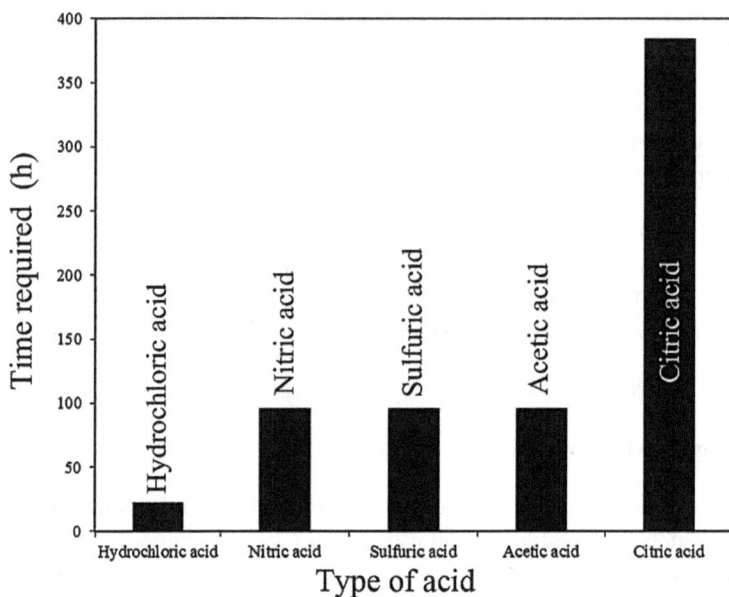

Figure 7.7: Leaching time required for various lixiviants to recover metal from printed circuit boards. Reproduced with permission from Ref. [25].

recovery, which are not able to fully remove metals from the PCBs pieces even after longer incubation periods. Furthermore, HCl is more cost effective than HNO_3, H_2SO_4 or $C_6H_8O_7$ for metal recovery from large PCBs pieces. HCl reacts immediately with Zn, Sn, Ni, Fe, Al, and Ag metals forming metal chlorides as per following reactions [47, 48]:

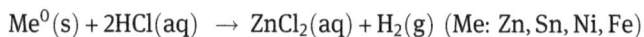

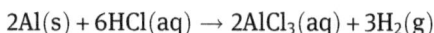

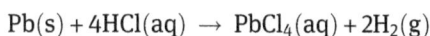

$$Me^0(s) + 2HCl(aq) \rightarrow ZnCl_2(aq) + H_2(g) \text{ (Me: Zn, Sn, Ni, Fe)}$$

$$2Al(s) + 6HCl(aq) \rightarrow 2AlCl_3(aq) + 3H_2(g)$$

$$2Ag(s) + 2HCl(aq) \rightarrow 2AgCl(aq) + H_2(g)$$

$$Pb(s) + 4HCl(aq) \rightarrow PbCl_4(aq) + 2H_2(g)$$

Due to the nonoxidizing nature of HCl, Cu, Pd, and Au can dissolve only in HCl when oxidants present such as oxygen in air [49–51]. Highly acidic conditions and halide anions are required to dissolve metal Au in HCl in addition to the presence of oxidant [52]:

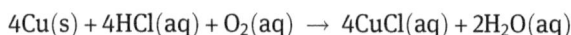

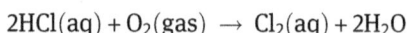

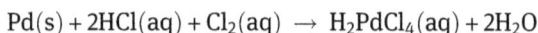

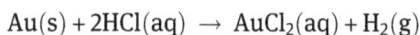

$$4Cu(s) + 4HCl(aq) + O_2(aq) \rightarrow 4CuCl(aq) + 2H_2O(aq)$$

$$2HCl(aq) + O_2(gas) \rightarrow Cl_2(aq) + 2H_2O$$

$$Pd(s) + 2HCl(aq) + Cl_2(aq) \rightarrow H_2PdCl_4(aq) + 2H_2O$$

$$Au(s) + 2HCl(aq) \rightarrow AuCl_2(aq) + H_2(g)$$

Apart from the nonoxidizing HCl acid, oxidizing HNO_3 acid is also used alone (for leaching of Cu and Pb) or together with HCl acid forming aqua regia (for leaching of Au) [53]:

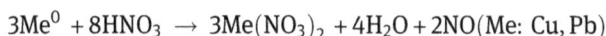

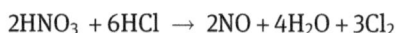

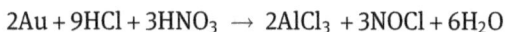

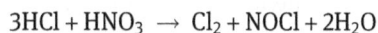

$$3Me^0 + 8HNO_3 \rightarrow 3Me(NO_3)_2 + 4H_2O + 2NO \text{(Me: Cu, Pb)}$$

$$2HNO_3 + 6HCl \rightarrow 2NO + 4H_2O + 3Cl_2$$

$$2Au + 9HCl + 3HNO_3 \rightarrow 2AlCl_3 + 3NOCl + 6H_2O$$

$$3HCl + HNO_3 \rightarrow Cl_2 + NOCl + 2H_2O$$

Moreover, sulfuric acid (H_2SO_4) with oxidant H_2O_2 results in better copper recovery at 80 °C as per following reaction [54]:

$$Cu^0 + H_2O_2 + H_2SO_4 \rightarrow CuSO_4 + 2H_2O$$

Apart from the aforesaid acid-based leaching, iodine leaching with 1% of hydrogen peroxide (H_2O_2) can achieve high yield of Au leaching (95%) as per following reaction:

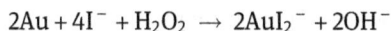

$$2Au + 4I^- + H_2O_2 \rightarrow 2AuI_2^- + 2OH^-$$

The above leaching processes using halide anions are known as halide leaching, which have high leaching rate. However, halide leaching involves strong corrosive acids and oxidizing conditions requiring corrosion-resistant equipment [55].

Metal extraction from leach liquor

After leaching processes, metals need to be further extracted from leaching liquor by various techniques such as electrowinning or chemical reduction processes. Chemical reduction routes with low-energy consumption, such as cementation, solvent extraction, and carbon adsorption, are promising for large-scale metal recovery as seen in Figure 7.8. As gold is the most valuable metal in e-waste, gold recovery from leach liquor will be focused in this section to discuss the different chemical reduction techniques for metal extraction.

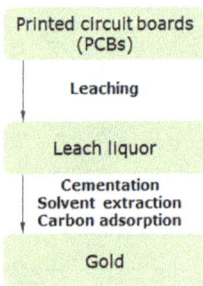

Printed circuit boards
(PCBs)

Leaching

Leach liquor

Cementation
Solvent extraction
Carbon adsorption

Gold

Figure 7.8: A simple flowchart of gold recovery from printed circuit boards.

Cementation

Zinc cementation has been widely used since 1890 to extract gold medal from leached liquor of gold cyanide solution. After adding zinc dust or shavings into the solution, gold will precipitate due to a higher affinity of zinc for cyanide iron than gold. The zinc cementation is an electrochemical reaction taking place on the surface of zinc, where the cathodic and anodic reactions are for gold deposition and zinc corrosion, respectively [56]:

$$\text{Cathodic reaction: } 4Au(CN)_2- + 2e^- \rightarrow Au + 4CN^-$$

$$\text{Anodic reaction: } Zn + 4CN^- \rightarrow Zn(CN)_4^{2-} + 2e^-$$

In thiocyanate solutions, iron powder can be used for gold cementation as represented by the bellowing reaction [57]. Notably, Fe^{3+} can reduce gold precipitation efficiency and lower the grade of gold cement:

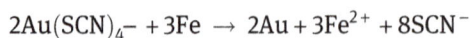

$$2Au(SCN)_4- + 3Fe \rightarrow 2Au + 3Fe^{2+} + 8SCN^-$$

Solvent extraction

Solvent extraction employs organic solvents to extract metals like gold from leach liquor into organic phases (Table 7.3). The gold-loaded organic phase can readily transfer back to the aqueous phase by stripping step using water or KCN solutions. Commercially, reagents such as LIX-79 and Cyanex 921 are used to extract gold from gold cyanide media [58, 59]. MIBK, DBC, and 2-EH are utilized to extract gold in the form of metalate $AuCl_4^-$ from aqueous HCl solution [60]. Solvent extraction can possess high selectivity by designing a reagent that preferably transfers a single metal into an organic phase from an aqueous leach liquor with mixed metals [61]. It is particularly desirable when only one metal like gold is targeted to be recovered [62]. Doidge et al. propose the use of a simple primary amide **L** (Figure 7.9a) to selectively extract gold into toluene from an aqueous HCl solution [28]. The percentage of gold extraction using amide **L** reaches the maximum for HCl concentration of 4 M as seen in Figure 7.9b, showing high selectivity of gold metalate extraction. Furthermore, 88% of the gold metalate $AlCl_4^-$ in the organic phase can be transferred back into the aqueous phase by sampling adding water. Combined experimental and simulation

Table 7.3: Typical commercial solvents for gold extraction from cyanide or chloride media.

Solvent	Molecular formula	Structure
LIX-79 (*N,N'*-bis(2-ethylhexyl)guanidine)	$C_{17}H_{37}N_3$	
Cyanex 921 (trioctylphosphine oxide)	$C_{24}H_{51}OP/$ $OP(C_8H_{17})_3$	
MIBK (methyl isobutyl ketone)	$C_6H_{12}O/$ $CH_3COCH_2CH(CH_3)_2$	
DBC (dibutyl carbitol)	$C_{12}H_{26}O_3/$ $[CH_3(CH_2)_3OCH_2CH_2]_2O$	
2-EH (2-ethylhexanal)	$C_8H_{16}O/$ $C_4H_9CH(C_2H_5)CHO$	

(a)

(b)   **(c)**

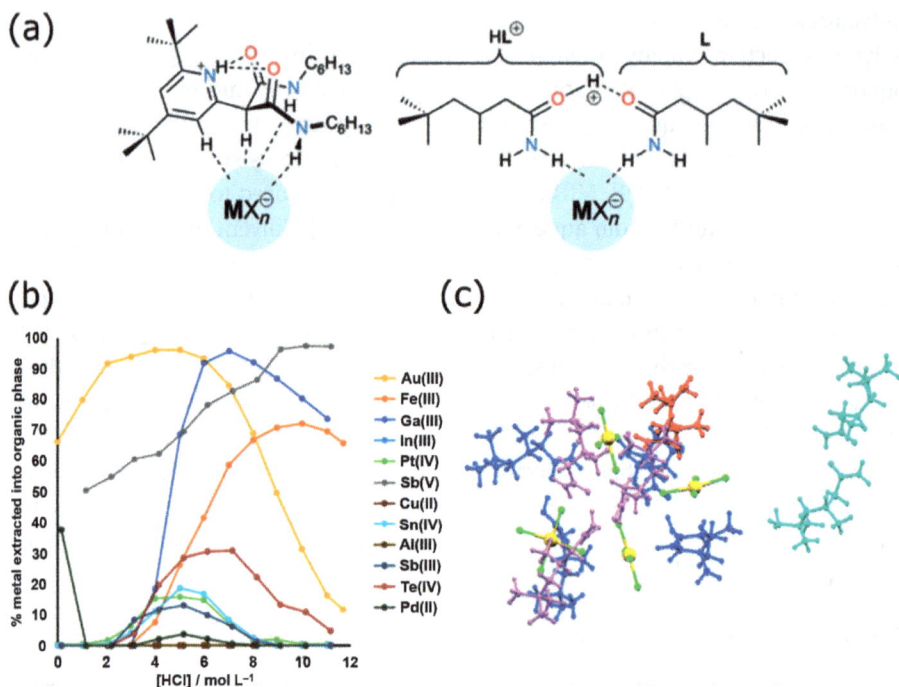

Figure 7.9: (a) Metalate extraction by protonated receptors. (b) Extraction of metals in HCl solution into toluene using amide L. (c) Simulated structures of L/H[AuCl$_4$] aggregates. Red: [(HL)(AuCl$_4$)$_2$]$^-$, dark blue: [(HL$_2$)(AuCl$_4$)$_2$]$^-$, magenta: [(HL$_3$)(AuCl$_4$)$_2$]$^-$, and cyan: (L)$_2$. Reproduced with permission from Ref. [28].

results (Figure 7.9c) show that the gold-loaded organic supramolecular clusters are spontaneously formed by bridging hydrogen bonding interactions of **L** and **LH$^+$** with AuCl$_4^-$ ions [28].

Carbon materials for metal extraction

Metal adsorption from leach liquor is one of the most widely used methods to re-cover valuable metals. In typical adsorption experiments, the metal adsorption capacity q_e (mg/g) of sorbent is calculated using the following equation,

$$q_e = \frac{(C_0 - C_e)V}{m}$$

where C_0 (mmol/L) and C_e (mmol/L) are the initial and equilibrium adsorbate concentration in the solution, respectively; V (L) and m (g) are the volume of solution and mass of the sorbent. The percentage of metal adsorbed by the sorbent can be obtained by $\frac{C_0 - C_f}{C_0} \times 100\%$. Typically, solid–liquid adsorption processes can be represented by two different isotherms: Langmuir and isotherms as described [63, 64]:

Langmuir isotherm:

$$\frac{C_e}{q_e} = \frac{1}{q_{max}b} + \frac{C_e}{q_{max}}$$

Freundlich isotherm:

$$\log q_e = \log K_f + \frac{1}{n}\log C_e$$

where q_{max} (mmol/g) and b (L/mmol) are the maximum adsorption capacity and adsorption equilibrium constant, respectively. K_f and n are Freundlich isotherm constants. Therefore, linear fitting of C_e/q_e vs C_e can be used to determine q_{max} and b for Langmuir isotherm and linear fitting log q_e vs log C_e can be applied to obtain K_f and n for Freundlich isotherm. The Langmuir isotherm and Freundlich isotherm assume monolayer and multilayer adsorption, which are applicable to adsorption on homogeneous and heterogeneous surfaces, respectively [65–67]. All sites on a homogeneous surface have the same affinity for adsorbate, so that adsorbate will not transmigrate in the surface plane [68, 69]. On a heterogeneous surface, the distribution of adsorption heat and affinities over the surface sites are nonuniform for adsorbate [70].

Moreover, three different models including pseudo-first-order, pseudo-second-order, and intraparticle diffusion models can be adapted to analyze the time-dependent adsorption amounts of Au(III) as shown [71–73].

Pseudo-first-order equation:

$$q_t = q_e \times \left(1 - e^{-k_1 t}\right)$$

Pseudo-second-order equation:

$$q_t = \frac{k_2 \times q_e^2 \times t}{1 + k_2 \times q_e \times t}$$

Intraparticle diffusion equation:

$$q_t = k_p \times t^{0.5} + I$$

where q_t is the adsorption amount of adsorbate at time t (h), k_1 (h^{-1}), k_2 (g·mmol^{-1}·h^{-1}), and k_p (mmol·g^{-1}·h$^{-0.5}$) are rate constants of the corresponding models, and I (mmol·g^{-1}) is a constant. The pseudo-first-order and pseudo-second-order adsorption signify that the rate of the adsorption is limited by the concentration of the adsorbate in the solution and amount of the sorption sites on the surface of sorbent, respectively [74, 75]. The intraparticle diffusion model signifies that the movement of adsorbate into the interior of sorbent is the rate-limiting step [76].

Carbon-based materials are economical, nontoxic, metal selective, and regenerative, and thus are regarded as promising sorbents for metal extraction. Activated carbon was first used to adsorb gold cyanide complexes from aqueous solution in 1951 [77]. Furthermore, carbon nanotube, carbon spherules, or amorphous carbon have been extensively investigated for PM recovery. Pang et al. proposed the gold adsorption by multi-walled carbon nanotubes (MWCNTs) from the acidic $AuCl_4^-$ solutions forming metallic gold on the MWCNTs as seen in Figure 7.10a [29]. $AuCl_4^-$ was reduced to Au via an intermediate $AuCl_2^-$, leading to elemental gold deposition on MWCNTs. The combined adsorption and reduction process to directly obtain Au metals demonstrates great potential to simplified leaching and purification steps for metal recovery. Wang et al. synthesized hydrothermal carbon spherules (HCSs) as

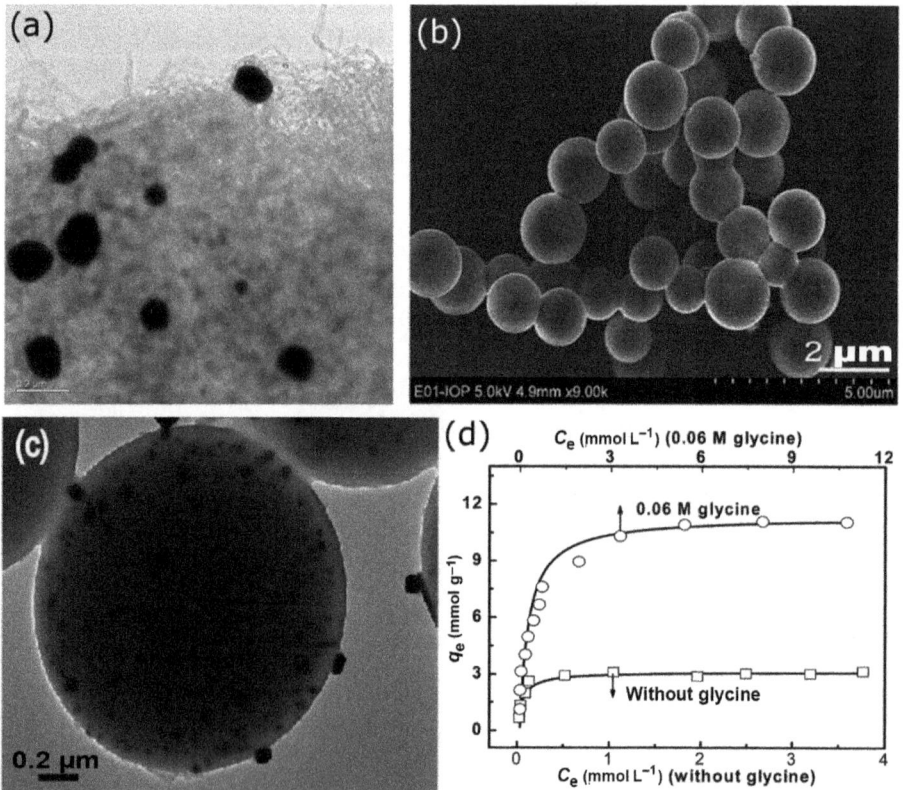

Figure 7.10: (a) HRTEM image of multiwalled carbon nanotubes (MWCNTs) loaded with Au nanoparticles. Reproduced with permission from Ref. [29]. (b) SEM image of as-synthesized hydrothermal carbon spherules (HCSs). (c) TEM image of Au-loaded HCSs. (d) Adsorption isotherms of Au(III) by HCSs with and without glycine. Reproduced with permission from Ref. [30].

seen in Figure 7.10b and c, demonstrating high selectivity for Au(III) adsorption against Pd(II), Pt(VI), Rh(III), Fe(III), Co(II), Cu(II), and Ni(II). The adsorption capacities, selectivity, and rate can be enhanced in the presence of extra reductant glycine as seen in Figure 7.10d.

In addition to selectivity of PM adsorption, efficient separation from the leach liquor is highly favorable. Gunawan et al. reported the fabrication of magnetic 1D γ-Fe_2O_3@C core-shell nanostructures for selective gold recovery and magnetic separation [31]. First, 1D α-Fe_2O_3 nanorods were prepared by microwave heating of a mixed solution containing iron(III) nitrite and TEAOH as seen in Figure 7.11a. Thereafter, the 1D α-Fe_2O_3 nanorods and glucose were mixed in water forming a homogeneous mixture. Amorphous carbon layer was then coated onto the 1D nanorods by hydrothermal heating of the mixture in a Teflon-lined autoclave. Finally, the nanorods were calcined in nitrogen gas to form 1D γ-Fe_2O_3@C core-shell structures. The microstructure and magnetic property of the core-shell nanorods are stable in ambient conditions.

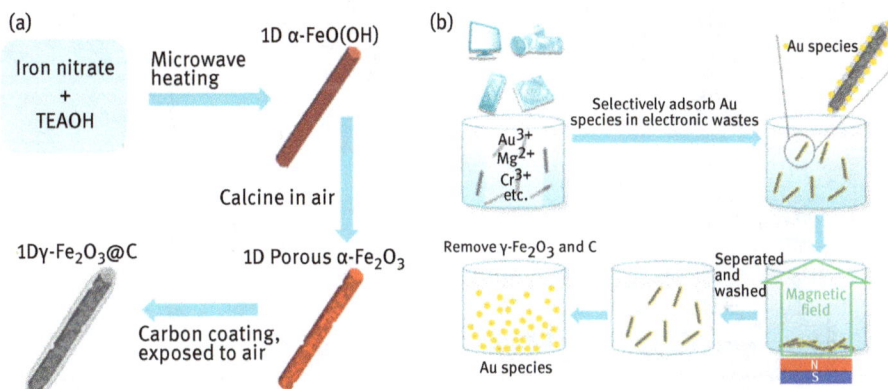

Figure 7.11: (a) Synthesis of 1D carbon coated γ-Fe_2O_3 nanorods and (b) selective adsorption of Au from mixed metal salts solution. Reproduced with permission from Ref. [31].

Gold recovery performance of the 1D γ-Fe_2O_3@C core-shell nanorods was examined using a mixed solution of cations including Au^{3+}, Cr^{3+}, Cu^{2+}, Mg^{2+}, Ni^{2+}, and Zn^{2+} cations. The sorbent possesses high adsorption selectivity toward Au^{3+} with the highest removal efficiency of 98%. The high selectivity may be attributed to the high affinity of the carbon layer to Au ions. Moreover, nitrogen calcination can result in defects on carbon surface, which could interact with O_2 and H_2O forming negatively charged species and functional groups. The negatively charged species on the carbon surface can adsorb and reduce Au^{3+} cations into a metallic phase. Furthermore, the 1D γ-Fe_2O_3@C core-shell structures demonstrate ultrahigh maximum adsorption capacity of ~600 mg/g sorbent or 1,132 mg/g carbon depositing a large amount of metallic Au particles on the carbon layers as seen in Figure 7.12. Most importantly, the Au-loaded

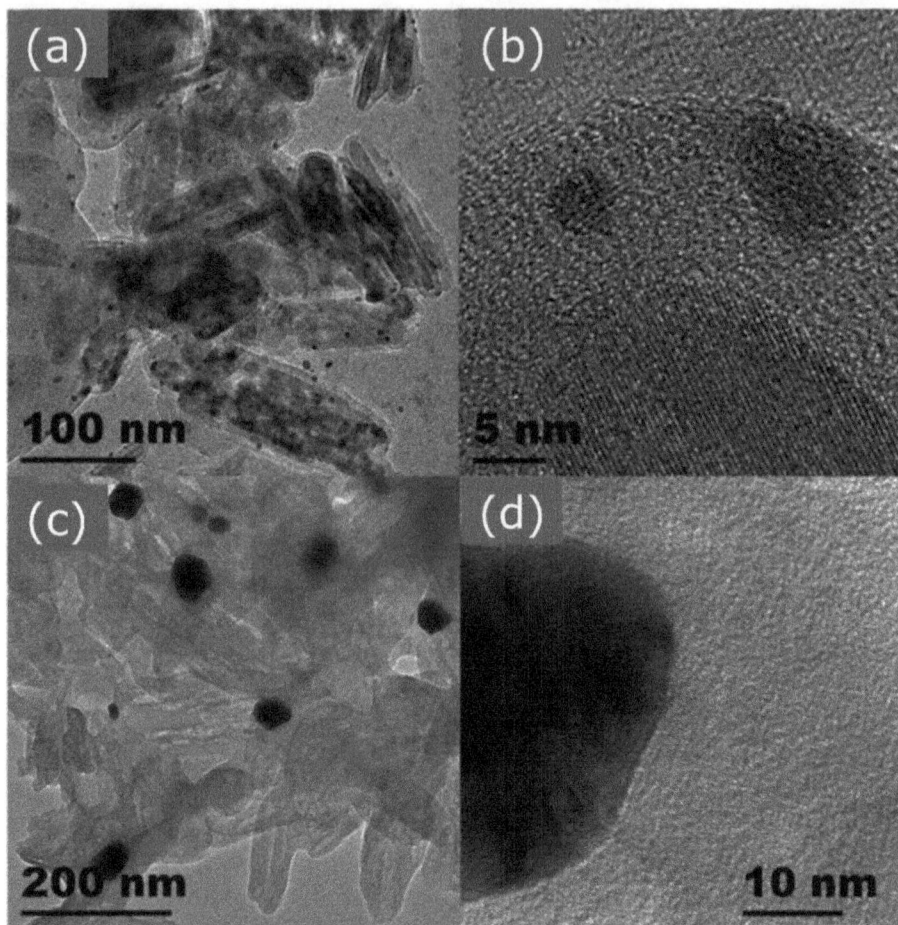

Figure 7.12: TEM images of 1D γ-Fe$_2$O$_3$@C adsorbed with Au from (a–b) 1,000 mg/L and (c–d) 2,500 mg/L Au aqueous solutions. Au nanoparticles of (a–b) <5 nm and (c–d) 10–60 nm in size form on the 1D γ-Fe$_2$O$_3$@C structure in 1,000 mg/L and 2,500 mg/L Au aqueous solutions, respectively. It shows that high selectivity toward Au adsorption of the 1D γ-Fe$_2$O$_3$@C structure is related to the high reducibility of Au cations. The tunable size of Au nanoparticles formed in different concentrations of Au aqueous solutions demonstrates wide applications in catalysis. Reproduced with permission from Ref. [31].

magnetic core-shell structures can be readily separated from the aqueous solution by simply applying a magnetic field as seen in Figure 7.11b. High-purity Au particles can be finally retrieved by combusting the Au-loaded structures in the air to remove the carbon layer and dissolving the γ-Fe$_2$O$_3$ core in dilute HCl solution. The combined gold adsorption and magnetic separation route provides a possible cyanide-free solution for fast and selective gold recovery from e-waste.

Apart from metal recovery via extraction, separation, and purification, PMs extracted by adsorption or reduction of nanomaterials can also be directly used as catalysts for other reactions. Teng et al. dispersed uniform nanoscale zero-valent iron (nZVI) nanoparticles (~16 nm) in ordered mesoporous carbon matrix for rapid and efficient gold reduction schematically shown in Figure 7.13 [78]. The nZVI possesses high performance toward metal removal from low-concentration sources [79]. The mesoporous carbon framework prevents magnetic aggregation and oxidation of nZVI and precisely confines the nZVI nanoparticles in the carbon channels. The highly ordered mesopores (~5.2 nm) of the mesoporous carbon support leads to a high surface area (~500 m^2/g) of the nZVI@C composite and high exposure of nZVI surface active sites. The nZVI@C composite is first utilized to recover gold from gold aqueous solution of low concentrations of 0.01 mg/L and 1 mg/L. Figure 7.14a shows that the nZVI@C composite extracts 96% of gold within 1 min and 99% removal of gold after 5 min demonstrating high extraction efficiency. It can be seen in Figure 7.14b that the iron concentration of the solution increases to 0.8 mg/L after the gold extracting, indicating only ~4% of the nZVI nanoparticles are dissolved. Due to the rapid in situ reduction and immobilization of gold, small gold nanoparticles (~6 nm) form and are uniformly distributed over the nZVI@C composite. Furthermore, the nZVI@C composite with reclaimed gold nanoparticles can be readily collected using a magnet due to the embedded magnetic iron nanoparticles

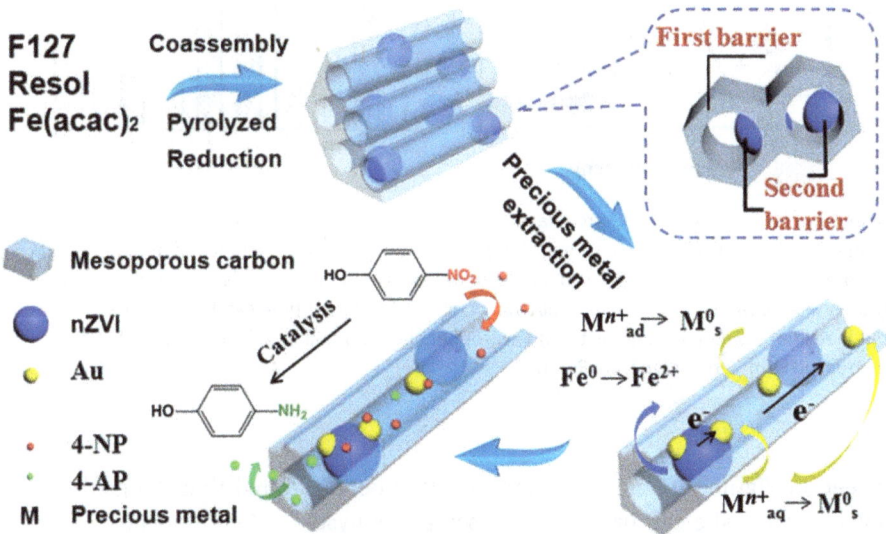

Figure 7.13: Schematic diagram of nanoscale zero-valent iron (nZVI) in mesoporous carbon (nZVI@C) being used for metal extraction (e.g., gold) and catalytic reduction of 4-nitrophenol. Reproduced with permission from Ref. [78].

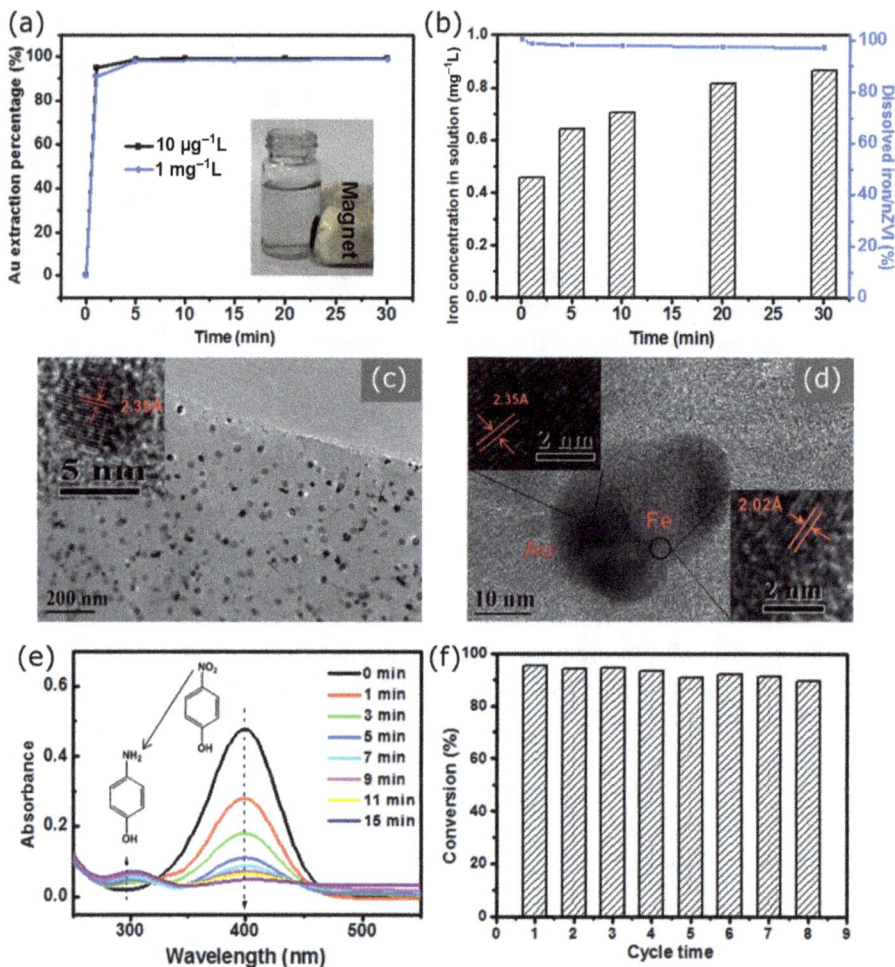

Figure 7.14: (a) Gold extraction percentage as a function time from gold aqueous solution of low initial concentrations of (0.01 and 1 mg/L). (b) Concentration of ion as a function of time. (c) TEM and (d) HRTEM images of nZVI@C composite loaded with gold nanoparticles via gold extraction. (e) UV–vis spectra of 4-nitrophenol reduction reaction at different time catalyzed by the gold-loaded nZVI@C nanocomposite. (f) 4-Nitrophenol conversion as a function of cycle time. Reproduced with permission from Ref. [78].

as seen in Figure 7.14c and d. Moreover, it is demonstrated that the gold-loaded nZVI@C composite can be directly reused as catalysts for 4-nitrophenol reduction as seen in Figure 7.14e. The gold-loaded nZVI@C catalyst shows close to 95% conversion of 4-nitrophenol reduction suggesting high catalytic activity as seen in Figure 7.14f.

Coprecipitation

Apart from the cementation, solvent extraction, and adsorption methods, Liu et al. proposed a coprecipitation method for gold extraction based on host–guest chemistry [32].

First, concentrated HBr and HNO_3 (3/1) was prepared to dissolve gold-bearing materials resulting in a solution containing HAuBr4. Thereafter, KOH was used to neutralize the dissolved solution with a pH value of 4–6 converting the $HAuBr_4$ to $KAuBr_4$. After removing the insoluble bromide, α-CD (CD: cyclodextrin) was added to the solution and $KAuBr_4 \cdot (a\text{-CD})_2$ **(α·Br)** rapidly coprecipitated by reaction between α-CD and $KAuBr_4$ as shown schematically in Figure 7.15a. The solid α·Br precipitate containing gold was then filtered and separated. The α·Br complex is a 1D coaxial core-shell superstructure with a high aspect ratio, possessing length and diameter of a few hundreds of nanometers and several tens of micrometers, respectively, as seen in Figure 7.15b. By dispersing the solid α·Br into water and reducing with $Na_2S_2O_5$, gold was further recovered from the α·Br and precipitates in metallic phase. After decanting off the aqueous solution containing liquid α-CD, the recovered gold metal

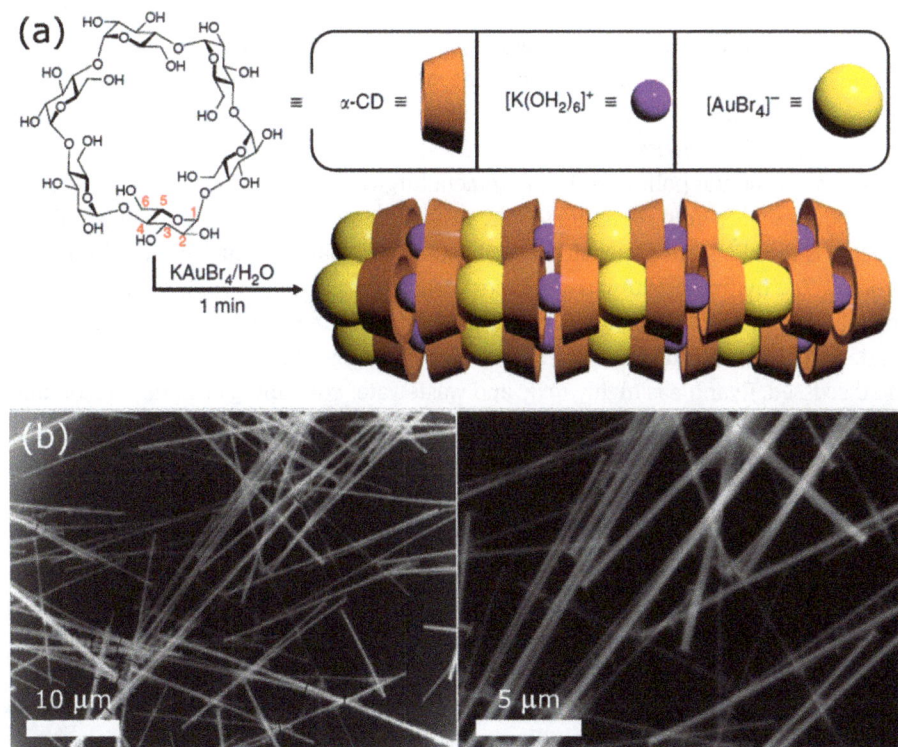

Figure 7.15: (a) Schematic spontaneous self-assembly of α·Br and (b) TEM images of crystalline α·Br. Reproduced with.permission from Ref. [32].

was collected. The liquid α-CD can then be recycled from the aqueous solution by recrystallization. The extracted gold metal has high purity (>95%), indicating high selectivity of the coprecipitation process toward gold recovery. This process has the advantages of being eco-friendly, economical, fast, and feasible.

7.3 Conclusions

The rapid development of today's advanced technologies, such as physical–digital integration, Internet-of-things, automation and robotics, and so on, lead to dramatically increased demand for EEE and as a consequence high consumption of PMs and rare earth elements. The shortened lifespan of electronic products further accelerates the generation of e-waste and make PMs severely scarce. Therefore, recovering valuable materials and metals from the secondary source – e-waste, is imperative considering economic value, environmental protection, and energy efficiency.

In this chapter, the background of e-waste generation and current status of e-waste recycling are first overviewed. The general process of e-waste recycling is introduced including collection, separation, and pyro-/hydrometallurgical recovery. The pyrometallurgical process is economical and has a good recovery rate of PMs. However, installing the recycling plants needs a large investment of valuable metals and environmental protection. Heating e-waste containing halogenated flame retardants results in the release of hazardous gases, which requires off-gas treatment to reduce environmental pollution. The pyrometallurgical process is also energy intensive and requires high-grade feed materials. Another drawback of the pyrometallurgical process is that PMs are segregated into BMs (copper, lead, and nickel) in the smelting process. The hydrometallurgical route for metal recovery is especially focused including leaching processes and metal extraction methods. Leaching processes such as cyanide leaching, thiosulfate leaching, thiourea leaching, and acid and halide leaching are discussed. Cyanide is highly toxic and wastewater containing cyanide can contaminate rivers and seawater, posing serious health risks to the inhabitants. In addition, the cyanidation process is long due to slow leaching rate. Therefore, noncyanide leaching processes have attracted great interest recently. Thiosulfate leaching has high selectivity, nontoxic, and noncorrosive. However, the process is slow and consumes a large amount of reagent, which is uneconomical. Thiourea lixiviant is currently limited to gold leaching due to the high cost and consumption of thiourea. Halide leaching has high leaching rate. However, it involves strong corrosive acids and oxidizing conditions requiring corrosion-resistant equipment.

Moreover, metal extraction methods to recover metals in particular gold from leach liquor are presented, including cementation, solvent extraction, adsorption, and coprecipitation. Examples include various nanomaterials for gold recovery such as MWCNTs, HCSs, 1D γ-Fe_2O_3/carbon core-shell nanorods, composites of nanoscale zero-valent iron on mesoporous carbon and 1D coaxial core-shell α·Br superstructure. Carbon materials for metal recovery are of particular interest and theories of adsorption isotherms and kinetics are reviewed. Magnetic separation processes using the magnetic γ-Fe_2O_3/carbon nanorods or iron nanoparticles are highlighted. This chapter illustrates a comprehensive picture on the recovery of metals from e-waste including not only materials and processes but also theories and mechanisms.

References

[1]   Akcil A., Erust C., Gahan C.S., Ozgun M., Sahin M., Tuncuk A. Precious metal recovery from waste printed circuit boards using cyanide and non-cyanide lixiviants–a review, Waste Manage., 2015, 45, 258–271.

[2]   Advancing Sustainable Materials Management: Facts and Figures 2013, 2015. https://www.epa.gov/sites/production/files/2015-09/documents/2013_advncng_smm_rpt.pdf

[3]   Tiwary C., Kishore S., Vasireddi R., Mahapatra D., Ajayan P., Chattopadhyay K. Electronic waste recycling via cryo-milling and nanoparticle beneficiation, Mater. Today, 2017, 20, 67–73.

[4]   Cui J., Zhang L. Metallurgical recovery of metals from electronic waste: a review, J. Hazard. Mater., 2008, 158, 228–256.

[5]   Schluep M., Hagelueken C., Kuehr R., Magalini F., Maurer C., Meskers C., Mueller E., Wang F. Sustainable innovation and technology transfer industrial sector studies: Recycling–from e-waste to resources, 2009. http://www.unep.fr/shared/publications/pdf/DTIx1192xPA-Recycling%20from%20ewaste%20to%20Resources.pdf

[6]   Heacock M, Kelly CB, Asante KA, Birnbaum LS, Bergman ÅL, Bruné M-N., Buka I., Carpenter D.O., Chen A., Huo X. E-waste and harm to vulnerable populations: a growing global problem, Environ. Health Perspect., 2016, 124, 550.

[7]   Reuter M., Hudson C., Van Schaik A., Heiskanen K., Meskers C., Hagelüken C. Metal recycling: Opportunities, limits, infrastructure, in A Report of the Working Group on the Global Metal Flows to the International Resource Panel, 2013. https://wedocs.unep.org/handle/20.500.11822/8423

[8]   Goosey M., Kellner R. Recycling technologies for the treatment of end of life printed circuit boards (PCBs), Circuit World, 2003, 29, 33–37.

[9]   Kang H-Y., Schoenung J.M. Electronic waste recycling: a review of US infrastructure and technology options, Resour. Conserv. Recycl., 2005, 45, 368–400.

[10]   Balde C., Kuehr R., Blumenthal K., Gill S.F., Kern M., Micheli P., Magpantay E., Huisman J. E-waste statistics: Guidelines on classifications, reporting and indicators, 2015. https://i.unu.edu/media/ias.unu.edu-en/project/2238/E-waste-Guidelines_Partnership_2015.pdf

[11]   Zhang B., Guan D. Take responsibility for electronic-waste disposal, Nature, 2016, 536, 4.

[12]   Zhang K., Schnoor JL., Zeng EY. E-waste recycling: where does it go from here?, Environ. Sci. Technol. 2012, 46, 20, 10861–10867.

[13]   Daum K., Stoler J., Grant RJ. Toward a more sustainable trajectory for e-waste policy: a review of a decade of e-waste research in Accra, Ghana, Int. J. Environ. Res. Publ. Health, 2017, 14, 135.

[14]   Meskers C., Hagelüken C., Salhofer S., Spitzbart M. Impact of pre-processing routes on precious metal recovery from PCs. in Proceedings of EMC. 2009.

[15] Khaliq A., Rhamdhani M.A., Brooks G., Masood S. Metal extraction processes for electronic waste and existing industrial routes: a review and Australian perspective, Resources, 2014, 3, 152–179.

[16] Guo J., Guo J., Xu Z. Recycling of non-metallic fractions from waste printed circuit boards: a review, J. Hazard. Mater., 2009, 168, 567–590.

[17] Guo J., Cao B., Guo J., Xu Z. A plate produced by nonmetallic materials of pulverized waste printed circuit boards, Environ. Sci. Technol., 2008, 42, 5267–5271.

[18] Jin Y., Li G., He W-Z. Preparation of a Composite plate Using Nonmetallic Materials Powder from the Waste Printed Circuit Boards. in 4th International Conference on Bioinformatics and Biomedical Engineering (iCBBE). 2010: IEEE.

[19] Antrekowitsch H., Potesser M., Spruzina W., Prior F. Metallurgical recycling of electronic scrap. in EPD Congress. 2006.

[20] Anindya A., Swinbourne D., Reuter M., Matusewicz R. Distribution of elements between copper and FeOx–CaO–SiO$_2$ slags during pyrometallurgical processing of WEEE: Part 1–Tin, Miner. Process. Extr. Metall., 2013, 122, 165–173.

[21] Anindya A. Minor elements distribution during the smelting of WEEE with copper scrap, 2012.

[22] Hagelüken C. Recycling of electronic scrap at Umicore's integrated metals smelter and refinery, Erzmetall, 2006, 59, 152–161.

[23] Hagelüken C. Recycling of electronic scrap at Umicore precious metals refining, Acta Metallurgica Slovaca, 2006, 12, 111–120.

[24] Rocchetti L., Vegliò F., Kopacek B., Beolchini F. Environmental impact assessment of hydrometallurgical processes for metal recovery from WEEE residues using a portable prototype plant, Environ Sci. Technol., 2013, 47, 1581–1588.

[25] Jadhav U., Hocheng H. Hydrometallurgical recovery of metals from large printed circuit board pieces, Sci. Rep., 2015, 5.

[26] Park Y.J., Fray D.J. Recovery of high purity precious metals from printed circuit boards, J. Hazard. Mater., 2009, 164, 1152–1158.

[27] Anand A., Jha M.K., Kumar V., Sahu R. Recycling of Precious Metal Gold from Waste Electrical and Electronic Equipments (WEEE): a review. in Proceedings of the XIII International Seminar on Mineral Processing Technology. 2013.

[28] Doidge E.D., Carson I., Tasker P.A., Ellis R.J., Morrison C.A., Love J.B. A simple primary amide for the selective recovery of gold from secondary resources, Angew. Chem. Int. Ed., 2016, 55, 12436–12439.

[29] Pang S-K., Yung K-C. Prerequisites for achieving gold adsorption by multiwalled carbon nanotubes in gold recovery, Chem. Eng. Sci., 2014, 107, 58–65.

[30] Wang F., Zhao J., Zhu M., Yu J, Hu Y-S., Liu H. Selective adsorption–deposition of gold nanoparticles onto monodispersed hydrothermal carbon spherules: a reduction–deposition coupled mechanism, J. Mater. Chem. A, 2015, 3, 1666–1674.

[31] Gunawan P., Xiao W., Chua M.W.H, Tan CP-C, Ding J., Zhong Z. One-dimensional fossil-like γ-Fe2O3@ carbon nanostructure: preparation, structural characterization and application as adsorbent for fast and selective recovery of gold ions from aqueous solution, Nanotechnology, 2016, 27, 415701.

[32] Liu Z., Frasconi M., Lei J., Brown Z.J, Zhu Z., Cao D., Iehl J., Liu G., Fahrenbach A.C., Botros Y.Y. Selective isolation of gold facilitated by second-sphere coordination with α-cyclodextrin, Nat. Commun., 2013, 4, 1855.

[33] Tuncuk A., Stazi V., Akcil A., Yazici E.Y., Deveci H. Aqueous metal recovery techniques from e-scrap: hydrometallurgy in recycling, Miner. Eng., 2012, 25, 28–37.

[34] Tue N.M., Takahashi S., Subramanian A., Sakai S., Tanabe S. Environmental contamination and human exposure to dioxin-related compounds in e-waste recycling sites of developing countries, Environ. Sci. Process. Impact, 2013, 15, 1326–1331.
[35] Zhang Y., Liu S., Xie H., Zeng X., Li J. Current status on leaching precious metals from waste printed circuit boards, Procedia Environ. Sci., 2012, 16, 560–568.
[36] Andrews D., Raychaudhuri A., Frias C. Environmentally sound technologies for recycling secondary lead, J. Power Sources, 2000, 88, 124–129.
[37] Hilson G., Monhemius A. Alternatives to cyanide in the gold mining industry: what prospects for the future?, J. Clean Prod., 2006, 14, 1158–1167.
[38] Akcil A. A new global approach of cyanide management: international cyanide management code for the manufacture, transport, and use of cyanide in the production of gold, Miner. Process. Extr. Metall. Rev., 2010, 31, 135–149.
[39] Feng D., Van Deventer J. Leaching behaviour of sulphides in ammoniacal thiosulphate systems, Hydrometallurgy, 2002, 63, 189–200.
[40] Jeffrey M., Breuer P., Choo W. A kinetic study that compares the leaching of gold in the cyanide, thiosulfate, and chloride systems, Metall. Mater. Trans. B, 2001, 32, 979–986.
[41] Byerley J.J., Fouda S.A., Rempel G.L. Kinetics and mechanism of the oxidation of thiosulphate ions by copper (II) ions in aqueous ammonia solution, J. Chem. Soc., Dalton Trans., 1973, 889–893.
[42] Byerley J., Fouda S., Rempel G. The oxidation of thiosulfate in aqueous ammonia by copper (II) oxygen complexes, Inorg. Nucl. Chem. Lett., 1973, 9, 879–883.
[43] Kai T., Hagiwara T., Haseba H., Takahashi T. Reduction of thiourea consumption in gold extraction by acid thiourea solutions, Ind. Eng. Chem. Res., 1997, 36, 2757–2759.
[44] Murthy D., Kumar V., Rao K. Extraction of gold from an Indian low-grade refractory gold ore through physical beneficiation and thiourea leaching, Hydrometallurgy, 2003, 68, 125–130.
[45] Gönen N., Körpe E., Yıldırım M., Selengil U. Leaching and CIL processes in gold recovery from refractory ore with thiourea solutions, Miner. Eng., 2007, 20, 559–565.
[46] Wu J., Chen L., Qiu L., Chen D. Study on selectively leaching gold from waste printed circuit boards with thiourea, Gold, 2008, 29, 55–58.
[47] Xiu F-R., Qi Y., Zhang F-S. Recovery of metals from waste printed circuit boards by supercritical water pre-treatment combined with acid leaching process, Waste Manage., 2013, 33, 1251–1257.
[48] Zhang R.L., Zhang X.F., Tang S.Z., Huang A.D. Ultrasound-assisted HCl–NaCl leaching of lead-rich and antimony-rich oxidizing slag, Ultrason. Sonochem., 2015, 27, 187–191.
[49] Havlik T., Orac D., Petranikova M., Miskufova A., Kukurugya F., Takacova Z. Leaching of copper and tin from used printed circuit boards after thermal treatment, J. Hazard. Mater., 2010, 183, 866–873.
[50] Behnamfard A., Salarirad M.M., Veglio F. Process development for recovery of copper and precious metals from waste printed circuit boards with emphasize on palladium and gold leaching and precipitation, Waste Manage., 2013, 33, 2354–2363.
[51] Barakat M., Mahmoud M., Mahrous Y. Recovery and separation of palladium from spent catalyst, Appl. Catal., A., 2006, 301, 182–186.
[52] Stanley R.W., Harris G.B., Monette S. Process for the recovery of gold from a precious metal bearing sludge concentrate, 1987, US4670052A.
[53] Mecucci A., Scott K. Leaching and electrochemical recovery of copper, lead and tin from scrap printed circuit boards, J. Chem. Technol. Biotechnol., 2002, 77, 449–457.
[54] Quinet P., Proost J., Van Lierde A. Recovery of precious metals from electronic scrap by hydrometallurgical processing routes, Miner. Metall. Process., 2005, 22, 17–22.

[55] Samina M., Karim A., Venkatachalam A. Corrosion study of iron and copper metals and brass alloy in different medium, J. Chem., 2011, 8, S344–S348.
[56] Fleming C. Hydrometallurgy of precious metals recovery, Hydrometallurgy, 1992, 30, 127–162.
[57] Wang Z., Chen D., Chen L. Gold cementation from thiocyanate solutions by iron powder, Miner. Eng., 2007, 20, 581–590.
[58] Sastre A.M., Madi A., Cortina J.L., Alguacil F.J. Solvent extraction of gold by LIX 79: experimental equilibrium study, J. Chem. Technol. Biotechnol., 1999, 74, 310–314.
[59] Alguacil F., Caravaca C., Cobo A., Martinez S. The extraction of gold (I) from cyanide solutions by the phosphine oxide Cyanex 921, Hydrometallurgy, 1994, 35, 41–52.
[60] Musikas C., Choppin G.R., Rydberg J. Principles and practices of solvent extraction, 1992.
[61] Wilson A.M., Bailey P.J., Tasker P.A., Turkington J.R., Grant R.A., Love J.B. Solvent extraction: the coordination chemistry behind extractive metallurgy, Chem. Soc. Rev., 2014, 43, 123–134.
[62] Izatt R.M., Izatt S.R., Bruening R.L., Izatt N.E., Moyer B.A. Challenges to achievement of metal sustainability in our high-tech society, Chem. Soc. Rev., 2014, 43, 2451–2475.
[63] Langmuir I. Kinetic model for the sorption of dye aqueous solution by clay-wood sawdust mixture, J. Am. Chem. Soc, 1916, 38, 2221–2295.
[64] Freundlich H. Über die adsorption in lösungen, Zeitschrift für physikalische Chemie, 1907, 57, 385–470.
[65] El-Khaiary M.I., Malash G.F. Common data analysis errors in batch adsorption studies, Hydrometallurgy, 2011, 105, 314–320.
[66] Alberti G., Amendola V., Pesavento M., Biesuz R. Beyond the synthesis of novel solid phases: review on modelling of sorption phenomena, Coord. Chem. Rev., 2012, 256, 28–45.
[67] Foo K., Hameed B. Insights into the modeling of adsorption isotherm systems, Chem. Eng. J., 2010, 156, 2–10.
[68] Kundu S., Gupta A.K. Arsenic adsorption onto iron oxide-coated cement (IOCC): regression analysis of equilibrium data with several isotherm models and their optimization, Chem. Eng. J., 2006, 122, 93–106.
[69] Pérez-Marín A.B., Zapata V.M., Ortuño J.F., Aguilar M., Sáez J., Lloréns M. Removal of cadmium from aqueous solutions by adsorption onto orange waste, J. Hazard. Mater., 2007, 139, 122–131.
[70] Adamson A.W., Gast A.P. Physical chemistry of surfaces, 1997.
[71] Lagergren S. About the theory of so-called adsorption of soluble substances, 1898.
[72] Ho Y-S. Absorption of heavy metals from waste streams by peat, 1995.
[73] Weber W.J., Morris J.C. Kinetics of adsorption on carbon from solution, J. Sanit. Eng. Div., 1963, 89, 31–60.
[74] Ho Y-S., McKay G. Pseudo-second order model for sorption processes, Process Biochem., 1999, 34, 451–465.
[75] Ho Y-S. Review of second-order models for adsorption systems, J. Hazard. Mater., 2006, 136, 681–689.
[76] Gupta S.S., Bhattacharyya K.G. Kinetics of adsorption of metal ions on inorganic materials: a review, Adv. Colloid Interface Sci., 2011, 162, 39–58.
[77] Mcquiston Jr Frank W. Recovery of gold or silver, 1951, US2545239A.
[78] Teng W., Fan J., Wang W., Bai N., Liu R., Liu Y., Deng Y., Kong B., Yang J., Zhao D. Nanoscale zero-valent iron in mesoporous carbon (nZVI@C): stable nanoparticles for metal extraction and catalysis, J. Mater. Chem. A, 2017, 5, 4478–4485.
[79] Stefaniuk M., Oleszczuk P., Ok YS. Review on nano zerovalent iron (nZVI): from synthesis to environmental applications, Chem. Eng. J., 2016, 287, 618–632.

Index

https://doi.org/10.1515/9783110544183-008

www.ingramcontent.com/pod-product-compliance
Lightning Source LLC
Chambersburg PA
CBHW061353210326
41598CB00035B/5968